Spatial Audio

Special Issue Editors

Woon Seng Gan

Jung-Woo Choi

MDPI • Basel • Beijing • Wuhan • Barcelona • Belgrade

MDPI

Special Issue Editors
Woon Seng Gan
Nanyang Technological University
Singapore

Jung-Woo Choi
Advanced Institute of Science
and Technology
Korea

Editorial Office
MDPI AG
St. Alban-Anlage 66
Basel, Switzerland

This edition is a reprint of the Special Issue published online in the open access journal *Applied Sciences* (ISSN 2076-3417) in 2017 (available at: http://www.mdpi.com/journal/applsci/special_issues/spatial_audio).

For citation purposes, cite each article independently as indicated on the article page online and as indicated below:

Author 1; Author 2. Article title. *Journal Name* **Year**, *Article number*, page range.

First Edition 2017

ISBN 978-3-03842-585-4 (Pbk)
ISBN 978-3-03842-586-1 (PDF)

Table of Contents

About the Special Issue Editors

Woon Seng Gan received his BEng (1st Class Hons) and PhD degrees, both in Electrical and Electronic Engineering from the University of Strathclyde, UK in 1989 and 1993, respectively. He is currently a Professor of Audio Engineering and the Director of the Centre for Infocomm Technology in the School of Electrical and Electronic Engineering in Nanyang Technological University. He also served as the Head of the Information Engineering Division in the School of Electrical and Electronic Engineering in Nanyang Technological University from 2011–2014, and as the Deputy Director of the Centre for Signal Processing from 2007–2008. His research regarding the connections between the physical world, signal processing and sound control resulted in the practical demonstration and licensing of spatial audio algorithms, directional sound beams, and active noise control for headphones. He has published more than 300 papers in international refereed journals and conferences, and has translated his research into six granted patents.

Jung-Woo Choi received his B.Sc., M.Sc., and Ph.D. degrees in Mechanical Engineering from the Korean Institute of Science and Technology (KAIST), in 1999, 2001, and 2005, respectively. He is currently an Assistant Professor in the School of Electrical Engineering, KAIST, Korea. Prior to joining KAIST, he was as an audio engineer at the Samsung Advanced Institute of Technology (SAIT) in Korea. His primary research area includes sound field control and array signal processing technologies for 3D audio, personal sound zone, and noise control. He is an author of more than 70 international journal and conference papers. He is also an inventor of 15 granted patents and the co-author of the book: Sound Visualization and Manipulation, Wiley, published in 2013. He is currently serving as an editorial board member of Mechanical Systems and Signal Processing (MSSP).

Preface to "Spatial Audio"

Three-dimensional (or spatial) audio is a growing research field that plays a key role in realizing immersive communication in many of today's applications for teleconferencing, entertainment, gaming, navigation guidance, and virtual reality (VR)/augmented reality (AR). Technologies in spatial sound capture and binaural recording are becoming an add-on module to our mobile devices to capture the surrounding soundscape, pickup directional and ambient cues, and create an immersive 3D audio media for playback. We are seeing a surge in research activities and applications that rely on digital spatial audio processing and rendering over loudspeakers (stereo, wave field synthesis, ambisonics) and headphones, and seeing new emerging fields of mobile spatial audio, personal assisted listening, and spatial audio for VR/AR. New developments in graphical processing units and multi-core processors are accelerating the pace for real-time spatial audio processing, and new techniques that can lead to high quality and immersive spatial audio reproduction.

Woon Seng Gan
Special Issue Editor

applied
sciences

MDPI

Editorial

Guest Editors' Note—Special Issue on Spatial Audio

Woon-Seng Gan [1],* and Jung-Woo Choi [2] (iD)

[1]   Digital Signal Processing Lab, School of Electrical and Electronic Engineering,
     Nanyang Technological University, Singapore 639798, Singapore
[2]   School of Electrical Engineering, Korea Advanced Institute of Science and Technology,
     Daejeon 34141, Korea; jwoo@kaist.ac.kr
*   Correspondence: ewsgan@ntu.edu.sg; Tel.: +65-6790-4538

Received: 1 August 2017; Accepted: 2 August 2017; Published: 3 August 2017

1. Introduction

Three-dimensional (or spatial) audio is a growing research field that plays a key role in realizing immersive communication in many of today's applications for teleconferencing, entertainment, gaming, navigation guidance, and virtual reality (VR)/augmented reality (AR). Technologies in spatial sound capture and binaural recording are becoming an add-on module to our mobile devices to capture the surrounding soundscape, pickup directional and ambient cues, and create an immersive 3D audio media for playback. In total, eight research papers and two review papers are published in this special issue. The research papers reported on new research techniques that resulted in higher quality and more immersive spatial audio reproduction; while the two review papers account for the state-of-the-art spatial audio recording and reproduction for playback in VR and AR headsets. A detailed accounts on these papers are summarized as follows.

2. Advanced Signal Processing Technologies for Spatial Audio

The current state-of-the-art technologies involved with spatial audio are summarized in two review papers [1,2]. The review paper written by Zhang, Samarasinghe, Chen, and Abhayapala [1] delivers a broad overview of existing and emerging spatial audio technologies involved with spatial audio. The paper begins with a summary of binaural technologies based on the head-related transfer function and covers sound field recording/reproduction techniques utilizing multichannel microphones and loudspeakers. The ending of the paper is devoted to the multi-zone sound reproduction problem, which aims to deliver multiple audio programs over multiple spatial regions.

The paper by Hong, He, Lam, Gupta, and Gan [2] puts a strong emphasis on the use of signal processing tools for the design of soundscapes. The review paper discusses the sound recording and reproduction technologies to render auditory sceneries resulting from the interaction of sound objects and surrounding environments. Beyond the simple reproduction of existing auditory scenery, soundscape design problems to improve the existing poor acoustic conditions are presented. The augmented reality in audio is especially highlighted as a means to provide an improved listening experience.

Proper localization of a sound source has been an important issue in the spatial audio for a long time, and it has been realized in various ways for stereo, discrete multichannel, and sound field control systems. Some new aspects of the localization problem are dealt with in this special issue, especially for proper recording, reproduction, perception, and evaluation.

In the paper written by Gößwein, Grosse, and van de Par [3], the authors propose a stereoscopic recording technique for enhancing the direct sound field in a reverberant environment. They employ two crossed linear microphone arrays combined with a super-directive endfire beam pattern. It is shown that the array recording can reduce the reverberation, while keeping compatibility with the amplitude panning technique.

The perception of sound localization has been studied over a decade, but the localization of elevated sound sources still remains a challenging problem. Wallis and Lee [4] study the influence of the interchannel time and level differences between two loudspeaker layers of different heights on the localization threshold. The results show the dependence of the localization threshold on the interchannel time difference. The required directivity of recording microphone in height direction is also discussed based on the identified localization threshold.

Objective evaluation of the reproduced sound field is another important issue. Mean squared error (MSE) has been popularly employed as a measure of similarity between target and reproduced sound fields. However, in their work [5], Chang and Jeong propose to use beamforming powers derived from given sound fields as a new measure. The primary reasoning behind the proposed measure is the weakness of MSE against room reflections and for 2.5-D reproduction techniques, such as wave field synthesis, that inevitably exhibit amplitude bias along the distance. The beam-power measure is expected to provide a more robust means to evaluate the directional cue or the direct component of a sound field.

The study by Mieth and Zölzer [6] deals with the objective evaluation problem for the pairwise panning-based upmix algorithm. To access the sound quality of upmix algorithms without subjective evaluations, they propose detailed procedures and measures regarding the direction of a virtual sound source, the amount of residual direct sound, loudness, and correlations in the frontal and surround channels.

The localization of sound is not only involved with the directional cue. Wendt, Zotter, Frank, and Höldrich [7] investigated the way to control the perceived distance through the variation of source directivities. The influence of the auralized room, source-listener distance, signal, and single-channel reverberation are considered together to build a model predicting the perceived distance. They tested various third-order beam patterns in a real room, which demonstrate that the distance perception caused by the source directivity is coupled with the sense of apparent source width.

For spatial audio, there are many auditory impressions to be carefully controlled along with the localization cues. Auditory sceneries deliver various spatial impressions such as stage width and ambience. The synthesis of late reverberation is studied in the paper written by Välimäki, Holm-Rasmussen, Alary, and Lehtonen [8]. They segmented the late part of a room impulse response and approximated the segments as filtered velvet noises that are very sparse in time but sound smoother. It is demonstrated that filtering with velvet noises greatly reduces the computational cost, only resulting in minor subjective differences for transient sounds.

The paper authored by Bai, Chung, Wu, Chiang, and Yang [9] proposes a general strategy to tackle the inverse problem that is often encountered in solving the source identification and separation problems for the spatial audio signal processing. Various inverse problem solvers, for both underdetermined and overdetermined problems, are investigated and compared in terms of PESQ and segSNR. Guidelines for choosing the right algorithm and regularization parameter are provided, with detailed examples of sound field analysis and synthesis problems.

Another inverse problem discussed by Gómez, Astley, and Fazi [10], is for the interactive auralization of sound fields in a low-frequency region. They utilized the finite element method to simulate a sound field in a room and then transformed the result using a plane wave expansion technique. Plane wave expansion has been popularly used for its simplicity in realizing the interactive sound rendering system that requires translation and rotation of sound fields. The transform of a sound field using plane wave expansion is a typical inverse problem, in which the determination of a regularization parameter is important to prevent singularity problems. The effect of regularization on the sound field representation is discussed, in view of plane waves' energy density and the size of sweet spot.

Appl. Sci. **2017**, *7*, 788

3. Summary

The trends reflected in the above papers stress that the signal processing technologies for spatial audio are heading towards a more natural reproduction of auditory impressions. The whole signal processing chain from recording to reproduction and evaluation is being revisited to cope with emerging applications, such as VR/AR and to render new auditory impressions. Although there is still a long way to go for the complete understanding of human listening and perfect control of auditory sceneries, the directions presented in this special issue demonstrate that a great deal of improvement can be made through the combination of perceptual and physical sides of spatial audio.

Acknowledgments: The editors would like to thank the strong administration support rendered by the MDPI editorial team, which include Managing Editor, Xiaoyan Chen; and Assistant Editors, Alice Zhang, Daria Shi, Candice Zhuo, Sydni Sun, and Jennifer Li. Their effort and quick responses in handling paper reviews and responding to our many questions has resulted in the fast appearance of this Special Issue on Spatial Audio.

Conflicts of Interest: The authors declare no conflict of interest.

References

1. Zhang, W.; Samarasinghe, P.N.; Chen, H.; Abhayapala, T.D. Surround by Sound: A Review of Spatial Audio Recording and Reproduction. *Appl. Sci.* **2017**, *7*, 532. [CrossRef]
2. Hong, J.Y.; He, J.; Lam, B.; Gupta, R.; Gan, W.-S. Spatial Audio for Soundscape Design: Recording and Reproduction. *Appl. Sci.* **2017**, *7*, 627. [CrossRef]
3. Gößwein, J.A.; Grosse, J.; van de Par, S. Stereophonic Microphone Array for the Recording of the Direct Sound Field in a Reverberant Environment. *Appl. Sci.* **2017**, *7*, 541.
4. Wallis, R.; Lee, H. The Reduction of Vertical Interchannel Crosstalk: The Analysis of Localisation Thresholds for Natural Sound Sources. *Appl. Sci.* **2017**, *7*, 278. [CrossRef]
5. Chang, J.-H.; Jeong, C.-H. A Measure Based on Beamforming Power for Evaluation of Sound Field Reproduction Performance. *Appl. Sci.* **2017**, *7*, 249. [CrossRef]
6. Mieth, M.; Zölzer, U. Objective Evaluation Techniques for Pairwise Panning-Based Stereo Upmix Algorithms for Spatial Audio. *Appl. Sci.* **2017**, *7*, 374. [CrossRef]
7. Wendt, F.; Zotter, F.; Frank, M.; Höldrich, R. Auditory Distance Control Using a Variable-Directivity Loudspeaker. *Appl. Sci.* **2017**, *7*, 666. [CrossRef]
8. Välimäki, V.; Holm-Rasmussen, B.; Alary, B.; Lehtonen, H.-M. Late Reverberation Synthesis Using Filtered Velvet Noise. *Appl. Sci.* **2017**, *7*, 483.
9. Bai, M.R.; Chung, C.; Wu, P.-C.; Chiang, Y.-H.; Yang, C.-M. Solution Strategies for Linear Inverse Problems in Spatial Audio Signal Processing. *Appl. Sci.* **2017**, *7*, 582. [CrossRef]
10. Gómez, D.M.M.; Astley, J.; Fazi, F.M. Low Frequency Interactive Auralization Based on a Plane Wave Expansion. *Appl. Sci.* **2017**, *7*, 558.

applied
sciences

MDPI

Review

Surround by Sound: A Review of Spatial Audio Recording and Reproduction

Wen Zhang [1,2,*], Parasanga N. Samarasinghe [1], Hanchi Chen [1] and Thushara D. Abhayapala [1]

[1] Research School of Engineering, College of Engineering and Computer Science, The Australian National University, Canberra 2601 ACT, Australia; prasanga.samarasinghe@anu.edu.au(P.N.S.); hanchi.chen@anu.edu.au (H.C.); thushara.abhayapala@anu.edu.au (T.D.A.)
[2] Center of Intelligent Acoustics and Immersive Communications, Northwestern Polytechnical University, 127 Youyi West Road, Xi'an 710072, Shaanxi, China
*   Correspondence: wen.zhang@anu.edu.au; Tel.: +61-2-6125-1438

Academic Editors: Woon-Seng Gan and Jung-Woo Choi
Received: 14 March 2017; Accepted: 11 May 2017; Published: 20 May 2017

Abstract: In this article, a systematic overview of various recording and reproduction techniques for spatial audio is presented. While binaural recording and rendering is designed to resemble the human two-ear auditory system and reproduce sounds specifically for a listener's two ears, soundfield recording and reproduction using a large number of microphones and loudspeakers replicate an acoustic scene within a region. These two fundamentally different types of techniques are discussed in the paper. A recent popular area, multi-zone reproduction, is also briefly reviewed in the paper. The paper is concluded with a discussion of the current state of the field and open problems.

Keywords: spatial audio; binaural recording; binaural rendering; soundfield recording; soundfield reproduction; multi-zone reproduction

---

1. Introduction

Spatial audio aims to replicate a complete acoustic environment, or to synthesize realistic new ones. Sound recording and sound reproduction are two important aspects in spatial audio, where not only the audio content but also the spatial properties of the sound source/acoustic environment are preserved and reproduced to create an immersive experience.

Binaural recording and rendering refer specifically to recording and reproducing sounds in two ears [1,2]. It is designed to resemble the human two-ear auditory system and normally works with headphones [3,4] or a few loudspeakers [5,6], i.e., the stereo speakers. However, it is a complicated process, as a range of localization cues, including the static individual cues captured by the individualized HRTF (head-related transfer function), dynamic cues, due to the motion of the listener, and environmental scattering cues, should be produced in an accurate and effective way for creating realistic perception of the sound in 3D space [7,8]. In gaming and personal entertainment, binaural rendering has been widely applied in Augmented Reality (AR)/Virtual Reality (VR) products like the Oculus Rift [9] and Playstation VR [10].

Soundfield recording and reproduction adopts physically-based models to analyze and synthesize acoustic wave fields [11–14]. It is normally designed to work with many microphones and a multi-channel speaker setup. With conventional stereo techniques, creating an illusion of location for a sound source is limited to the space between the left and right speakers. However, with more channels included and advanced signal processing techniques adopted, current soundfield reproduction systems can create a 3D (full-sphere) sound field within an extended region of space. Dolby Atmos (Hollywood, CA, USA) [15] and Auro 3D (Mol, Belgium) [16] are two well-known examples in this area, mainly used in commercial cinema and home theater applications.

This article gives an overview of various recording and reproduction techniques in spatial audio. Two fundamentally different types of rendering techniques are covered, i.e., binaural recording/rendering and soundfield recording/reproduction. We review both early and recent methods in terms of apparatus design and signal processing for spatial audio. We conclude with a discussion of the current state of the field and open problems.

2. Binaural Recording and Rendering

Humans have only two ears to perceive sound in a 3D space. Hence, it is intuitive to use two locally separated microphones to record audio as it is heard; and when played back through headphones or a stereo dipole, a 3D sound sensation is created for the listener. This is known as binaural recording, a specific approach of the two-channel stereo recording, where two microphones are placed at two ears either on a human head (known as *listening subject recording*) or an artificial head (known as *dummy head recording*).

The first binaural audio system was demonstrated in 1881 on a device called a Theatrophone introduced by a French engineer Clément Ader. However, there was not too much interest in this technology. It was not until 1974 that the first clip-in binaural microphones for a human subject was offered by Sennheiser and the first completely in-ear binaural microphones were offered by Sound Professionals in 1999. Nowadays, with the widespread availability of headphones and cheaper methods of recording, there has been a renewed interest in binaural spatial sound.

The sounds received in two ears are scattered and shaped by the human head, torso and ear geometry, resulting in spatial cues for binaural hearing being made available. This is fully captured by the HRTF, a filter defined in the spatial frequency domain that describes sound propagation from a specific point to the listener's ear, and a pair of HRTFs for two ears include all of the localization cues, such as the inter-aural time difference (ITD), inter-aural level difference (ILD), and spectral cues [17].

To generate spatial audio effects in binaural rendering, we can either use the recordings directly to simulate real objects or generate virtual objects by convolving a mono signal with the HRTFs corresponding to the virtual source positions. These are common practices used in Augmented Reality (AR) [3] and Virtual Reality (VR) [18].

2.1. Binaural Recording and HRTF Measurement

In both binaural recording and the HRTF measurement, a typical setup is to record acoustic properties of the dummy head or head and torso simulator (HATS), which are designed based on the average dimension of a human head/torso and have two high fidelity microphones inserted within each ear to record the two-ear signals. Some widely used binaural recording packages include (a) Dummy head: Neumann KU-100 (Berlin, Germany) [19], (b) head-and-torso simulator (HATS): Brüel & Kjær 4128D/C (Nærum, Denmark) [20], and (c) 3Dio Free Space Binaural Microphone (Vancouver, WA, USA) [21], as shown in Figure 1.

The dummy head recordings are supposed to achieve a good trade-off between individual variability and good spatial perception. However, as each human body has a unique size and shape features, it could cause confusion, especially in terms of elevation localization using the dummy head recordings directly for everyone's ears [22]. This problem is also known as one of the HRTF characteristics—that HRTFs differ greatly from person to person.

To understand the perceptual cues to spatial hearing, HRTFs are measured on both the HATS and listening subjects, and further analyzed for its characteristic variation as a function of source positions. The source positions are typically sampled at a predefined set of azimuths and elevations on a spherical surface with a radius of 1–2 m, which is when the HRTFs are assumed to be distance independent [23,24]. The whole measurement process, which involves the measurement system design/setup, is logistically complicated and takes at least 15–20 min to complete [25].

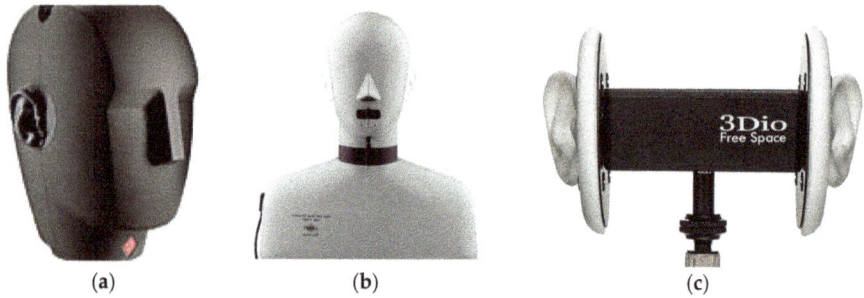

Figure 1. Binaural recording packages. (**a**) dummy head: Neumann KU-100; (**b**) head-and-torso simulator (HATS): Brüel Kjær 4128D/C; (**c**) binaural microphones: 3Dio Free Space.

2.1.1. Measurement Resolution

One problem for the HRTF measurement is that there is a lack of recognized standards, especially in terms of the measurement resolution, resulting in the fact that databases with varying sample positions and the number of samples [26]. Even though interpolation is widely used to generate HRTFs at non-measured positions [27], the spatial resolution for HRTF measurement greatly influences the success and quality of binaural rendering.

For azimuth sampling, it is shown that an angular spacing of 5° or less is necessary to reconstruct the HRTF data up to 20 kHz in the horizontal plane (the plane perpendicular to a vertical direction and at the same height of a person's ear) [28]. Zhong and Xie [29] further studied the maximal azimuthal resolution, or equivalently the minimum number of azimuthal measurements (MNAM), at each elevation plane. The MNAM increases with increasing frequency, i.e., from five measurements at the lowest frequency to more than 60 measurements at 20 kHz in the horizontal plane, while, for a fixed frequency, it decreases as the elevation deviates from the horizontal plane. Minnaar et al. [30] investigated the directional resolution of HRTF measurement in the horizontal, frontal, and median plane. A resolution of 8° over the sphere was deemed to be enough without introducing audible interpolation errors in binaural rendering, but this resolution threshold was largely based on listening experiments.

The Spherical Harmonics, an orthogonal basis function on the sphere, has been widely used for HRTF data representation [31,32], based on which the HRTF dimensionality [33], measurement resolution [34], and sampling scheme [35] are proposed. The Spherical Harmonics based analysis shows that, in order to capture the HRTF variations for the entire audible frequency range (i.e., 20 Hz to 20 kHz), the measurement grid should have a spatial resolution of 3–4° in elevation and azimuth, with fewer measurement points required towards polar directions.

2.1.2. Test Signal and Post-Processing

Various test signals with corresponding signal processing methods have been used for HRTF measurement, such as exponential swept-sine, maximum length sequences (MLS), Golay codes, etc. Post-processing is necessary to remove the responses of the measurement system as well as reflections from the measurement apparatus or reverberation for measurements performed in a semi-anechoic chamber. The swept-sine based method with appropriate time windowing has been shown effective to remove reflections [36]; in addition, a common practice is to divide the measured transfer function by the spatially averaged mean response, i.e., the Common Transfer Function, which is irrelevant to direction, in order to extract the so-called Directional Transfer Function [1]. Another major focus is to improve the measurement efficiency, in order to minimize the recording time but without reducing the spatial resolution. More specifically, the proposed approaches include the measurement via the reciprocity method [37], the multiple exponential sweep method [25], the continuous measurement method [38], and the HRTF acquisition with unconstrained movements of human subjects [39].

An ongoing project, the "Club Fitz", was initiated in 2004 with the aim to compare HRTF databases from laboratories across the world [27]. The database consists of physical measurements and numerical simulations performed on the same dummy head (Neumann KU-100), where HRTFs are converted to the widely-adopted open standard SOFA file format [40]. Investigations on 12 different HRTFs from 10 laboratories showed an observation of large ITD variations (up to 235 µs) and spectral magnitude variations (up to 12.5–23 dB) among these databases. This further demonstrates the profound impact of measurement systems and signal processing techniques on HRTF data acquisition.

2.2. Binaural Rendering

It requires rendering static, dynamic, and environmental cues within the audio stream to create an immersive spatial audio experience [41]. As stated earlier, the static cue is fully captured by the HRTF, which varies significantly between people [42]. Thus, it is normally claimed to use individualized HRTF for binaural rendering [4]. In addition, dynamic and reverberation cues must be added to generate a virtual audio scene with maximum fidelity—for example, with appropriate externalization and localization of the virtual source. Figure 2 shows a binaural rendering system for headphones with all of the signal processing techniques reviewed in this section.

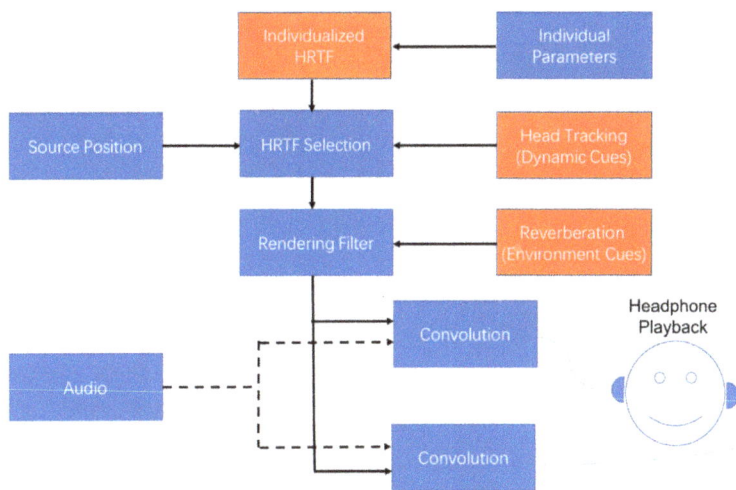

Figure 2. The signal flow of a binaural rendering system.

2.2.1. Individualized HRTF

The direct acoustic measurement on a listening subject is the most accurate and important way of obtaining the individualized HRTFs, which include all of the relevant static cues, such as ITD, ILD, and spectral cues, for sound localization in 3D, especially in terms of azimuth and elevation. However, these measurements are discrete over space and frequency, and thus should be used with interpolation for binaural rendering [43,44]. The methods can be classified as *local interpolation*, where HRTFs are computed from the measurements at its adjacent directions, i.e., bilinear interpolation [45] and inter-positional transfer function based interpolation [46], or *global interpolation*, where HRTFs are computed from all measurements based on a model using appropriate basis functions, i.e., the Spherical Harmonic based interpolation [26] and principle component analysis based interpolation [47]. Especially, the Spherical Harmonics, which are spatially continuous and orthogonal over sphere, are now widely used for HRTF representation [33,48]; based on this continuous representation, a novel spatial audio rendering technique for area and volumetric sources in VR was developed [18].

Due to the fact that HRTF individualization is strongly related to the anthropometry of a person, methods have been proposed for HRTF personalization by choosing a small set of anthropometry features with a pre-trained model [49–54]. The training was established based on a direct linear or nonlinear relationship between the anthropometric data and the HRTFs, where the first step is to reduce the HRTF data dimensionality. The effectiveness of this kind of model is heavily dependent on the selection of the anthropometry features [55]. For example, the study in [53] investigated the contribution of the external ear to the HRTF and proposed a relation between the spectra notches and pinna contours.

Recently, researchers from Microsoft proposed an indirect method for HRTF synthesis using sparse representation [56,57]. The main idea is firstly to learn a sparse representation of a person's anthropometric features from the training set and then apply this sparse representation directly for HRTF synthesis. Further work shows that, in this method, the pre-processing and post-processing are crucial for the performance of HRTF individualization [58].

Some other techniques used for HRTF individualization include perceptual feedback (tunning-based HRTF manipulation) [59], binaural synthesis using frontal projection headphones [60], and by finite element modelling of the human head and torso to simulate the source-ear acoustic path [61]. Especially, a significant amount of work has been done in HRTF numerical simulations, such as using boundary element method [62,63] and finite difference time domain simulation [64], or estimating key localization cues of the HRTF data from anthropometric parameters [65]. Recently, a collection of 61-subject HRTFs and the high-resolution surface meshes of these subjects obtained from magnetic resonance imaging (MRI) data are released in the SYMARE database [66]. The predictions using the fast-multiple boundary element method show a high correlation with the measurement ones, especially at frequencies below 10 kHz.

Notice that, due to the logistical challenges of obtaining the individualized HRTF, the database of non-individualized HRTFs are often used in the acoustic research [67,68]. In addition, the perception using non-individualized HRTFs can be strengthened if dynamic cues are appropriately included as shown in the following section.

2.2.2. Dynamic Cues

The dynamic cue arising from the motion of the listener, which changes the relative position of the source, can reinforce localization. Studies on this cue show that the well-known front-back confusion problem in static binaural rendering disappears when the listener can turn their head to assist localization [69]. It also demonstrates that the perception using non-individualized HRTFs can be strengthened if dynamic cues are appropriately included [42]. Thus, it is necessary in binaural rendering to have the simulated scene change with the listener movement. This is normally achieved using a low-latency head tracking system, based on computer vision techniques (such as regular cameras) to estimate the orientation of the listener's head. In smartphones, the low-cost sensors, such as accelerometers and gyroscopes, can be used for head-tracking.

2.2.3. Environment Cues

Using the static cue, i.e., the HRTF, along to render the binaural signals will have one big problem that the recreated sounds are not well externalized. Dummy head recordings performed in reverberant rooms reveal that reverberation effects are essential for an immersive 3D sensation. For example, the direct-to-reverberant energy ratio enables source distance estimation in the room [70]. Thus, it is important to incorporate environmental scattering cues to achieve good externalization or distance perception [70,71].

The environmental scattering is characterized by the room transfer function (RTF) or room impulse response (RIR), a function of the source and receiver positions that includes effects due to reflection at the boundaries, sound absorption, diffraction, and room resonance. The RIR can be separated into two parts, a small number of early strong reflections and very large numbers (hundreds or even thousands) of late weak reverberation. It is believed that the early reflections are helpful for source externalization, thus must be convolved with the appropriate HRTF for the corresponding image source direction [41]. The late reverberation, however, is directionless and thus can be approximated by a generic model for a given room geometry. It is clear that room reverberation has a strong impact on RIR simulation. Physically based reverberation models are widely used in virtual reality to reproduce the acoustics of a given real or virtual space. One example is the image source model for computing the early reflections of RIR in a rectangular room [72]. For a review on different artificial reverberation simulation methods, see, e.g., [73].

The rendering filter is normally constructed as a Finite-Impulse-Response (FIR) filter, where only the direct path and the first few reflections are computed in real time and the rest of the filter is computed once given the room geometry and boundary. For playback synthesis, simulation of multiple virtual sources at different positons can be performed in parallel, where the audio stream of each source is convolved with the corresponding FIR filter. The convolution can be performed either in the time domain to have a zero-processing lag but with high computational complexity or in the frequency domain for low computational complexity but with unavoidable latency. The synthesized results are mixed together for playback. To further reduce computational complexity, binaural rendering on headphones [74] and for interactive simulations [75] were implemented using a Graphical Processing Unit (GPU).

2.2.4. Rendering by Headphones or Loudspeakers

Headphones are the natural and most effective way to reproduce binaural sounds. Equalization is commonly applied to make them meet an ideal target response. The original reference was that of a frontal free field (i.e., free-field equalization), but a preference for a diffuse field with random incidence was later proposed (i.e., diffuse-field equalization) [76] and now has been widely adopted by the headphone manufactures. In terms of binaural rendering itself, the recent results show that only presenting the variation of the sound spectrum due to the source position changes can provide the localization cues from the perception point of view [2]. However, from a practical application point of view, compensation is necessary for accurate binaural synthesis over headphones, either through individual measurements [77] or non-individual recordings of the headphone transfer function (HpTF) [78]. The metrics for evaluating the perception of equalized HpTF was recently proposed [79].

Binaural rendering through loudspeakers is intrinsically affected by the crosstalk and thus requires a pre-processing crosstalk cancellation system (CCS) [5]. This system, however, is sensitive to the listener's head movement or misalignment given limited sweet spot. The research has focused on developing robust CCS for a two-speaker or multiple-speaker setup [6,80]. The optimal loudspeaker positions for CCS are of interest, among which an effective dual-speaker system called "stereo dipole" [81] and the optimal loudspeaker array configuration [82,83] were developed. Other recent work includes investigating the sound source localization performance provided by CCS [84] and designing CCS with a head tracker based on an online monitoring of ITD [85].

3. Soundfield Recording and Reproduction

The system that uses many loudspeakers for soundfield reproduction is principally different from the above mentioned binaural rendering. The soundfield reproduction system creates spatial audio effects (i.e., exact positions of the sound sources) within an extended region of space while a binaural audio system produces natural 3D sound at the listener's ears without an expensive set of loudspeakers. The recording and reproduction techniques for soundfield reproduction are described in the following.

3.1. Soundfield Representation

For a soundfield within a source free region, the sound pressure at any point can be expressed in the 3D spherical coordinate as [86]

$$P(x, w) \approx \sum_{n=0}^{N} \sum_{m=-N}^{N} \alpha_n^m(w) j_n(kr) Y_n^m(\hat{x}) \tag{1}$$

where $\alpha_n^m(w)$ are soundfield coefficients corresponding to the mode index (n, m), $j_n(kr)$ are Spherical Bessel functions of the first kind representing the mode amplitude at radius r, $Y_n^m(\hat{x})$ are Spherical Harmonics that are functions of the angular variables. Spherical Harmonics and Spherical Bessel functions together represent the propagation modes. Especially, due to the low-pass characteristics of the Spherical Bessel function, given the radius of a region of interest (ROI) r_0 and the wavenumber k, the truncation number $N \approx \lceil kr_0 \rceil$ [87]. This means that the soundfield within the ROI can be represented by a finite number of, i.e., $D = (N + 1)^2$, coefficients.

In soundfield recording and analysis, normally, the soundfield coefficients $\alpha_n^m(w)$ and the Spherical Bessel functions $j_n(kr)$ are integrated as one component. Equation (1) becomes an expansion with respect to Spherical Harmonics solely. The expansion order N is also the order of the system. For example, when $N = 1$, the system is called the first-order system (or the widely known Ambisonics); and when $N \geq 2$, the system is called the higher-order system (or higher-order Ambisonics).

3.2. 3D Soundfield Recording

The soundfield microphone, arranged as a microphone unit composed of multiple microphone capsules and a signal processor, with 3D pick up capability is commonly used for soundfield recording. This microphone normally has a 3D geometry, such as the B-Format Ambisonic microphone [88] and EigenMike (a spherical microphone array) [89], as shown in Figure 3. The design of the microphone for soundfield recording is based on the decomposition of the 3D soundfield using Spherical Harmonics, i.e., Equation (1).

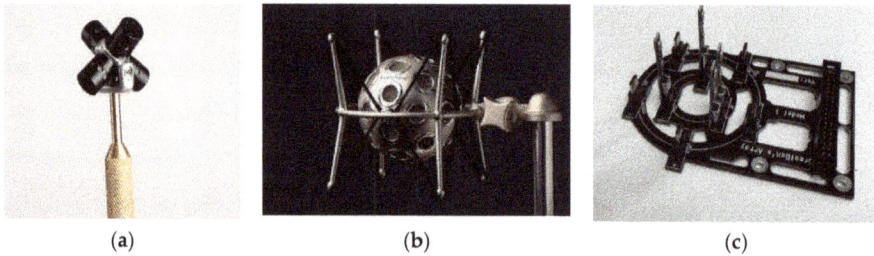

| (a) | (b) | (c) |

Figure 3. 3D soundfield recording microphones. (**a**) TetraMic; (**b**) EigenMike; (**c**) Planar Microphone Array.

3.2.1. 3D Microphone Array

The Ambisonic microphone was firstly designed by Dr. Jonathan Halliday at Nimbus Records (Monmouth, England) for recording the first-order Spherical Harmonics decomposition of a soundfield, i.e., B-format. The original design, known as a native or Nimbus/Halliday microphone array, has three coincident microphones, i.e., an omnidirectional microphone, one forward-facing, and one left-facing figure of eight microphone, to record the W, X, and Y components separately. Since it is impossible to build a perfectly coincident microphone array, Michael A. Gearzon developed an improved version, the tetrahedral microphone, which has four cardioid or sub-cardioid microphone capsules arranged in a tetrahedron and equalized for uniform diffuse-field response [90]. The recorded signals later are converted to B-format through a matrix operation. On the market, TetraMic developed in 2007 [88] (as shown in Figure 3a) the first portable single point, stereo and surround sound Ambisonic microphone.

Above the first-order, multiple microphones with very sophisticated digital signal processing are required to obtain the higher-order expansion components directly. An ideal higher order microphone would be comprised of a continuous spherical microphone array, which can decompose the measured 3D soundfield into its spherical harmonic coefficients. In practice, spherical microphone arrays are implemented with a discrete array that uniformly samples a spherical surface [11,12]. To record an Nth order soundfield, the minimum sampling requirement along the spherical surface is $(N + 1)^2$. Note that, since the soundfield order $N \approx \lceil kr_0 \rceil$ is proportional to the frequency and the ROI size (or radius), the number of microphones required in the array is also proportional to those two quantities. The Eigenmike [89], as shown in Figure 3b, is an example for a commercially available 4th order microphone array, which consists of 32 microphones uniformly spaced on a spherical baffle of radius 4 cm. Instead of using a spherical geometry, a novel array structure consisting of a set of parallel circular microphone arrays to decompose a wavefield into spherical harmonic components was proposed recently [91].

3.2.2. Plannar Microphone Array

Planar microphone arrays are widely used for applications such as beamforming and circular harmonic analysis of 2D spatial sound. However, generally speaking, due to the limitation of the 2D geometry, a single planar array cannot capture full 3D spatial sound.

A special planar microphone array configuration that is capable of recording 3D spatial sound was proposed by Chen et al. [92], as shown in Figure 3c. In this configuration, the combined use of omni-directional and vertically placed first order (cardioid or differential) microphones enables detection of the acoustic particle velocity in the vertical direction, which can be used to solve for the spatial soundfield components that are normally "invisible" to planar microphone arrays. It is shown that this planar microphone array offers the same capability as a spherical microphone array of the same radius, in terms of spatial sound recording.

3.2.3. Array of Higher Order Microphones

While higher order soundfield recording can be conveniently achieved via the microphone designs discussed above, with increasing size of the desired spatial region and increasing frequency, the array's minimum microphone requirement increases to impractical numbers. A recently proposed method to overcome this limitation is via utilizing an array of higher order microphones [14,93]. For example, a distributed array of fourth order Eigenmike/planar higher order microphones can replace a spherical array of 121 omnidirectional microphones. Such a design is highly suitable for spatial soundfield recording over large regions because it significantly reduces the implementation complexity of the array (reduced amount of cabling, reduced spatial samples, etc.) at the expense of increased complexity at each microphone unit.

3.3. Soundfield Reproduction

Spatial soundfield reproduction aims to create an immersive soundfield over a predefined spatial region so that the listener inside the region can experience a realistic but virtual replication of the original soundfield. This is achieved by controlling the placement of a set of loudspeakers usually put on the boundary that encloses the spatial region of interest and deriving the signals emitted from the loudspeakers. Loudspeaker array design and audio processing are two key aspects to control sound radiation and to deal with the complexity and uncertainty associated with soundfield reproduction.

3.3.1. Reproduction Methods

Given multi-channel (or multi-microphone) recordings of an acoustic scene, the *channel-based reproduction* plays these recordings with certain down/up mixing to replicate the scene. The *object-based reproduction*, on the other hand, is based on the location information of loudspeakers and the objects (or virtual source) in order to determine which loudspeakers are used and their driving signals for playing back the object's audio. The object-based reproduction formed on the panning law can render the audio objects in real time. Two well-known object-based pieces of audio equipment are stereo systems that consist of two channels, left and right, and surround sound systems that consist of multiple channels surrounding the listener. The stereo can provide good imaging (the perceived spatial location of the sound source) in the front quadrant while the surround sound can offer imaging around the listener.

Based on the amplitude panning principle [94,95], the audio object can be positioned in an arbitrary 2D or 3D setup using the vector base amplitude panning (VBAP) or distance-based amplitude panning (DBAP). The VBAP is based on a triplet-wise panning law, where three loudspeakers are arranged in a triangle layout to generate sound imaging [96]. The DBAP only takes the positions of the virtual source and loudspeakers into account for reproduction and thus is well suited to irregular loudspeaker layouts [97]. In object-based reproduction, the listener normally is required to be at the centre of the speaker array (i.e., the sweep spot position) where the generated audio effect is best; DBAP, however, is a noticeable exception to this restriction.

Table 1. Summary of soundfield reproduction methods and commercial systems.

Method	Typical Systems	Characteristics
Stereo/Surround	Dolby Stereo, Dolby 5.1/7.1 Dolby Atmos, NHK 22.2 Auro 3D	■ Channel-based Object-based reproduction ■ Amplitude/phase encoding
VBAP, DBAP	Software Demo [98]	■ Object-based reproduction ■ Amplitude encoding
Ambisonics (B-format) HOA	Youtube 360° VR Audio Kit [99]	■ Model-based reproduction ■ Amplitude/phase encoding
WFS	IOSONO	■ Model-based reproduction ■ Amplitude/phase encoding
Inverse Filtering	–	■ Channel-based reproduction ■ Amplitude/phase encoding

VBAP: vector base amplitude panning; DBAP: distance-based amplitude panning; HOA: higher order Ambisonics; WFS: wave-field synthesis.

The first known demonstration of reproducing a soundfield within a given region of space was conducted by Camras at the Illinois Institute of Technology in 1967, where loudspeakers were distributed on the surface enclosing the selected region and the listeners can move freely within the region [100]. Later, the well-known Ambisonics was designed based on Huygen's principle for more advanced spatial soundfield reproduction over a large region of space [101,102]. The system is based on the zero and first order spherical harmonic decomposition of the original soundfield into four channels, i.e., Equation (1), and from a linear combination of these four channels to derive the loudspeaker driving signals. This low-order system is optimum at low frequencies but less accurate at high frequencies and when the listener is away from the center point. Higher Order Ambisonics (HOA) based on the higher order decomposition of a soundfield, such as cylindrical two-dimensional (2D or horizontal plane) harmonic [87,103,104] or spherical three-dimensional (3D or full sphere) harmonic [13,105,106] decomposition, was developed especially for high reproduction frequencies and large reproduction regions. Based on the same principle, soundfield reproduction using the plane wave decomposition approach was recently proposed [107–109].

Wave-Field Synthesis (WFS) is another well-known sound reproduction technique initially conceived by Berkhout [110,111]. The fundamental principle is based on the Kirchhoff–Helmholtz integral to represent a soundfield in the interior of a bounded region of the space by a continuous distribution of monopole and normally oriented dipole secondary sources, arranged on the boundary of that region [112]. An array of equally spaced loudspeakers is used to approximate the continuous distribution of secondary sources. Reproduction artifacts due to the finite size of the array and the spatial discretization of the ideally continuous distribution of secondary sources were investigated [113]. The WFS technique has been mainly implemented in 2D sound reproduction using linear and planar arrays [114,115], for which a 2.5D operator was proposed to replace the secondary line sources by point sources (known as 2.5D WFS) [112,116,117]. In WFS, a large number of closely spaced loudspeakers is necessary.

For arbitrarily placed loudspeakers, a simple approach in sound reproduction known as inverse filtering is to use multiple microphones as matching points based on the least-square match to derive the loudspeaker weights [118], with the knowledge of the acoustic channel between the loudspeakers and matching points, i.e., the RTF. Tikhonov regularization is the common method for obtaining loudspeaker weights with limited energy and also for improving the system robustness. In addition, the fluctuation of the RTF requires an accurate online estimation of the inverse filter coefficients in this method [119].

Table 1 summarises the above mentioned soundfield reproduction techniques. A comparison of different soundfield reproduction techniques especially from the perception point of view was presented in the review paper [120]. One of the current research interests is to further improve these techniques with a thorough perceptual assessment [121].

3.3.2. Listening Room Compensation

In an acoustic reverberant environment, the multi-path propagation effect introduces echoes and spectral distortions into the generated soundfield. Room equalization has been studied in theory and applied in practice to cancel this effect for sound reproduction in cinema halls, home theaters and teleconferencing applications [122]. An efficient method for correcting room reverberation is by using active room compensation, that is, by applying compensation signals to the loudspeaker input to make a reverberant room problem look like an anechoic room problem. This method, however, requires the knowledge of the underlying acoustic system, i.e., the RIR (room impulse response) or the RTF (room transfer function).

Compensation schemes based on RIR/RTF modelling are theoretically capable of good performance; however, imperfections in the modelling process generally lead to a reduction of dereverberation performance in real environments [123]. This is further exacerbated by non-static room conditions, e.g., time-variant room responses because of a temperature variation or source/receiver position variations [124,125]. For soundfield reproduction in time-varying environments, a multiple-loudspeaker-multiple-microphone setup is employed and the RIR/RTF of this acoustic system must be determined online in an adaptive manner. In addition, as the purpose here is to reproduce sound over a region of interest, room compensation should be achieved essentially within the entire region as well.

In massive multichannel soundfield reproduction systems, for which the number of loudspeakers and microphones are large, active room compensation can be solved computationally and efficiently by using a wave-domain approach [103,126]. The principle of wave-domain signal representation is to use fundamental solutions of the Helmholtz wave-equation as basis functions to express a wave field over a spatial region as shown in Label (1). Processing directly on the decomposition coefficients therefore controls sound within the region.

The wave-domain adaptive filtering (WDAF) approach transforms the signals at the microphones and the loudspeaker signals into the wave domain, and then adaptively calculates the loudspeaker compensations signals (Figure 4). Especially, in the wave domain, the compensation filter is forced to be diagonal, and each diagonal entry can be determined from the decoupled adaptive filters [127,128]. This technique results in parallel implementation and significantly reduces the complexity of the adaption process.

More complex approaches for room compensation are by using fixed or variable directivity higher-order loudspeakers to minimize the acoustic energy directed towards the walls of a room [129,130], or by exploiting room reflections to reproduce a desired soundfield [131–133].

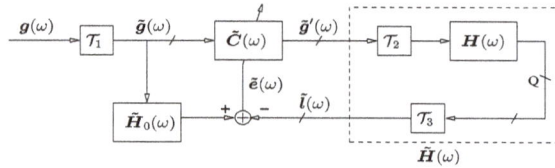

Figure 4. The listening room compensation using WDAF (wave-domain adaptive filtering). The free-field transformed loudspeaker signals $\tilde{g}$ are used in a reverberant room with the filter matrix $\tilde{C}$ to compensate the RTFs in matrix $\tilde{H}$, $\mathcal{T}_1$, $\mathcal{T}_3$ and $\mathcal{T}_2$ represent the forward and backward WDAF, respectively.

4. Multi-Zone Sound Reproduction

Multi-zone reproduction aims to extend spatial sound reproduction over multiple regions so that different listeners can enjoy their audio material simultaneously and independently of each other but without physical isolation or using headphones (Figure 5). The concept of multi-zone soundfield control has recently drawn attention due to a whole range of audio applications, such as controlling sound radiation from a personal audio device, creating independent sound zones in different kinds of enclosures (such as shared offices, private transportation vehicles, exhibition centres, etc.), and generating quiet zones in a noisy environment. A single array of loudspeakers is used, where sound zones can be placed at any desired location and the listener can freely move between zones; thus, the whole system provides significant freedom and flexibility.

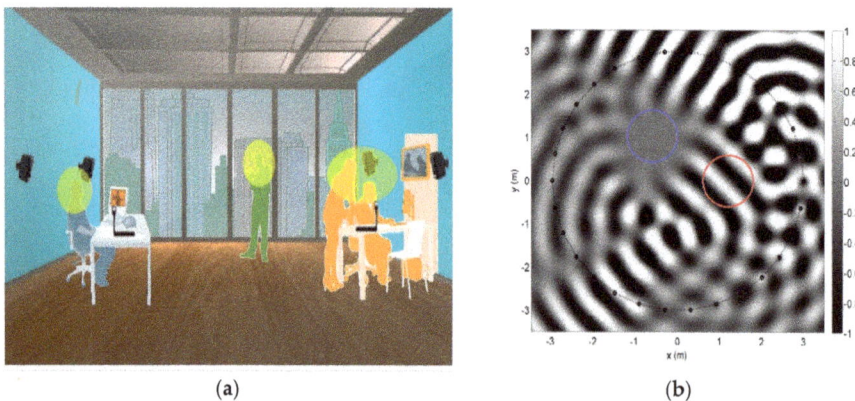

(a)

(b)

Figure 5. (a) an illustration of multi-zone sound reproduction in an office environment; (b) a plane wave of 500 Hz from 45° is reproduced in the bright zone (red circle) with a dark or quiet zone (blue circle) generated using a circular array of 30 loudspeakers [134].

The multi-zone reproduction was firstly formulated as creating two kinds of sound zones, the bright zone within which certain sounds with high acoustic energy are reproduced and the dark zone (or the quiet zone) within which the acoustic energy is kept at a low level [135]. The proposed method is to maximise the ratio of the average acoustic energy density in the bright zone to that in the dark zone, which is known as the acoustic contrast control (ACC) method. Since then, different forms of contrast control based on the same principle have been proposed, including an acoustic energy difference formulation [136], direct and indirect acoustic contrast formulations using the Lagrangian [137]. The technique has been implemented in different personal audio systems in an anechoic chamber [138,139] or in a car cabin [140]; over 19 dB contrast was achieved under the ideal condition, while, for real-time systems in the car cabin, the acoustic contrast was limited to a maximum value of 15 dB. This contrast control method, however, does not impose a constraint on the phase of the soundfield and thus cannot control the spatial aspects of the reproduced soundfield in the bright zone. A recent work by Coleman et al. proposed refining the cost function of the ACC with the aim of optimizing the extent to which the reproduced soundfield resembles a plane wave, thus optimising the spatial aspects of the soundfield [141]. Another issue in ACC is the self-cancellation problem, which results in a standing wave produced within the bright zone [142].

The pressure matching (PM) approach aims to reproduce a desired soundfield in the bright zone while producing silence in the dark zone [143]. The approach uses a sufficiently dense distribution of microphones within all the zones as the matching points and adopts the least-squares method to control the pressure at each point. A constraint on the loudspeaker weight energy (or the array effort) is added to control the sound leakage outside the sound zones and to ensure the implementation is robust against speaker positioning errors and changes in the acoustic environment [144]. When the desired soundfeld in the bright zone is due to a few virtual source directions, the multi-zone sound control problem can be solved using a compressive sensing idea dolwhere the loudspeaker weights are regularised with the L_1 norm. This results in only a few loudspeakers placed closely to the virtual source directions being activated for reproduction [145,146]. More recent works have been focusing on the combination of the ACC and PM formulations using the Lagrangian with a weighting factor to tune the trade-off between the two performance measures, i.e., the acoustic contrast and bright zone error (i.e., the reproduction error within the bright zone) [147–149]. The idea of performing time-domain filters for personal audio was recently investigated [150].

The multi-zone reproduction is formulated in the modal domain based on representing the soundfield within each zone through a spatial harmonic expansion, i.e., Equation (1). The local sound field coefficients are then transformed to an equivalent global soundfield coefficient using the harmonic translation theroem, from which the loudspeaker signals are obtained throung the mode matching [151]. The modal-domain approach can provide theoretical insights into the multi-zone problem. For example, through the modal domain analysis, a theoretical basis is established for creating two sound zones with no interference [152]. Modal-domain sparsity analysis shows that a significantly reduced number of microphone points could be used quite effectively for multi-zone reproduction over a wide frequency range [146]. The synthesis of soundfields with distributed modal constraints and quiet zones having an arbitrary predefined shape have also been investigated [153,154]. Based on modal-domain analysis, a parameter, the coefficient of realisability, is developed to indicate the achievable reproduction performance given the sound zone geometry and the desired soundfield in the bright zone [155].

5. Conclusions

In this article, we presented the recording and reproduction techniques for spatial audio. The techniques that have been explored include binaural recording and rendering, soundfield recording and reproduction, and multi-zone reproduction. Binaural audio that works with a pair of headphones or a few loudspeakers has the advantage of being easily incorporated into the personal audio products, while soundfield reproduction techniques that rely on many loudspeakers to control sounds within a

Appl. Sci. **2017**, *7*, 532

region are mainly used in commercial and professional audio applications. Both techniques strive for the best immersive experience; however, in soundfield reproduction, the spatial audio effect is not restricted to a single user or some spatial points. Therefore, a wide range of emerging research directions has appeared in this area, such as multi-zone reproduction and active noise control over space.

In binaural techniques, generating individualized dynamic auditory scenes are still challenging problems. Future research directions include adaptation to an individualized HRTF based on the available datasets, real-time room acoustic simulations for a natural AR/VR listening experience, and sound source separation for augmented audio reality. In soundfield reproduction, interference mitigation and room compensation robust to acoustic environment changes remain as the major challenges. Further opportunities exist in higher-order surround sound using an array of directional sources and wave-domain active room compensation to perform sound field reproduction in reverberant rooms.

Acknowledgments: The authors acknowledge National Natural Science Foundation of China (NSFC) No. 61671380 and Australian Research Council Discovery Scheme DE 150100363.

Conflicts of Interest: The authors declare no conflict of interest.

References

1. Hammershoi, D.; Møller, H. Methods for binaural recording and reproduction. *Acta Acust. United Acust.* **2002**, *88*, 303–311.
2. Møller, H. Fundamentals of binaural technology. *Appl. Acoust.* **1992**, *36*, 171–218. [CrossRef]
3. Ranjan, R.; Gan, W.-S. Natural listening over headphones in augmented reality using adaptive filtering techniques. *IEEE/ACM Trans. Audio Speech Lang. Process.* **2015**, *23*, 1988–2002. [CrossRef]
4. Sunder, K.; He, J.; Tan, E.-L.; Gan, W.-S. Natural sound rending for headphones: Integration of signal processing techniques. *IEEE Signal Process. Mag.* **2015**, *23*, 100–114. [CrossRef]
5. Bauer, B.B. Stereophonic earphones and binaural loudspeakers. *J. Acoust. Soc. Am.* **1961**, *9*, 148–151.
6. Huang, Y.; Benesty, J.; Chen, J. On crosstalk cancellation and equalization with multiple loudspeakers for 3-D sound reproduction. *IEEE Signal Process. Lett.* **2007**, *14*, 649–652. [CrossRef]
7. Ahveninen, J.; Kopčo, N.K.; Jääskeläinen, I.P. Psychophysics and neuronal bases of sound localization in humans. *Hear. Res.* **2014**, *307*, 86–97. [CrossRef] [PubMed]
8. Kolarik, A.J.; Moore, B.C.J.; Zahorik, P.; Cirstea, S.; Pardhan, S. Auditory distance perception in humans: A review of cues, development, neuronal bases and effects of sensory loss. *Atten. Percept. Pyschophys.* **2016**, *78*, 373–395. [CrossRef] [PubMed]
9. Oculus Rift ǀ Oculus. Available online: https://www.oculus.com/rift/ (accessed on 26 April 2017).
10. PlayStation VR—Virtual Reality Headset for PS4. Available online: https://www.playstation.com/en-us/explore/playstation-vr/ (accessed on 26 April 2017).
11. Abhayapala, T.D.; Ward, D.B. Theory and design of high order sound field microphones using spherical microphone array. In Proceedings of the IEEE International Conference on Acoustics, Speech, and Signal Processing (ICASSP), Orlando, FL, USA; 2002; pp. 1949–1952.
12. Meyer, J.; Elko, G. A highly scalable spherical microphone array based on an orthonormal decomposition of the soundfield. In Proceedings of the IEEE International Conference on Acoustics, Speech, and Signal Processing (ICASSP), Orlando, FL, USA; 2002; pp. 1781–1784.
13. Poletti, M.A. Three-dimensional surround sound systems based on spherical harmonics. *J. Audio Eng. Soc.* **2005**, *53*, 1004–1025.
14. Samarasinghe, P.N.; Abhayapala, T.D.; Poletti, M.A. Spatial soundfield recording over a large area using distributed higher order microphones. In Proceedings of the IEEE Workshop on Applications of Signal Processing to Audio and Acoustics (WASPAA), New Paltz, NY, USA; 2011; pp. 221–224.
15. Dolby Atmos Audio Technology. Available online: https://www.dolby.com/us/en/brands/dolby-atmos.html (accessed on 26 April 2017).
16. Auro-3D/Auro Technologies: Three-dimensional Sound. Available online: http://www.auro-3d.com/ (accessed on 26 April 2017).

17. Cheng, C.I.; Wakefield, G.H. Introduction to Head-Related Transfer Functions (HRTFs): Representations of HRTFs in time, frequency and space. *J. Audio Eng. Soc.* **2001**, *49*, 231–249.

18. Schissler, C.; Nicholls, A.; Mehra, R. Efficient HRTF-based spatial audio for area and volumetric sources. *IEEE Trans. Vis. Comput. Gr.* **2016**, *22*, 1356–1366. [CrossRef] [PubMed]

19. Neumann—Current Microphones, Dummy Head KU-100 Description. Available online: http://www.neumann.com/?lang=en&id=current_microphones&cid=ku100_description (accessed on 10 March 2017).

20. Brüel & Kjær—4128C, Head and Torso Simulator HATS. Available online: http://www.bksv.com/Products/transducers/ear-simulators/head-and-torso/hats-type-4128c?tab=overview (accessed on 10 March 2017).

21. 3Dio—The Free Space Binaural Microphone. Available online: http://3diosound.com/index.php?main_page=product_info&cPath=33&products_id=45 (accessed on 10 March 2017).

22. Wenzel, E.M.; Arruda, M.; Kistler, D.J.; Wightman, F.L. Localization using nonindividualized head-related transfer functions. *J. Acoust. Soc. Am.* **1993**, *94*, 111–123. [CrossRef] [PubMed]

23. Brungart, D.S. Near-field virtual audio displays. *Presence Teleoper. Virtual Environ.* **2002**, *11*, 93–106. [CrossRef]

24. Otani, M.; Hirahara, T.; Ise, S. Numerical study on source distance dependency of head-related transfer functions. *J. Acoust. Soc. Am.* **2009**, *125*, 3253–3261. [CrossRef] [PubMed]

25. Majdak, P.; Balazs, P.; Laback, B. Multiple exponential sweep method for fast measurment of head related transfer functions. *J. Audio Eng. Soc.* **2007**, *55*, 623–630.

26. Andreopoulou, A.; Begault, D.R.; Katz, B.F.G. Inter-laboratory round robin HRTF measurement comparison. *IEEE J. Sel. Top. Signal Process.* **2015**, *9*, 895–906. [CrossRef]

27. Duraiswami, R.; Zotkin, D.N.; Gumerov, N.A. Interpolation and range extrapolation of HRTFs. In Proceedings of the IEEE International Conference on Acoustics, Speech, and Signal Processing (ICASSP), Montreal, QC, Canada; 2004; pp. 45–48.

28. Ajdler, T.; Faller, C.; Sbaiz, L.; Vetterli, M. Sound field analysis along a circle and its applications to HRTF interpolation. *J. Audio Eng. Soc.* **2008**, *56*, 156–175.

29. Zhong, X.L.; Xie, B.S. Maximal azimuthal resolution needed in measurements of head-related transfer functions. *J. Acoust. Soc. Am.* **2009**, *125*, 2209–2220. [CrossRef] [PubMed]

30. Minnaar, P.; Plogsties, J.; Christensen, F. Directional resolution of head-related transfer functions required in binaural synthesis. *J. Audio Eng. Soc.* **2005**, *53*, 919–929.

31. Zhang, W.; Abhayapala, T.D.; Kennedy, R.A.; Duraiswami, R. Modal expansion of HRTFs: Continuous representation in frequency-range-angle. In Proceedings of the IEEE International Conference on Acoustics, Speech, and Signal Processing (ICASSP), Taipei, Taiwan, 19–24 April 2009; pp. 285–288.

32. Zhang, M.; Kennedy, R.A.; Abhayapala, T.D. Empirical determination of frequency representation in spherical harmonics-based HRTF functional modeling. *IEEE/ACM Trans. Audio Speech Lang. Process.* **2015**, *23*, 351–360. [CrossRef]

33. Zhang, W.; Abhayapala, T.D.; Kennedy, R.A.; Duraiswami, R. Insights into head-related transfer function: Spatial dimensionality and continuous representation. *J. Acoust. Soc. Am.* **2010**, *127*, 2347–2357. [CrossRef] [PubMed]

34. Zhang, W.; Zhang, M.; Kennedy, R.A.; Abhayapala, T.D. On high-resolution head-related transfer function measurements: An efficient sampling scheme. *IEEE Trans. Audio Speech Lang. Process.* **2012**, *20*, 575–584. [CrossRef]

35. Bates, A.P.; Khalid, Z.; Kennedy, R.A. Novel sampling scheme on the sphere for head-related transfer function measurements. *IEEE/ACM Trans. Audio Speech Lang. Process.* **2015**, *23*, 1068–1081. [CrossRef]

36. Muller, A.; Massarani, P. Transfer-function measurement with sweeps. *J. Audio Eng. Soc.* **2001**, *49*, 443–471.

37. Zotkin, D.N.; Duraiswami, R.; Grassi, E.; Gumerov, N.A. Fast head-related transfer function measurement via reciprocity. *J. Acoust. Soc. Am.* **2006**, *120*, 2202–2215. [CrossRef] [PubMed]

38. Fukudome, K.; Suetsugu, T.; Ueshin, T.; Idegami, R.; Takeya, K. The fast measurment of head related impulse responses for all azimuthal directions using the continuous measurement method with a servoswiveled chair. *Appl. Acoust.* **2007**, *68*, 864–884. [CrossRef]

39. He, J.; Ranjan, R.; Gan, W.-S. Fast continuous HRTF acquisition with unconstrained movements of human subjects. In Proceedings of the IEEE International Conference on Acoustics, Speech, and Signal Processing (ICASSP), Shanghai, China, 20–25 March 2016; pp. 321–325.

40. Majdak, P.; Iwaya, Y.; Carpentier, T. Spatially oriented format for acoustics: A data exchange format representing head-related transfer functions. In Proceedings of the 134th Audio Engineering Society Convention, Rome, Italy, 4–7 May 2013; pp. 1–11.
41. Zotkin, D.N.; Duraiswami, R.; Davis, L.S. Rendering localized spatial audio in a virtual auditory scene. *IEEE Trans. Multimedia* **2004**, *6*, 553–563. [CrossRef]
42. Xie, B. *Head-Related Transfer Function and Virtual Auditory Display*; J Ross Publishing: Plantation, FL, USA, 2013.
43. Gamper, H. Head-related transfer function interpolation in azimuth, elevation, and distance. *J. Acoust. Soc. Am.* **2013**, *134*. [CrossRef] [PubMed]
44. Queiroz, M.; de Sousa, G.H.M.A. Efficient binaural rendering of moving sound sources using HRTF interpolation. *J. New Music Res.* **2011**, *40*, 239–252. [CrossRef]
45. Savioja, L.; Huopaniemi, J.; Lokki, T.; Väänänen, R. Creating interactive virtual acoustic environments. *J. Audio Eng. Soc.* **1999**, *47*, 675–705.
46. Freeland, F.P.; Biscinho, L.W.P.; Diniz, P.S.R. Efficient HRTF interpolation in 3D moving sound. In Proceedings of the 22nd AES International Conference: Virtual, Synthetic, and Entertainment Audio, Espoo, Finland, 15–17 June 2002; pp. 1–9.
47. Kistler, D.J.; Wightman, F.L.L. A model of HeadRelated Transfer Functions based on Principal Components Analysis and Minimum-Phase reconstruction. *J. Acoust. Soc. Am.* **1992**, *91*, 1637–1647. [CrossRef] [PubMed]
48. Romigh, G.D.; Brungart, D.S.; Stern, R.M.; Simpson, B.D. Efficient real spherical harmonic representation of head-related transfer function. *IEEE J. Sel. Top. Signal Process.* **2015**, *9*, 921–930. [CrossRef]
49. Zotkin, D.N.; Hwang, J.; Duraiswami, R.; Davis, L.S. HRTF personalization using anthropometric measurements. In Proceedings of the IEEE Workshop on Applications of Signal Processing to Audio and Acoustics (WASPAA), New Paltz, NY, USA, 19–22 October 2003; pp. 157–160.
50. Hu, H.; Zhou, L.; Ma, H.; Wu, Z. HRTF personalization based on airtificial neural network in individual virtual auditory space. *Appl. Acoust.* **2008**, *69*, 163–172. [CrossRef]
51. Li, L.; Huang, Q. HRTF personalization modeling based on RBF neural network. In Proceedings of the IEEE International Conference on Acoustics, Speech, and Signal Processing (ICASSP), Vancouver, BC, Canada, 26–31 May 2013; pp. 3707–3710.
52. Grindlay, G.; Vasilescu, M.A.O. A multilinear (tensor) framework for HRTF analysis and synthesis. In Proceedings of the IEEE International Conference on Acoustics, Speech, and Signal Processing (ICASSP), Honolulu, HI, USA, 16–20 April 2007; pp. 161–164.
53. Spagnol, S.; Geronazzo, M.; Avanzini, F. On the relation between pinna reflection patterns and head-related transfer functon features. *IEEE Trans. Audio Speech Lang. Process.* **2013**, *21*, 508–519. [CrossRef]
54. Geronazzo, M.; Spagnol, S.; Bedin, A.; Avanzini, F. Enhancing vertical localization with image-guided selection of non-individual head-related transfer functions. In Proceedings of the IEEE International Conference on Acoustics, Speech, and Signal Processing (ICASSP), Florence, Italy, 4–9 May 2014; pp. 4496–4500.
55. Zhang, M.; Kennedy, R.A.; Abhayapala, T.D.; Zhang, W. Statistical method to identify key anthropometric parameters in HRTF individualization. In Proceedings of the Hands-free Speech Communication and Microphone Arrays (HSCMA), Edinburgh, UK, 30 May–1 June 2011; pp. 213–218.
56. Bilinski, P.; Ahrens, J.; Thomas, M.R.P.; Tasheve, I.J.; Platt, J.C. HRTF magnitude synthesis via sparse representation of anthropometric features. In Proceedings of the IEEE International Conference on Acoustics, Speech, and Signal Processing (ICASSP), Florence, Italy, 4–9 May 2014; pp. 4468–4472.
57. Tasheve, I.J. HRTF phase synthesis via sparse representation of anthropometric features. In Proceedings of the Information Theory and Applications Workshop (ITA), San Diego, CA, USA, 9–14 February 2014; pp. 1–5.
58. He, J.; Gan, W.-S.; Tan, E.-L. On the preprocessing and postprocessing of HRTF individualizaion based on sparse representation of anthropometric features. In Proceedings of the IEEE International Conference on Acoustics, Speech, and Signal Processing (ICASSP), Brisbane, Australia, 19–24 April 2015; pp. 639–643.
59. Fink, K.J.; Ray, L. Individualization of head related transfer functions using principal component analysis. *Appl. Acoust.* **2015**, *87*, 162–173. [CrossRef]
60. Sunder, K.; Tan, E.-L.; Gan, W.-S. Individualization of binaural synthesis using frontal projection headphones. *J. Audio Eng. Soc.* **2013**, *61*, 989–1000.

61. Cai, T.; Rakerd, B.; Hartmann, W.M. Computing interaural differences through finite element modeling of idealized human heads. *J. Acoust. Soc. Am.* **2015**, *138*, 1549–1560. [CrossRef] [PubMed]

62. Katz, B.F.G. Boundary element method calculation of individual head-related transfer function. I. Rigid model calculation. *J. Acoust. Soc. Am.* **2001**, *110*, 2440–2448. [CrossRef] [PubMed]

63. Otani, M.; Ise, S. Fast calculation system specialized for head-related transfer function based on boundary element method. *J. Acoust. Soc. Am.* **2006**, *119*, 2589–2598. [CrossRef] [PubMed]

64. Prepeliță, S.; Geronazzo, M.; Avanzini, F.; Savioja, L. Influence of voxelization on finite difference time domain simulations of head-related transfer functions. *J. Acoust. Soc. Am.* **2016**, *139*, 2489–2504. [CrossRef] [PubMed]

65. Mokhtari, P.; Takemoto, H.; Nishimura, R.; Kato, H. Preliminary estimation of the first peak of HRTFs from pinna anthropometry for personalized 3D audio. In Proceedings of the 5th International Conference on Three Dimensional Systems and Applications, Osaka, Japan, 26–28 June 2013; p. 3.

66. Jin, C.T.; Guillon, P.; Epain, N.; Zolfaghari, R.; van Schaik, A.; Tew, A.I.; Hetherington, C.; Thorpe, J. Creating the sydney york morphological and acoustic recordings of ears database. *IEEE Trans. Multimedia* **2014**, *16*, 37–46. [CrossRef]

67. Voss, P.; Lepore, F.; Gougoux, F.; Zatorre, R.J. Relevance of spectral cues for auditory spatial processing in the occipital cortex of the blind. *Front. Psychol.* **2011**, *2*, 48. [CrossRef] [PubMed]

68. Kolarik, A.J.; Cirstea, S.; Pardhan, S. Discrimination of virtual auditory distance using level and direct-to-reverberant ratio cues. *J. Acoust. Soc. Am.* **2013**, *134*, 3395–3398. [CrossRef] [PubMed]

69. Wightman, F.L.; Kistler, D.J. Resolution of front-back ambiguity in spatial hearing by listener and source movement. *J. Acoust. Soc. Am.* **1999**, *102*, 2325–2332. [CrossRef]

70. Kolarik, A.J.; Cirstea, S.; Pardhan, S. Evidence for enhanced discrimination of virtual auditory distance among blind listeners using level and direct-to-reverberant cues. *Exp. Brain Res.* **2013**, *224*, 623–633. [CrossRef] [PubMed]

71. Shinn-Cunningham, B.G. Distance cues for virtual auditory space. In Proceedings of the IEEE Pacific Rim Conference (PRC) on Multimedia, Sydney, Australia, 26-28 August 2001; pp. 227–230.

72. Allen, J.B.; Berkeley, D.A. Image method for efficiently simulating small-room acoustics. *J. Acoust. Soc. Am.* **1979**, *75*, 943–950. [CrossRef]

73. Valimaki, V.; Parker, J.D.; Savioja, L.; Smith, J.O.; Abel, J.S. Fifty years of artificial reverberation. *IEEE Trans. Audio Speech Lang. Process.* **2012**, *20*, 1421–1448. [CrossRef]

74. Belloch, J.A.; Ferrer, M.; Gonzalez, A.; Martinez-Zaldivar, F.J.; Vidal, A.M. Headphone-based virtual spatialization of sound with a GPU accelerator. *J. Audio Eng. Soc.* **2013**, *61*, 546–561.

75. Taylor, M.; Chandak, A.; Mo, Q.; Lauterbach, C.; Schissler, C.; Manocha, D. Guided multiview ray tracing for fast auralization. *IEEE Trans. Vis. Comput. Gr.* **2012**, *18*, 1797–1810. [CrossRef] [PubMed]

76. Theile, G. On the standardization of the frequency response of high-quality studio headphones. *J. Audio Eng. Soc.* **1986**, *34*, 959–969.

77. Hiipakka, M.; Kinnari, T.; Pulkki, V. Estimating head-related transfer functions of human subjects from pressure-velocity measurements. *J. Acoust. Soc. Am.* **2012**, *13*, 4051–4061. [CrossRef] [PubMed]

78. Lindau, A.; Brinkmann, F. Perceptual evaluation of headphone compensation in binaural synthesis based on non individual recordings. *J. Audio Eng. Soc.* **2012**, *60*, 54–62.

79. Boren, B.; Geronazzo, M.; Brinkmann, F.; Choueiri, E. Coloration metrics for headphone equalization. In Proceedings of the 21st International Conference on Auditory Display, Graz, Austria, 6–10 July 2015; pp. 29–34.

80. Takeuchi, T.; Nelson, P.A.; Hamada, H. Robustness to head misalignment of virtual sound imaging system. *J. Acoust. Soc. Am.* **2001**, *109*, 958–971. [CrossRef] [PubMed]

81. Kirkeby, O.; Nelson, P.A.; Hamada, H. Local sound field reproduction using two closely spaced loudspeakers. *J. Acoust. Soc. Am.* **1998**, *104*, 1973–1981. [CrossRef]

82. Takeuchi, T.; Nelson, P.A. Optimal source distribution for binaural synthesis over loudspeakers. *J. Acoust. Soc. Am.* **2002**, *112*, 2786–2797. [CrossRef] [PubMed]

83. Bai, M.R.; Tung, W.W.; Lee, C.C. Optimal design of loudspeaker arrays for robust cross-talk cancellation using the Taguchi method and the generic algorithm. *J. Acoust. Soc. Am.* **2005**, *117*, 2802–2813. [CrossRef] [PubMed]

84. Majdak, P.; Masiero, B.; Fels, J. Sound localization in individualized and non-individualized crosstalk cancellation systems. *J. Acoust. Soc. Am.* **2013**, *133*, 2055–2068. [CrossRef] [PubMed]
85. Lacouture-Parodi, Y.; Habets, E.A. Crosstalk cancellation system using a head tracker based on interaural time differences. In Proceedings of the International Workshop on Acoustic Signal Enahcancement, Aachen, Germany, 4–6 September 2012; pp. 1–4.
86. Williams, E.G. *Fourier Acoustics: Sound Radiation and Nearfield Acoustical Holography*; Academic Press: San Diego, CA, USA, 1999.
87. Ward, D.B.; Abhayapala, T.D. Reproduction of a plane-wave sound field using an array of loudspeakers. *IEEE Trans. Speech Audio Process.* **2001**, *9*, 697–707. [CrossRef]
88. Core Sound TetraMic. Available online: http://www.core-sound.com/TetraMic/1.php (accessed on 10 March 2017).
89. Eigenmike Microphone. Available online: https://www.mhacoustics.com/products#eigenmike1 (accessed on 10 March 2017).
90. Gerzon, M.A. The design of precisely conincident microphone arrays for stereo and surround sound. In Proceedings of the 50th Audio Engineering Society Covention, London, UK, 4–7 March 1975; pp. 1–5.
91. Abhayapala, T.D.; Gupta, A. Spherical harmonic analysis of wavefields using multiple circular sensor arrays. *IEEE Trans. Audio Speech Lang. Process.* **2010**, *18*, 1655–1666. [CrossRef]
92. Chen, H.; Abhayapala, T.D.; Zhang, W. Theory and design of compact hybrid microphone arrays on two-dimensional planes for three-dimensional soundfield analysis. *J. Acoust. Soc. Am.* **2015**, *138*, 3081–3092. [CrossRef] [PubMed]
93. Samarasinghe, P.N.; Abhayapala, T.D.; Poletti, M.A. Wavefield analysis over large areas using distributed higher order microphones. *IEEE/ACM Trans. Audio Speech Lang. Process.* **2014**, *22*, 647–658. [CrossRef]
94. Pulkki, V.; Karjalainen, M. Localization of amplitude-panned virtual sources, Part 1: Stereophonic panning. *J. Audio Eng. Soc.* **2001**, *49*, 739–752.
95. Pulkki, V. Localization of amplitude-panned virtual sources, Part 2: Two and three dimensional panning. *J. Audio Eng. Soc.* **2001**, *49*, 753–767.
96. Pulkki, V. Virtual sound source positioning using vector base amplitude panning. *J. Audio Eng. Soc.* **1997**, *45*, 456–466.
97. Lossius, T.; Baltazar, P.; de la Hogue, T. DBAP—distance-based amplitude panning. In Proceedings of the 2009 International Computer Music Conference, Montreal, QC, Canada, 16–21 August 2009; pp. 1–4.
98. VBAP Demo. Available online: http://legacy.spa.aalto.fi/software/vbap/VBAP_demo/ (accessed on 26 April 2017).
99. Developers—3D Sound Labs. Available online: http://www.3dsoundlabs.com/category/developers/ (accessed on 26 April 2017).
100. Cameras, M. Approach to recreating a sound field. *J. Acoust. Soc. Am.* **1967**, *43*, 1425–1431. [CrossRef]
101. Gerzon, M.A. Periphony: With-height sound reproduction. *J. Audio Eng. Soc.* **1973**, *21*, 2–10.
102. Gerzon, M.A. Ambisonics in multichannel broadcasting video. *J. Audio Eng. Soc.* **1985**, *33*, 859–871.
103. Betlehem, T.; Abhayapala, T.D. Theory and design of sound field reproduction in reverberant rooms. *J. Acoust. Soc. Am.* **2005**, *117*, 2100–2111. [CrossRef] [PubMed]
104. Wu, Y.; Abhayapala, T.D. Theory and design of soundfield reproducion using continuous loudspeakers concept. *IEEE Trans. Audio Speech Lang. Process.* **2009**, *17*, 107–116. [CrossRef]
105. Daniel, J. Spatial sound encoding including near field effect: Introducing distance coding filters and a viable, new ambisonic format. In Proceedings of the 23rd AES International Conference: Signal Processing in Audio Recording and Reproduction, Copenhagen, Denmark, 23–25 May 2003.
106. Ahrens, J.; Spors, S. Applying the ambisonics approach to planar and linear distributions of secondary sources and combinations thereof. *Acta Acust. United Acust.* **2012**, *98*, 28–36. [CrossRef]
107. Ahrens, J.; Spors, S. Wave field synthesis of a sound field described by spherical harmonics expansion coefficients. *J. Acoust. Soc. Am.* **2012**, *131*, 2190–2199. [CrossRef] [PubMed]
108. Bianchi, L.; Antonacci, F.; Sarti, A.; Turbaro, S. Model-based acoustic rendering based on plane wave decomposition. *Appl. Acoust.* **2016**, *104*, 127–134. [CrossRef]
109. Okamoto, T. 2.5D higher-order Ambisonics for a sound field described by angular spectrum coefficients. In Proceedings of the IEEE International Conference on Acoustics, Speech, and Signal Processing (ICASSP), Shanghai, China, 20–25 March 2016; pp. 326–330.

110. Berkhout, A.J. A holographic approach to acoustic control. *J. Audio Eng. Soc.* **1988**, *36*, 977–995.
111. Berkhout, A.J.; de Vries, D.; Vogel, P. Acoustic control by wave field synthesis. *J. Acoust. Soc. Am.* **1993**, *93*, 2764–2778. [CrossRef]
112. Spors, S.; Rabenstein, R.; Ahrens, J. The theory of wave field synthesis revisited. In Proceedings of the 124th Audio Engineering Society Convention, Amsterdam, The Netherlands, 17–20 May 2008.
113. Spors, S.; Rabenstein, R. Spatial aliasing aritifacts produced by linear and circular loudspeaker arrays used for wave field synthesis. In Proceedings of the 120th Audio Engineering Society Convention, Paris, France, 20–23 May 2006.
114. Boone, M.M.; Verheijen, E.N.G.; Tol, P.F.V. Spatial sound-field reproduction by wave-field synthesis. *J. Audio Eng. Soc.* **1995**, *43*, 1003–1012. [CrossRef]
115. Boone, M.M. Multi-actuator panels (MAPs) as loudspeaker arrays for wave field synthesis. *J. Audio Eng. Soc.* **2004**, *52*, 712–723.
116. Spors, S.; Ahrens, J. Analysis and improvement of pre-equalization in 2.5-dimensional wave field synthesis. In Proceedings of the 128 Audio Engineering Society Convention, London, UK, 23–25 May 2010.
117. Firtha, G.; Fiala, P.; Schultz, F.; Spors, S. Improved referencing schemes for 2.5D wave field synthesis driving functions. *IEEE/ACM Trans. Audio Speech Lang. Process.* **2017**, *25*, 1117–1127. [CrossRef]
118. Kirkeby, O.; Nelson, P.A. Reproduction of plane wave sound fields. *J. Acoust. Soc. Am.* **1993**, *94*, 2992–3000. [CrossRef]
119. Tatekura, Y.; Urata, S.; Saruwatari, H.; Shikano, K. On-line relaxation algorithm applicable to acoustic fluctuation for inverse filter in multichannel sound reproduction system. *IEICE Trans. Fundam. Electron. Commun. Comput. Sci.* **2005**, *E88-A*, 1747–1756. [CrossRef]
120. Spors, S.; Wierstorf, H.; Raake, A.; Melchior, F.; Frank, M.; Zotter, F. Spatial sound with loudspeakers and its perception: A review of the current state. *Proc. IEEE* **2013**, *101*, 1920–1938. [CrossRef]
121. Wierstorf, H. *Perceptual Assessment of Sound Field Synthesis*; Technical University of Berlin: Berlin, Germany, 2014.
122. Bharitkar, S.; Kyriakakis, C. *Immersive Audio Signal Processing*; Springer: New York, NY, USA, 2006.
123. Corteel, E.; Nicol, R. Listening room compensation for wave field sysnthesis. What can be done? In Proceedings of the 23rd Audio Engineering Society Convention, Copenhagen, Denmark, 23–25 May 2003.
124. Mourjopoulos, J.N. On the variation and invertibility of room impulse response functions. *J. Sound Vib.* **1985**, *102*, 217–228. [CrossRef]
125. Hatziantoniou, P.D.; Mourjopoulos, J.N. Erros in real-time room acoustics dereverberation. *J. Audio Eng. Soc.* **2004**, *52*, 883–889.
126. Spors, S.; Buchner, H.; Rabenstein, R.; Herbordt, W. Active listening room compensation for massive multichannel sound reproduction systems. *J. Acoust. Soc. Am.* **2007**, *122*, 354–369. [CrossRef] [PubMed]
127. Talagala, D.; Zhang, W.; Abhayapala, T.D. Efficient multichannel adaptive room compensation for spatial soundfield reproduction using a modal decomposition. *IEEE Trans. Audio Speech Lang. Process.* **2014**, *22*, 1522–1532. [CrossRef]
128. Schneider, M.; Kellermann, W. Multichannel acoustic echo cancellation in the wave domain with increased robustness to nonuniqueness. *IEEE Trans. Audio Speech Lang. Process.* **2016**, *24*, 518–529. [CrossRef]
129. Poletti, M.A.; Fazi, F.M.; Nelson, P.A. Sound-field reproduction systems using fixed-directivity loudspeakers. *J. Acoust. Soc. Am.* **2010**, *127*, 3590–3601. [CrossRef] [PubMed]
130. Poletti, M.A.; Abhayapala, T.D.; Samarasinghe, P.N. Interior and exterior sound field control using two dimensional higher-order variable-directivity sources. *J. Acoust. Soc. Am.* **2012**, *131*, 3814–3823. [CrossRef] [PubMed]
131. Betlehem, T.; Poletti, M.A. Two dimensional sound field reproduction using higher-order sources to exploit room reflections. *J. Acoust. Soc. Am.* **2014**, *135*, 1820–1833. [CrossRef] [PubMed]
132. Canclini, A.; Markovic, D.; Antonacci, F.; Sarti, A.; Tubaro, S. A room-compensated virtual surround sound system exploiting early reflections in a reverberant room. In Proceedings of the 20th European Signal Processing Conference (EUSIPCO), Bucharest, Romania, 27–31 August 2012; pp. 1029–1033.
133. Samarasinghe, P.N.; Abhayapala, T.D.; Poletti, M.A. Room reflections assisted spatial sound field reproduction. In Proceedings of the European Signal Processing Conference (EUSIPCO), Lisbon, Portugal, 1–5 September 2014; pp. 1352–1356.
134. Betlehem, T.; Zhang, W.; Poletti, M.A.; Abhayapala, T.D. Personal sound zones: Delivering interface-free audio to multiple listeners. *IEEE Signal Process. Mag.* **2015**, *32*, 81–91. [CrossRef]

135. Choi, J.-W.; Kim, Y.-H. Generation of an acoustically bright zone with an illuminated region using multiple sources. *J. Acoust. Soc. Am.* **2002**, *111*, 1695–1700. [CrossRef] [PubMed]

136. Shin, M.; Lee, S.Q.; Fazi, F.M.; Nelson, P.A.; Kim, D.; Wang, S.; Park, K.H.; Seo, J. Maximization of acoustic energy difference between two spaces. *J. Acoust. Soc. Am.* **2010**, *128*, 121–131. [CrossRef] [PubMed]

137. Elliott, S.J.; Cheer, J.; Choi, J.-W.; Kim, Y.-H. Robustness and regularization of personal audio systems. *IEEE Trans. Audio Speech Lang. Process.* **2012**, *20*, 2123–2133. [CrossRef]

138. Chang, J.-H.; Lee, C.-H.; Park, J.-Y.; Kim, Y.-H. A realization of sound focused personal audio system using acoustic contrast control. *J. Acoust. Soc. Am.* **2009**, *125*, 2091–2097. [CrossRef] [PubMed]

139. Okamoto, T.; Sakaguchi, A. Experimental validation of spatial Fourier transform-based multiple sound zone generation with a linear loudspeaker array. *J. Acoust. Soc. Am.* **2017**, *141*, 1769–1780. [CrossRef] [PubMed]

140. Cheer, J.; Elliott, S.J.; Gálvez, M.F.S. Design and implementation of a car cabin personal audio system. *J. Audio Eng. Soc.* **2013**, *61*, 414–424.

141. Coleman, P.; Jackson, P.; Olik, M.; Pederson, J.A. Personal audio with a planar bright zone. *J. Acoust. Soc. Am.* **2014**, *136*, 1725–1735. [CrossRef] [PubMed]

142. Coleman, P.; Jackson, P.; Olik, M.; M′øller, M.; Olsen, M.; Pederson, J.A. Acoustic contrast, planarity and robustness of sound zone methods using a circular loudspeaker array. *J. Acoust. Soc. Am.* **2014**, *135*, 1029–1940. [CrossRef] [PubMed]

143. Poletti, M.A. An investigation of 2D multizone surround sound systems. In Proceedings of the 125th Audio Engineering Society Convention, San Francisco, CA, USA, 2–5 October 2008.

144. Betlehem, T.; Withers, C. Sound field reproduction with energy constraint on loudspeaker weights. *IEEE Trans. Audio Speech Lang. Process.* **2012**, *20*, 2388–2392. [CrossRef]

145. Radmanesh, N.; Burnett, I.S. Generation of isolated wideband soundfield using a combined two-stage Lasso-LS algorithm. *IEEE Trans. Audio Speech Lang. Process.* **2013**, *21*, 378–387. [CrossRef]

146. Jin, W.; Kleijn, W.B. Theory and design of multizone soundfield reproduction using sparse methods. *IEEE/ACM Trans. Audio Speech Lang. Process.* **2015**, *23*, 2343–2355.

147. Chang, J.-H.; Jacobsen, F. Sound field control with a circular double-layer array of loudspeakers. *J. Acoust. Soc. Am.* **2012**, *131*, 4518–4525. [CrossRef] [PubMed]

148. Chang, J.-H.; Jacobsen, F. Experimental validation of sound field control with a circular double-layer array of loudspeakers. *J. Acoust. Soc. Am.* **2013**, *133*, 2046–2054. [CrossRef] [PubMed]

149. Cai, Y.; Wu, M.; Yang, J. Sound reproduction in personal audio systems using the least-squares approach with acoustic contrast control constraint. *J. Acoust. Soc. Am.* **2014**, *135*, 734–741. [CrossRef] [PubMed]

150. Gálvez, S.; Marcos, F.; Elliott, S.J.; Jordan, C. Time domain optimisation of filters used in a loudspeaker array for personal audio. *IEEE/ACM Trans. Audio Speech Lang. Process.* **2015**, *23*, 1869–1878. [CrossRef]

151. Wu, Y.J.; Abhayapala, T.D. Spatial multizone soundfield reproduction: Theory and design. *IEEE Trans. Audio Speech Lang. Process.* **2011**, *19*, 1711–1720. [CrossRef]

152. Poletti, M.A.; Fazi, F.M. An approach to generating two zones of silence with application to personal sound systems. *J. Acoust. Soc. Am.* **2015**, *137*, 1711–1720. [CrossRef] [PubMed]

153. Menzies, D. Sound field synthesis with distributed modal constraints. *Acta Acust. United Acust.* **2012**, *98*, 15–27. [CrossRef]

154. Helwani, K.; Spors, S.; Buchner, H. The synthesis of sound figures. *Multidimens. Syst. Signal Process.* **2014**, *25*, 379–403. [CrossRef]

155. Zhang, W.; Abhayapala, T.D.; Betlehem, T.; Fazi, F.M. Analysis and control of multi-zone sound field reproduction using modal-domain approach. *J. Acoust. Soc. Am.* **2016**, *140*, 2134–2144. [CrossRef] [PubMed]

applied
sciences

MDPI

Review

Spatial Audio for Soundscape Design: Recording and Reproduction

Joo Young Hong [1,*], Jianjun He [2], Bhan Lam [1], Rishabh Gupta [1] and Woon-Seng Gan [1]

[1] School of Electrical and Electronic Engineering, Nanyang Technological University, Singapore 639798, Singapore; blam002@e.ntu.edu.sg (B.L.); grishabh@ntu.edu.sg (R.G.); ewsgan@ntu.edu.sg (W.-S.G.)
[2] Maxim Integrated Products Inc., San Jose, CA 95129, USA; jianjun.he@maximintegrated.com
[*] Correspondence: jyhong@ntu.edu.sg; Tel.: +65-8429-7512

Academic Editor: Gino Iannace
Received: 29 March 2017; Accepted: 9 June 2017; Published: 16 June 2017

Featured Application: This review introduces the concept of spatial audio in the perspective of soundscape practitioners. A selection guide based on the spatial fidelity and degree of perceptual accuracy of the mentioned spatial audio recording and reproduction techniques is also provided.

Abstract: With the advancement of spatial audio technologies, in both recording and reproduction, we are seeing more applications that incorporate 3D sound to create an immersive aural experience. Soundscape design and evaluation for urban planning can now tap into the extensive spatial audio tools for sound capture and 3D sound rendering over headphones and speaker arrays. In this paper, we outline a list of available state-of-the-art spatial audio recording techniques and devices, spatial audio physical and perceptual reproduction techniques, emerging spatial audio techniques for virtual and augmented reality, followed by a discussion on the degree of perceptual accuracy of recording and reproduction techniques in representing the acoustic environment.

Keywords: soundscape; spatial audio; recording; reproduction; virtual reality; augmented reality

1. Introduction

Urban acoustic environments consist of multiple types of sound sources (e.g., traffic sounds, biological sounds, geophysical sounds, human sounds) [1]. The different types of sound sources have different acoustical characteristics, meanings and values [2–4]. Conventionally, environmental noise policies are primarily centered on the energetic reduction of sound pressure levels (SPL). However, SPL indicators provide limited information on perceived acoustic comfort as it involves higher cognitive processes. To address the limitations of traditional noise management, the notion of a soundscape has been applied as a new paradigm. Schafer, a Canadian composer and music educator, introduced the term soundscape to encompass a holistic acoustic environment as a macrocosmic musical composition [5]. In this context, soundscape considers sound as a resource rather than waste and focuses on people's contextual perception of the acoustic environment [6].

Due to rising prominence of the soundscape approach, ISO TC43 SC1 WG 54 was started with the aim of standardizing the perceptual assessment of soundscapes. As defined in ISO 12913-1: definition and conceptual framework, acoustic environment is "sound from all sound sources as modified by the environment" and the modification by the environment includes effects of various physical factors (e.g., meteorological conditions, absorption, diffraction, reverberation, and reflection) on sound propagation. This implies that soundscape is a perceptual construct related to physical acoustic environment in a place.

According to ISO 12913-1, the context plays a critical role in the perception of a soundscape, as it affects the auditory sensation, the interpretation of auditory sensation, and the responses to the acoustic environment [7]. The context includes all other non-acoustic components of the place (e.g., physical as well as previous experience of the individual). Herranz-Pascul et al. [8] proposed a people-activity-place framework for contexts based on four clusters: person, place, person-place interaction and activity. The suggested framework shows interrelationships between person and activity and place, which may influence a person's experience of the acoustic environment.

It is evident from [8] that the human auditory process is deeply entwined in the person-activity-place framework. Thus, sufficient insight about the human auditory process is imperative to record and reproduce an acoustic environment with sufficient perceptual accuracy for proper analysis. For simplicity, Zwicker and Fastl has grouped the human auditory processing system into two stages: (1) the preprocessing in the peripheral system, and (2) information processing in the auditory system [9].

Current reproduction methods (e.g., loudspeakers, headphones, etc.) have sufficient fidelity for proper interpretation of frequency and temporal characteristics in the peripheral system (stage 1). As the spatial dimension of sound is only interpreted in stage 2, where there is significant exchange of information between both ears, complexity of reproduction is greatly increased. For instance, to locate a sound source in space, the characteristics of intensity, phase, and latency must be presented accurately to both ears. Hence, spatial audio recording and reproduction techniques should be reviewed in the technological perspective with a focus on their relationship with soundscape perception—the main goal of this paper.

As recording and reproduction techniques are heavily employed in the soundscape design process, a brief discussion will shed light on the areas where such techniques are most commonly used. The soundscape design process can be summarized into three stages, as illustrated in Figure 1 [10]. Stage 1 aims to define and analyze existing soundscapes. In this stage, soundscape researchers and planners are required to evaluate and identify 'wanted sounds' to be preserved or added, and 'unwanted sounds' to be removed or reduced based on the context. In Stage 2, soundscape planning and design scenarios are proposed based on the analysis of the existing soundscapes. The soundscape design proposals will then be simulated to be objectively and subjectively evaluated by stakeholders to determine the final soundscape design. In Stage 3, the final soundscape design will be implemented in situ. After implementation, validation of the soundscape design will be performed with iteration of Stage 1 for analysis of implemented soundscape design.

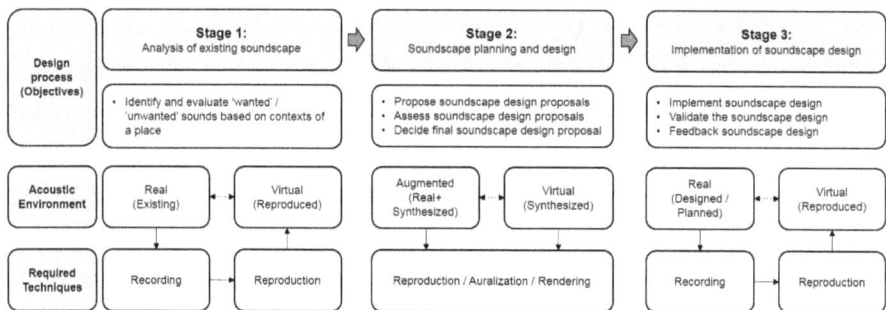

Figure 1. Schematic illustration of soundscape design process, types of acoustic environment and required techniques.

Throughout the soundscape design process, soundscapes can be evaluated based on the real acoustic environment or virtual (reproduced or synthesized) environment [6,11]. Similarly, in Stage 1 (analysis of existing soundscape) and Stage 3 (implementation of soundscape design), soundscapes can be assessed in real or reproduced acoustic environments. The reproduced acoustic environment is constructed from audio recordings in the real world. If the physical location for soundscape implementation is not available (i.e., not built), future soundscape design scenarios might also be evaluated using synthesized (or simulated) acoustic environment in Stage 2 (soundscape planning and design). Moreover, augmented acoustic technologies consisting of synthesized and real sounds can also be applied in soundscape design stage if the physical locations are available.

A real acoustic environment is evaluated in situ by means of field study methods, such as soundwalk and behavior observations, which generally focus on the short-term in situ experience [6,11]. Narrative interview and social survey usually deals with long-term responses or recalled experiences based on the real acoustic environment [12–14]. In contrast, the virtual acoustic environment, created by recording followed by reproduction, are usually evaluated in a laboratory conditions [6,11]. Aletta et al. [11] reviewed the advantages and limitations of such methods for soundscape assessment. Evaluation methods in the real acoustic environments are advantageous as they provide the most realistic representation of the real-life settings, which can guarantee high ecological validity. However, those methods are subjected to uncontrolled factors, such as temperature, humidity, daylight and wind speed. Accordingly, such methods may yield results that are difficult to generalized as they are limited to a specific context. These methods have been primarily applied to collect and characterize the existing acoustic environment at the early stages in urban soundscape planning and to validate the soundscape design at the implementation stages [6,10,15].

On the other hand, the virtual acoustic environment, are usually evaluated under controlled laboratory conditions. Researchers can design and control the experimental variables, such that it enables us to investigate various casual relationships or correlations. A laboratory-based experiment can obtain results that minimize the effects of other distracting environmental factors. Regardless of the strengths of the laboratory experiments based on virtual acoustic environment, criticisms on discrepancies between virtual and real environments, such as the absence of physical contact and possible perceptual alterations, have been raised. In general, the more control researchers exert in a laboratory experiment, the less ecological validity the findings have, which can be less generalized to real-life scenarios [11,16]. Despite this limitation, acoustic simulation and reproduction techniques can provide powerful tools to evaluate the performance of different soundscape design approaches in Stage 2 before they are implemented in the real world [11].

Currently, the ISO WG54 is working on the Part 2 of the soundscape assessment standard to provide information on minimum reporting requirement for soundscape studies and applications [17]. The reporting requirement includes physical (recording and reproduction techniques) and perceptual data collections both on-site (e.g., soundwalk) and off-site (e.g., laboratory experiments). In particular, recording and reproducing techniques play a critical role to achieve high ecological validity of soundscape research conducted in the virtual acoustic environment. This is because soundscape is the human perception of the reproduced acoustic environment, and the perception is dependent on the perceptual accuracy of the recording and reproduction techniques. Perceptual accuracy is the basis of auralization [18], and is largely influenced by the spatial properties of sounds such as source distance, source location, and reverberation [19,20]. Several laboratory experiments on ecological validation of the reproduction systems have been conducted based on the psycholinguistic measure, such as semantic differential method [16,21,22]. Those studies compared the subjective responses of verbal descriptors on soundscape among in situ acoustic environments and the reproduced acoustic environments created by different techniques in laboratory conditions to explore the ecological validity of reproduction systems.

Recently, virtual reality (VR) [23–25] and augmented reality (AR) [26–28] technologies have been increasingly adopted in soundscape studies due to their potential in creating a perceptually accurate audio-visual scene. To create virtual acoustic environments with high ecological validity, it is essential to have a holistic understanding of recording and reproducing techniques with respect to soundscape evaluation methodologies. Even though more studies on the standardization of soundscape recording and reproduction in an ecologically valid way that reflect context of real acoustic environment are necessary, relatively few studies have been done on these technical aspects of soundscape research. This paper attempts to formulate a framework for the appropriate use of audio recording and reproduction techniques in soundscape design. Sections 2 and 3 systemically reviews the acoustic recording and reproduction techniques that can be adopted in soundscape research, respectively. In Section 4, application of spatial audio in virtual and augmented Reality for soundscape design is discussed. Lastly, the degree of perceptual accuracy of the recording and reproduction techniques in representing acoustic environment is addressed in the discussion.

2. Spatial Audio Recording Techniques for Soundscape Design

Recording of the acoustic environment builds the foundation in soundscape research through analyzing and understanding the soundscape in a real or virtual environment. The recordings, therefore, requires sufficient characteristics of the acoustic environment to be captured for perceptual accuracy. To achieve this, it is necessary to consider two aspects of recording: timbre and spatial qualities. In general, timbre qualities of the recordings are largely dependent on the electrical and acoustic properties of the microphones, such as frequency response, directionality, impedance, sensitivity, equivalent noise level, total harmonic distortion, maximal SPL, etc. [29]. In terms of the spatial aspects, it is common to use multiple microphones for capturing better spatial characteristics of the acoustic environment. In this section, we will review the audio recording techniques, and discuss how they can be applied in the field of soundscape. A comparison of the advantages and disadvantages of these recording techniques are also presented.

2.1. Recording Techniques

In this subsection, we review various recording techniques that are commonly used in audio and acoustics in academic studies and commercial applications.

2.1.1. Stereo and Surround Recording

Sound recording can be conducted in the form of one, two or any number in microphone array. Compared to a single microphone recording, stereo recordings could provide more spatial information on the sound field, including sense of direction, distance, ensemble of the sound stage and ambience. There are basically four types of stereo recording configurations: coincident pair, spaced pair, near-coincident pair, and baffled pair. With the evolution of sound playback system from stereo to surround sound, it is straightforward to record the sound field in surround format, even though most of the surround sound comes from mixing process. Most surround recordings are to be played (directly or after some mixing) via 5.1/7.1 surround sound system. There are various types of surround sound microphone settings. Examples of such surround microphone system include Optimized Cardioid Triangle (OCT) surround, which employs five cardioid mics with one facing front, two facing the sides, and two facing rear; IRT surround, which employs four cardioid mics placed in a square with 90 degree angles; Hamasaki square, which employs four bidirectional mics facing the sides; and a few other configurations. For more practical guidelines on microphone recordings please refer to [30,31].

2.1.2. Microphone Array Recording

To better capture the spatial sound field, more microphones or an array of microphones are required. This idea dated back to the 1930s, where Steinberg and Snow introduced an "acoustic curtain" system that consists of a wall of microphones, which are directly connected to a matching loudspeaker array on the playback end [32]. With a good matching (closely positioned) between the microphone array and loudspeaker array, the recreated sound field is more realistic.

However, such well-matched recording and playback system is hardly practical in today's applications, where the configurations of the playback system (including headphone, number and positions of loudspeakers) vary greatly in different room settings. As a result, microphone recordings are no longer used directly without processing, but require real-time or post-processing to analyze the characteristics of the sound field. In general, a microphone array with signal processing could provide three types of the information about the sound field [33]: (i) the locations of the sound sources; (ii) the corresponding ("clean") sound source signal; and (iii) the characteristics of the sound environment, including the reflections and ambience. Therefore, microphone arrays are widely used in sound source localization, noise reduction, speech enhancement, echo cancellation, source separation, room impulse response estimation, and sound reproduction [34].

Microphone arrays are classified into two categories based on whether they are moving or stationary during the recording. Moving microphone arrays could be used for static or moving sound events. For static sound events, the movements of the microphone array create multiple virtual arrays or enlarges the array size so that more precise information on the sound field could be obtained. A facility enabling the regular movements of the array equipment (e.g., a motor) is required for the moving arrays system. In this case, the assumption that the sound field is spatially static, is very critical. In addition, the effect of recording system movement on the recordings (including the time difference, Doppler effect) should be accounted for. Furthermore, if the movement pattern of the array could be recorded, moving arrays could also be useful in recording non-static sound events. For example, the array could follow the main sound event to give it more focus.

The other category of microphone arrays is the static array. For practical reasons, static arrays are widely adopted as it is easier to set up a static microphone array for recording than moving arrays. Since microphones are to simulate human ears, it is reasonable to establish static microphone arrays as sound reproduction normally assumes the listeners to be static. The most commonly-used static microphone array is the uniform linear array (ULA), which is regularly deployed in beamforming applications. Multiple linear arrays can be arranged in a plane to realize a planar array. Circular array is another commonly used geometric configuration of static microphone array. Compared to linear arrays, circular array provides a two-dimensional scan (360° in the horizontal plane) with uniform performance. Extensions to the circular array include multiple circular arrays at the same plane with different radius, or multiple circular arrays at different planes to form a sphere (i.e., spherical microphone array). Spherical microphone arrays are attractive as they can be decomposed using spherical harmonics, which facilitates the sound field representation and interpretation. Spherical arrays with a large radius can be configured to surround the sound event, creating a third-person perspective, while compact spherical arrays are used to record sound events from the listener's perspective (i.e., first-person perspective). Theoretical foundations of spherical array processing are discussed in-depth in [35] As shown in Figure 2, examples of circular or spherical array include B&K [36], Nokia OZO [37], MTB [38], Eigenmike microphone from MHAcoustics [39], and the VisiSonics [40] microphone array. An illustration of various microphone array geometries can be found in [41].

Figure 2. Examples of spherical microphone array (from left to right) from Nokia OZO with 8 microphones, Dysonics Randomic with 8 channels, Eigenmike with 32 channels, B&K with 36 channels, and VisiSonics with 64 channels.

For interactive sound reproduction applications, such as virtual reality, the listening positions might be changing from time to time. In this case, the way of using moving array or static array might be different. Under this circumstance, the moving array should be used to follow the main sound event [42]. The static array can also be used with advanced signal processing applied so that the reproduced sound adapts to sound event or user movements [29].

2.1.3. Binaural Recording

Binaural recording is an extended form of stereo recording. Ideally, binaural recording only captures the sound received at the left and right ear positions (eardrum for the dummy head and blocked or open entrance of the ear canal for human listeners). Therefore, binaural recording is the format that is the closest to the human hearing, when it is played through calibrated headphones. By recording sound at the eardrum positions, it automatically embeds all the cues needed for sound localization in the 3D space and the natural alteration of the sound timbre due to propagation, reflection, and scattering of sound as it interacts with our body.

The different binaural recording setups can be classified into four main types: (1) binaural microphones inserted in the ears of the human listeners; (2) microphones in the dummy head with torso, head and ears; (3) microphones in the simulator with head and ears; (4) microphones in the simulator with only ears.

First, a miniature microphone is usually placed at the entrance of the ear canal of human beings since it is impractical and difficult to be placed inside the ear canal near the eardrum. Examples of binaural microphones vary from professional, calibrated types (e.g., from Brüel & Kjær [43], shown in Figure 3) or low-cost types. Some other binaural recording microphones are integrated into a headset. An appropriate recording device is required to obtain the binaural recording.

In addition to the use of human beings for binaural recording, dummy heads are also commonly used in academia and the industry to obtain consistent and comparable results. A dummy head usually consists of a torso, head and ears, which are made up of special materials whose acoustic properties are similar to human body and its anthropometry generally follows closely with the average of the whole or a certain part of the population [44]. Several dummy heads that consist of a torso, head and ears include KEMAR (Knowles Electronics Manikin for Acoustic Research) [45], Brüel & Kjær 4128 HATS (head and torso simulator) [46], Head acoustics HMS IV [47], as shown in Figure 4. An example for dummy heads with only a head and two ears is the Neumann KU-100 [48]. As shown in Figure 5, the 3Dio binaural microphones [45] offer superb portability for binaural recording when compared to the bulky dummy heads, but at the expense of lacking head shadowing effects due to the absence of a head. Furthermore, the 3Dio omni binaural microphone records sound from four different head orientations, which could probably be more useful in interactive spatial audio in VR applications.

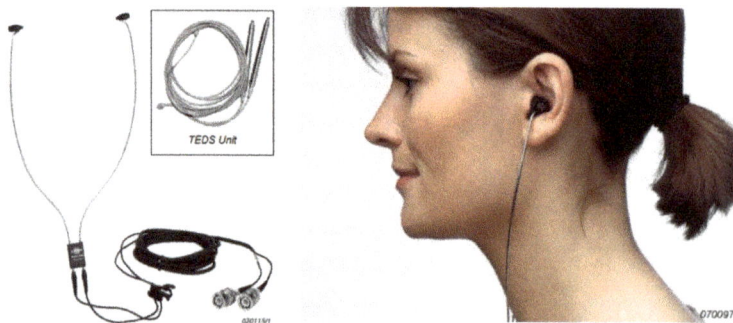

Figure 3. Brüel & Kjær 4101 Binaural microphone worn on the ear.

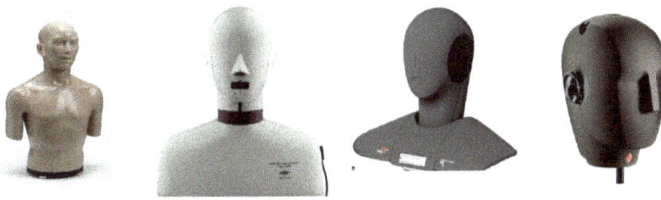

Figure 4. Dummy heads (from left to right): KEMAR, Brüel & Kjær 4128 HATS, Head Acoustics HMS III, and Neumann KU-100. Note that Neumann KU-100 dummy head is torso-free.

Figure 5. Binaural microphones from 3Dio, Free space binaural microphones.

2.1.4. Ambisonic Recording

Ambisonics is a method of recording and reproducing audio of a sound field in full-sphere surround. Essentially, ambisonics is a multi-channel surround sound format that does not contain any information or requirement for the playback configuration as demanded by other surround sound recording formats. This implies that the ambisonics recorded signals can be used in any playback setups. It is capable of full surround, including height and depth from a single point source in space [49].

Ambisonics format, or more precisely first order ambisonics (FOA), which is widely known as B-format surround signal, can be achieved by a special type of microphone in the shape of tetrahedron using four nearly coincident capsules. In other words, ambisonics can be simply understood as an extension of XY stereo recording with addition of the two other dimensions. The four components are labelled as W, X, Y and Z, where W is corresponding to an omnidirectional microphone and X, Y, Z corresponds to three spatial directions, i.e., front-back, left-right, and up-down, respectively, captured using figure-of-eight microphone capsules. For a given source signal, S with azimuth angle θ and elevation angle ϕ, ambisonics pans the desired four components as:

$$W = \frac{S}{\sqrt{2}}$$
$$X = S \cdot \cos \theta \cdot \cos \phi$$
$$Y = S \cdot \sin \theta \cdot \cos \phi$$
$$Z = S \cdot \sin \phi$$

The resulting four-channel signal can then be transcoded into outputs of various formats, from a single source in mono to multichannel surround sound arrays. The major advantage being that with the initial capture, you can use post processing to vary the pan, tilt, zoom and rotation of the sound field, which is hard to achieve with other systems. The limitation of the first order ambisonics is the limited spatial resolution, which affects the sound localization and is only effective in a relatively smaller sweet spot. To improve the performance of first order ambisonics, higher order ambisonics are employed by adding more microphones to record sound field at higher orders. As a result, the reproduction of higher order ambisonics also requires more loudspeakers.

Examples of ambisonics microphones in the market include: Sennheiser AMBEO VR Microphone [50], Core Sound TetraMic [51], SoundField SPS200 Software Controlled Microphone [52], as shown in the Figure 6. Moreover, the spherical microphone arrays described in Section 2.1.2 can also be converted into first or higher order ambisonics format. For example, the Eigenmike consists of 32 microphones, which supports up to 4th order ambisonics. A comprehensive objective and subjective comparative study regarding the use of different ambisonics microphones was conducted in [53,54].

Figure 6. Ambisonics microphones (from left to right): Sennheiser AMBEO, Core Sound TetraMic, and SoundField SPS200.

2.2. Application of Spatial Audio Recording in Soundscape Studies

The section discusses how we can adopt different recordings techniques for soundscape studies. As the application of soundscape is relatively new, there are no comprehensive standards in terms of how recordings should be conducted for studies or reports on soundscape. Soundscape researchers employ different recording techniques for capturing the acoustic environment depending on various factors, such as the fidelity of reproduction medium, ease of capture, cost, etc.

Binaural and ambisonics are the two most common recording techniques in soundscape studies. Many have used binaural measurement systems, such as an artificial head or binaural microphones to record acoustic environment and binaural recordings are used for the reproduction of acoustic environments in laboratory-based listening experiments [54–56]. Ambisonic recording methods have recently received much attention in soundscape studies, as they are not restricted by playback mediums. Boren et al. [28] captured the acoustic environment in different parts of New York with an ambisonic microphone. Davies et al. [22] also used a FOA ambisonics microphone to record urban areas in Manchester, UK.

Although the mentioned recording methods for soundscape applications are not extensive, some insights can be gained to better understand how different recording techniques can be used in soundscape studies. Let us consider the three stages of the soundscape design process as introduced in Section 1. In Stage 1, recordings are used to evaluate the existing soundscape. In Stage 2, recordings are used to design a better soundscape. Finally, recordings are captured to validate the design soundscape after its implementation. In all these stages, the most critical requirement for soundscape recording is that it must sufficiently represent the characteristics of the acoustic environment in question that would facilitate the reproduction of the acoustic environment with sufficient perceptual accuracy. It should be made clear that the degree of perceptual accuracy is dependent on the goal of the study. On this note, the recording techniques must also be chosen with consideration of the reproduction mediums. A detailed review of the spatial audio reproduction techniques can be found in Section 3.

While the community continues to develop recording techniques for spatial audio reproduction, there are some trends and early examples that we can learn from. The strengths and weaknesses of these recording techniques in terms of their applications in soundscape studies are summarized in Table 1. With its simplicity, it is evident that ambisonics is the leading recording technique for interactive spatial audio reproduction. On the other hand, conventional stereo and surround recording techniques can still be employed for specific applications, such as non-diegetic sound (or ambience) of the acoustic environment in soundscape studies. Microphone arrays are very useful to capture a more complete sound field (depending on the number of microphones used) and could be used in postprocessing to further emphasize certain sound components, such as, speech from noisy environment. Binaural recording still works great with static listening positions, but it does not allow interactions in soundscape studies. Even monophonic microphones are required for dry recordings that can be spatialized with post-processing, such as in auralization. Thus, it is clear that choosing the most suitable recording techniques depends on the reproduction techniques as well as the degree of perceptual accuracy required for soundscape studies. For high degree of perceptual accuracy, interactive spatial audio reproduction is required, and thus, ambisonic recordings might be a more suitable choice.

Table 1. Comparison of recording techniques for soundscape studies.

Recording Techniques	Strengths	Weaknesses	Remarks
Stereo and Surround recording	• Legacy recording methods; • Widely adopted in industry; • Direct playback over loudspeaker system.	• Does not capture 3D sound field (usually only covers 1D or 2D); • Not suitable for direct playback over headphones; • Does not support head movements.	• Limited spatial quality in soundscape studies.
Microphone arrays	• Ability to focus on certain sounds in different sound fields; • Support head movements.	• Requires large number of microphones for good performance; • Requires sophisticated signal processing to obtain desired sound from recording.	• A general method that could be used to record a more complete sound field for soundscape reproduction.
Ambisonics	• Records 3D sound fields with only 4 microphones; • Good mathematical foundations for recording and playback; • Efficient rendering for interactive applications; • Rapidly increasing popularity in industry.	• Not suitable for non-diegetic sound like music; • Better performance requires higher order ambisonics; • Absence of international standards.	• Well suited for interactive reproduction in soundscape studies.
Binaural recordings	• Closest to human hearing; • Direct playback over headphones.	• Specialized equipment needed, e.g., ear simulators, or wearable binaural microphones; • Lack of support for head movements; • Non-personalized rendering (e.g., dummy head recordings).	• Most commonly used recording technique for soundscape studies due to its simplicity; • Good spatial quality but limited interaction in soundscape studies; • Personalized rendering (i.e., from in-ear binaural microphones).

3. Spatial Audio Reproduction Techniques for Soundscape Design

Perceptually accurate reproduction of the acoustic environment is crucial to achieve high ecological validity for evaluation of soundscape in laboratory conditions. This requires synthesis and rendering of different sound sources to create an immersive playback system, for instance, in subjective listening tests with sufficient perceptual accuracy. Soundscape composition and acoustic reproduction was pioneered by Schafer and his group in 1970s when they published a record titled "The Vancouver soundscape" [57]. As the reproduction techniques for spatial audio become more advanced, the soundscape researchers started adopting them in their studies from stereophonic techniques, multi-channel setups [57], to ambisonics and wave field synthesis [58].

A rendering algorithm along with multiple transducers are often used for reproduction of spatial sound. The sound can be rendered using either a pair of headphones or an array of loudspeakers arranged in a specific configuration. The aim of the system is to reproduce the sound field in such a manner so as to give a sense of perception of spaciousness and directivity of sound objects located in 3D space [27]. The sounds generated in such a scenario are often referred to as virtual sound in contrast to real sound we commonly hear in everyday life [59].

The earliest attempt for reproduction of sound dates back to the phonograph invented in 1887. The first stereo system was introduced by Bluemin [60] in 1931 with its first commercial use in 1949. Dolby introduced the first surround sound system in 1970 and in the same year, Gerzon invented ambisonics [49]. In 1984, binaural reproduction technique is introduced by Sony and since 1989, the transaural technique gained popularity [61]. In 1993, another technique called the wave field synthesis (WFS) was proposed with its first major commercial application demonstrated in 2001 by Carrouso. In 2011, IOSONO came up with the world's first real-time spatial audio processor for WFS. In 2015, MPEG-H standard was announced, with the aim to ensure all types of sound formats are supported in any types of playback systems. The most recent attempts of spatial audio rendering were focused on VR/AR applications [27,62].

The above mentioned spatial audio reproduction techniques are primarily divided into two categories. The first category uses the technique of physical reconstruction of sound, which aims to synthesize the entire sound field in the listening area as close to the desired signal as possible. The second category is the perceptual reconstruction of sound, which employs psychoacoustic techniques to create a perception of spatial characteristics of sound [59]. The evolution of reproduction techniques is shown in Figure 7 along with their classification into physical and perceptual reproduction methods.

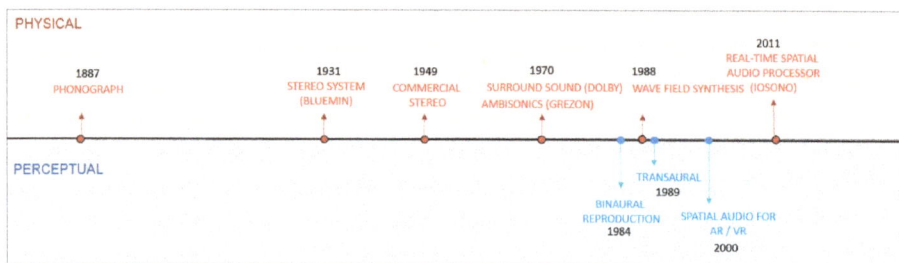

Figure 7. Evolution of spatial audio reproduction systems.

The primary aim of the section is to introduce the spatial audio techniques and to highlight the soundscape studies that have used these techniques. Moreover, since these techniques have different encoding and decoding formats, they require appropriate tools for rendering audio. Table A1 in Appendix A describes some of such tools that are available, where, many of these tools are open source and free to use. Usage of such tools in soundscape studies could enable the researchers to choose appropriate techniques according to the merits and demerits of each technique, which are described in the following Sections 3.1 and 3.2 and summarized in Table 2 in Section 3.3.

3.1. Physical Reconstruction of Sound

The aim of physical reconstruction of sound is to create the sound field in the listening area as close as possible to the desired sound field [59]. The oldest and one of the most popular methods of sound reproduction for soundscape uses two speakers in stereo configuration, i.e., placed at an angle of ±30° from the listener [60]. In soundscape studies, the stereo configuration was used by Schafer and his group. Vogel et al. [63] used stereophonic sound technique to study soundscape for French cities in 1997. Payne used a 2.1 channel system for testing the Perceived Restorativeness Soundscape Scale (PRSS) [64].

Multi-channel reproduction methods became popular in consumer devices from 1950. The international standard ITU-R BS.1116-1 establishes the recommendation for reproducing sound and assessment methods for multi-channel sound systems [65]. These methods have been widely used in the reproduction of acoustic environment in soundscapes. Gustavino et al. studied the subjective perception for playback through different multi-channel setups [66]. The speakers were arranged in different spatial configurations with a hexagon shaped prototype listening room created to test out different reproduction methods. The results of the experiments showed that while frontal perception was best for a 1-D configuration, the spatial definition of audio was best reproduced with 2-D configuration consisting of 6 speakers located at edges of the hexagonal room. In another article, Guastavino et al. [16] showed that multichannel reproduction is more suitable for processing complex auditory scenes, and playback of urban soundscapes in laboratory conditions, as compared to stereophonic reproduction techniques.

Wave field synthesis and ambisonics are the other two physical reconstruction techniques, which aim to create the same acoustical pressure field as present in the surrounding. Ambisonics, introduced by Gerzon et al. [49] is based on the decomposition of a sound field using spherical harmonics. Ambisonic reproduction is flexible to be applied in any sound playback configurations, which makes it very attractive to be used in a wide variety of applications [67]. Davis et al. [21] used the first order ambisonic microphones to record the background ambient soundscape and used a monophonic microphone to record the foreground sounds separately. The soundscape synthesis was carried out using a simulation tool. It allowed the sounds to be layered with each other and effects, like reverberation and reflections, to be added in real-time [68]. An eight-channel loudspeaker setup was used in a semi-anechoic chamber for playback. Boren et al. [58] described the usage of 16-channel audio configuration for playback of soundscape recording done in different parts of New York using sound field tetramic for recording. Moreover, there is increased support for ambisonics in VR/AR through leading industry players, including Google [69], BBC [70], and Facebook [71]. The ambisonics systems mentioned thus far are in the first-order configuration, with satisfactory but limited spatial resolution as compared to higher-order configurations, as discussed in Section 2.1.4. However, higher-order ambisonics systems are still pre-mature and costly today.

In a study for comparing the ecological validity among the different reproduction methods related to physical reconstruction of sound, Gustavino et al. [16] compared the stereophonic and ambisonic reproduction techniques. The study analyzed verbal data collected through questionnaires, and compared it to the field survey using semantic differential analyses for different sound samples. The study concluded that a "neutral visual environment" along with spatial immersion of recreated soundstage is essential to ensure high ecological validity.

3.2. Perceptual Reconstruction of Sound

Perceptual reconstruction techniques for spatial audio aim to replicate the natural listening experience by generating sufficient audio cues to represent the physical sound. Binaural technique, according to Blauert [72], is defined as "a body of methods that involve the acoustic input signals to both ears of the listener for achieving practical purposes, e.g., by recording, analyzing, synthesizing, processing, presenting and evaluating such signals". Reproduction through this technique has two parts: one is the synthesis and rendering portion of the signal, and the other is the playback system. Head related transfer functions (HRTFs) are used to describe the change in sound spectrum due to the interaction of sound waves with listener's body, head and pinna [59]. The synthesis and rendering of binaural audio is usually realized by convolving a dry sound source with the HRTF at a particular direction. For accurate reproduction, personalized binaural rendering using individualized HRTFs is required [73]. There are various techniques to accurately measure or synthesize the individualized HRTFs [74,75]. Recently, He et al. [76,77] proposed a fast and continuous HRTF acquisition system that incorporates the head-tracker to allow unconstrained head movements for human subjects. Binaural technique is used by soundscape researchers to playback the recorded sound with sufficient spatial fidelity. Axelsson et al. [54] used it for their study with the aim of finding the underlying and primary components of soundscape perception. They used an artificial head (Brüel & Kjær Type 4100) for binaural recording and headphone (Senheiser HD600) for playback.

Binaural reproduction can also be realized over a pair of loudspeakers. Transaural audio is the method used to deliver correct binaural signals to ears of a listener using speakers (mostly in a stereo setup). However, there are some challenges to this approach, namely effective crosstalk cancellation and limited sweet spot range. Several solutions to these problems have been described in the literature, e.g., [78]. Often, double transaural approaches [78] are used to make sure that the front back confusion is minimum. However, this method has rarely been used in soundscape studies. Gustavino et al. [79] compared the quality of spatial audio reproduction between transaural, ambisonics and stereo methods. They used the Ircam default decoder to playback different auditory scenes from both indoors and outdoors [66], such as road traffic noise, a car interior, music concert, etc. Two experiments were performed, one for overall spatial quality evaluation and the other for localization accuracy. The results from the study indicate that ambisonics provide good immersion but poor localization for a sound scene. Stereophony and transaural techniques, on the other hand could be useful in the case where precise localization is required but do not have good immersive spatial sound.

3.3. Comparison of Different Spatial Audio Reproduction Techniques for Soundscape Studies

It is crucial that soundscape researchers select the appropriate sound reproduction technique for their use to make sure that the playback sounds similar to the natural listening scenario. To this end, Table 2 highlights the strengths and weaknesses of common reproduction techniques mentioned above, extended from studies in [80].

Table 2. Strengths and weaknesses of reproduction techniques for soundscape studies.

Reproduction Techniques	Number of Channels	Strengths	Weaknesses
Perceptual Reconstruction			
Binaural	Two	• Enables creation of virtual source in 3D space with two channels. • Lower equipment cost as compared to other solutions.	• It is only suitable for single user. • Individualized HRTFs needed to avoid front back confusions and in-head localizations.
Transaural	Two or four	• Enhances spaciousness and realism of audio with limited number of speakers. • Accurate rendering of spatial images using fewer loudspeakers.	• Requires effective crosstalk cancellation. • It is only suitable for single user.
Physical Reconstruction			
Stereo	Two	• Legacy reproduction method. • Widely adopted in industry.	• Poor spatial effect. • The sound phantom image is always created at sweet spot.
Multichannel	Three or more	• Better spaciousness of 360° audio as compared to stereo setups. • Well adopted by industry.	• Large numbers of channels and speaker systems needed for spatial realism. • Unable to achieve accurate accurate 360 degree phantom images.
Ambisonics	$(N + 1)^2$ for Nth Ambisonic	• Can be used with any speaker arrangement. • Core technology is not patented and free to use.	• Listener movement can cause artifacts in sound. • Complicated setup. • Not popular in industry.
Wave Field Synthesis	More than 100 usually	• The sweet spot covers entire listening area. • The virtual sound sources independent of listener position	• High frequency sounds beyond aliasing frequency not well reproduced. • Large number of speakers needed.

In addition to the above techniques, Auralization techniques have recently become popular among soundscape researchers to achieve sufficient perceptual accuracy of acoustic reproduction. Auralization can be defined as a technique of creating audible sound files from numerical data [81]. It is not strictly a perceptual technique and can include the physical reproduction methods as well. The technique has been used in reproducing various acoustic environments, such as concert halls [82], auditoriums [83] and in room acoustics. The development of accurate models and efficient algorithms for accurate sound reproduction employs various techniques, like ray tracing [84], radiosity methods [85], finite element methods, etc. [86]. These models are used in conjugation with the reproduction setups to add audio effects and cues for increasing realism and spaciousness for spatial sound.

4. Spatial Audio in Virtual and Augmented Reality for Soundscape Design

Due to the importance of audio-visual interaction in soundscape perception, the audio techniques reviewed in this paper are commonly paired with a variety of visual mediums. Increasingly, virtual reality (VR) and augmented reality (AR) systems have been employed in soundscape research. Visuals recorded using omni-directional cameras, or designed in 3D modelling software, when rendered through head-mounted displays, are perceived to be immersive and realistic. To complement the immersive visuals, the accompanying audio has to possess a minimum degree-of-spatialness to achieve natural realism. This argument stems from the perceptual nature of soundscapes and thus, a disjoint in the spatialness of the audio and visual elements will degrade the ecological validity [87]. Hence, the general rule of thumb for audio with VR and AR should correlate with the receivers' movements (mostly head movements), to create an interactive and immersive perception of soundscape.

Although spatial audio techniques for VR are discussed in detail in various studies [88,89], sufficient care is needed when applying these techniques into soundscape research. With the above rule in mind, the spatial audio recording techniques that is suitable for VR (or virtual sound in AR) must be able to capture the acoustic environment from all directions. From this perspective, ambisonics recording is the most commonly used recording techniques for VR/AR, though general microphone arrays with suitable post-processing can also be applied.

The spatial audio reproduction for VR/AR is usually realized through binaural rendering with head-tracking, though multichannel playback systems with an ambisonics decoder is also possible. In the binaural rendering system, when the human head moves, these HRTFs must be updated accordingly to account for the changes of the sound directions, which help to create a more interactive and natural listening experience. Note that the binaural rendering of the ambisonics recording is implemented by virtual loudspeakers in the ambisonics reproduction setup. For instance, a stationary VR scene that allows for head-track views of the visual scene (e.g., user is standing still in the virtual environment but able to turn the head to "look" around) needs to be accompanied by head-tracked binaural reproduction. An example of a reproduced VR scene with head-tracked binaural reproduction over headphones is shown in the right half of Figure 8. Recently, this approach has been used in soundscape research since it allows more immersive reproduction of the acoustic environment. One good example that soundscape research has benefitted from the use of VR techniques is in the SONORUS project [90]. Maffei et al. used VR for evaluating the influence of visual characteristics of barriers and wind turbine on noise perception [23,91,92].

Along with VR, the rising popularity of AR devices can be attributed to its availability in the consumer space. Devices such as the Hololens and Meta 2 have the capability to render holograms through head mounted displays [93,94]. With the consumerization of AR systems, soundscape researchers and practitioners have access to tools that can virtually augment visual and audio elements in the real environment. The general rule for applying audio to AR should still be adhered to, for achieving the desired perceptual accuracy during auralization in the soundscape design process.

There are several ongoing research works to achieve a high degree of perceptual accuracy for AR audio. Härmä et al. [95] described an augmented reality audio headset using binaural microphones to assist the listener with pseudo-acoustic scenes. Tikander et al. [96] also developed a mixer for equalizing and mixing virtual objects with real environment. Ranjan et al. [27] proposed an augmented reality headset in which they use open-back headphones with pairs of external and internal microphones to achieve sound playback that is very similar to natural listening scenario.

These devices and upcoming innovations would be useful for projects that involve altering the soundscapes of existing locations. They could be used in soundscape studies to test different hypothesis in an immersive environment, which allows both virtual reproduced and real ambient sounds to be heard at the same time. AR devices have the advantage of including different elements of the perceptual construct of the soundscape, including meteorological conditions and other sensory factors [7]. Moreover, if accurate spatial sound is used with these AR devices, it would enable the soundscape researchers to fuse the virtual sound sources seamlessly with the real sound, thus enabling highly accurate interpretation of auditory sensation by the user. The viability of an augmented reality audio-visual system has been demonstrated with a consumer AR headgear and open-backed headphones with positional translation- and head-tracked spatial audio, as shown in the left half of Figure 8.

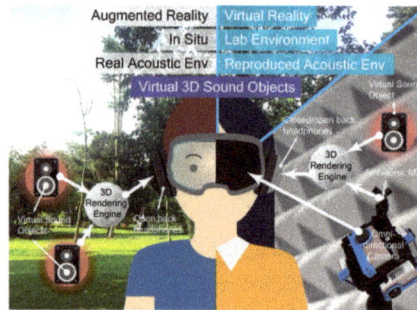

Figure 8. (**Left**) AR setup with spatial audio from rendered virtual sound objects for in situ environments. (**Right**) A lab-based VR setup from omni-directional camera recordings and a reproduced acoustic environment using spatial audio from ambisonic microphone recordings and rendered virtual sound objects [97].

In essence, VR systems with high fidelity spatial audio are well suited for an immersive acoustic environment with a good degree of control under laboratory conditions; whereas AR systems with high fidelity spatial audio are more immersive but are subject to a high degree of variability in the in situ environment. Hence, soundscape practitioners should exercise care in the selection of audio-visual mediums and should consider the complementary nature of VR and AR techniques mentioned.

5. Discussion

Soundscape design should be based on the relationship between human perceptual construct and physical phenomenon of the acoustic environment. As shown in Figure 1, acoustic recording and reproduction techniques are essentially adopted through every stage of soundscape design process. In particular, soundscape recording and reproduction techniques play a more critical role in Stage 2 for proposing and evaluating the soundscape designs before their implementations in situ. In Stage 2, various soundscape design approaches might be applied to improve the existing poor acoustic conditions, and those approaches should be assessed through subjective tests based on human perception of acoustic environment. In this context, soundscape design needs to consider the degree of ecological validity of the soundscape for reliable solutions, which largely depends on the adopted recording and reproduction techniques. Therefore, for clarity, the discussion will be focused on highlighting the degree of perceptual accuracy of the recording and reproduction techniques introduced in Sections 2 and 3 respectively, in representing the acoustic environment.

As stated in ISO 12913-1:2014, any evaluation of soundscape is based upon the human perception of the acoustic environment (actual or simulated). The technique of constructing an audible acoustic environment, or Auralisation, is commonly used for soundscape evaluation in controlled environments. According to Vorländer [81], the goal of auralisation also stems from the human perception of the acoustic environment, in which the synthesized acoustics only needs to be perceptually correct and not physically perfect. Perceptual accuracy, also called plausibility, of the virtual acoustic environment is defined by Lindau and Weinzierl as "a simulation in agreement with the listener's expectation towards an equivalent real acoustic event" [98]. The listener's expectation, however, is subjective and can vary depending on the intended tasks in the acoustic environment. Ultimately, the ecological validity of a virtual acoustic environment will be application specific. This task-specific criterion implies that there are different levels of perceptual accuracy that can create "a suitable reproduction of all required quality features for a given application" [87]. Out of the quality features suggested by Pellegrini, emphasis will be directed to features, which are attributed to recording and reproduction techniques.

Therefore, it would be beneficial to classify the characteristics of the acoustic environment, such that the appropriate techniques are selected to achieve the desired level of perceptual accuracy. The characteristics of the acoustic environment are summarized in Table 3.

Table 3. Recommended audio reproduction and recording techniques for virtualizing/augmenting acoustic environments.

Characteristics of the Acoustic Environment					Recommended Techniques		
Spatial Fideli [1]	Type of Environment [2]	Movements		Virtual Sound Source Localization [3]	Reproduction Techniques	Recording Techniques	Use Case(s) (Selected References, if Any)
		Listener Position [4]	Head				
Low	Virtual (R/S)	×	×	0D	Mono loudspeaker; stereo headphone	Mono	Masking road traffic noise with birdsongs [99]
	Virtual (R/S)	×	×	1D	Stereo/surround loudspeaker; stereo headphone	Stereo/surround	Reproduced acoustic environment [25]; Perceived restorative-ness soundscape scale [71]
	Virtual (R/S)	×	×	2D	Surround sound loudspeakers with height	Array	
Med	Virtual (R/S)	×	×		Ambisonics (2D)	Ambisonics	Perception of reproduced soundscapes [22]
	Virtual (R/S)	×	×	3D−	Ambisonics; Binaural	Ambisonics; Binaural;	Auralising noise mitigation measures [100]; Masking noise with water sounds [101,102]
	Virtual (R/S)	×	×	3D+	Personalized binaural (PB) [5]	Personalized binaural; Ambisonics [6]	
	Virtual (R/S)	×	✓	3D+	Binaural/PB with head tracking	Ambisonics	
High	Virtual(S)	✓	✓	3D+	WFS; Binaural/PB with positional & head tracking	Mono (anechoic); Ambisonics	LISTEN project [103]
	Real + Virtual(S) [7]	✓	✓	3D+	WFS; Binaural/PB with positional & head tracking	Mono (anechoic); Ambisonics	Augmented soundscape [27,97]

[1] Spatial fidelity refers to the quality of the perceived spatiality (location of sound in 3D space) of sound sources, which directly influences the auditory sensation. [2] Acoustic environments can be real, captured, and reproduced (R), or synthesized from existing sound tracks (S). [3] As described in Section 5. [4] Refers to listeners' translation movements. The virtual sound reproduction must be able to adapt the sound to translation movements. [5] Refers to binaural rendering with individualized HRTFs. [6] Requires convolution of down-mixed binaural with individualized HRTFs. [7] This refers to the case of recreating a virtual soundscape in a real acoustic environment. * Note that not all spatial audio recording and reproduction techniques reviewed have been used in soundscape studies, e.g., personalized binaural recording and rendering. The perceptual accuracy and ecological validity of these techniques need to be further examined.

The acoustic environment characteristics are organized firstly by its type, namely: (1) "Virtual(R)", a simulated environment reproduced from recordings at a real location; (2) "Virtual(S)", a simulated environment produced by auralization [81]; and (3) "Real", the actual physical environment.

Next, the characteristics based on interactivity of the listener in the acoustic environment is identified. The interactivity or 'Movement' characteristics in the table are extremely relevant for VR and AR applications, as described in Section 4 For instance, the sound has to be in sync with the receivers' positional and head movements as in the real environment (for AR) or in the virtual environment (in VR).

The ability to localize sound sources in the acoustic environment is also an important characteristic, which is also included and labelled as "Virtual sound source localization" in Table 3. The degree of localization is further categorized into 0D, 1D, 2D, 3D−, and 3D+. Virtual sound source localization in: (1) 0D refers to the perception with no spatial cues; (2) 1D refers to sound perception limited to left and right in horizontal plane; (3) 2D refers to sound perception with the inclusion of azimuthal and elevation dimensions; and (4) 3D refers to the sound perception with azimuth, elevation and distance, where − and + shows poor and good performance, respectively. Lastly, based on the degree of movement and localization characteristics, the spatial fidelity of the acoustic environment is simply classified into three grades for simplicity: Low, Medium, and High.

The description of the acoustic environment characteristics in Table 3, are accompanied by their respective reproduction and recording techniques. The limitations of the soundscape study, in terms of the spatial characteristics of the acoustic environment, can be decided by referring to Table 3. For reference, past soundscape evaluation studies are included in the last column of the table. It should be stressed that Table 3 does not suggest that all soundscape studies should employ high spatial fidelity audio, but instead provides a guide to construct the experimental conditions needed for soundscape research.

6. Conclusions

Recently, soundscape approaches have attracted more attention due to the increasing importance of evidence-based and perception-driven solutions to build better urban acoustic environments. Soundscape recording and reproducing techniques are essential tools for soundscape research and design. The present paper provides an overall picture of various spatial audio recording and reproduction techniques, which can be applied in soundscape studies and applications. Soundscape researchers should understand the strengths and weaknesses of these spatial audio recording and reproduction techniques, and apply the most appropriate techniques to suit their research purposes. Notably, the emerging VR/AR technologies, together with the advanced spatial audio recording and reproduction techniques, enable a more interactive and immersive auditory and visual perception, and would be a great fit for soundscape design with high ecological validity. Future research needs to focus more on spatial aspects of soundscape design elements for developing more accurate soundscape models.

Acknowledgments: Vectors modified under creative commons CC BY 2.0 for Figure 8 are designed by Vvstudio and Freepik. The background used in Figure 8 is modified for reuse under creative commons CC BY 3.0, was created by guillaumepaumier.com.

Conflicts of Interest: The authors declare no conflict of interest.

Appendix A

Table A1. Tools for rendering spatial audio for soundscape studies.

Tools (Type of Application)	Reproduction Format/Rendering Usage							
	BIN	TRA	AMB	WFS	STE	VAR	MUL	AUR
Soundscape Renderer (C++ application for rendering) [104]	×		×	×				
Transpan (Max/MSP application) [105]	×	×						
Simmetry 3D (Multi channel reproduction) [106]							×	
PSTD (Blender interface for auralization) [107]								×
UrbanRemix (Collaborative soundscape measurement) [108]					×			
CATT acoustics (Room acoustics and auralization) [109]	×							×
EARS (Indoor auralization) [110]	×				×			×
Urban street auralizer (Micro scale urban areas simulation) [111]	×					×		×
FB360 spatializer (Sound spatializer for VR) [71]	×		×		×	×	×	
Google VR (Spatial sound plugin) [112]					×	×		×
Slab 3D plugin (Rendering plugin) [113]	×							×
Hololens spatial audio plugin (AR based spatial sound rendering) [94]						×		

BIN: Binaural, TRA: Transaural, AMB: Ambisonic first order WFS: Wave field synthesis, STE: Stereo, VAR: VR/AR, MUL: Multi channel surround Sound, VBAP: Vector Base Amplitude Panning, and AUR: Auralization.

1. Pijanowski, B.C.; Farina, A.; Gage, S.H.; Dumyahn, S.L.; Krause, B.L. What is soundscape ecology? An introduction and overview of an emerging new science. *Landsc. Ecol.* **2011**, *26*, 1213–1232. [CrossRef]
2. Guastavino, C. The ideal urban soundscape: Investigating the sound quality of French cities. *Acta Acust. United Acust.* **2006**, *92*, 945–951.
3. Schulte-Fortkamp, B. Soundscapes and living spaces sociological and psychological aspects concerning acoustical environments. In Proceedings of the Forum Acusticum, Sevilla, Spain, 18 September 2002; p. 6.
4. Hong, J.Y.; Jeon, J.Y. Influence of urban contexts on soundscape perceptions: A structural equation modeling approach. *Landsc. Urban Plan.* **2015**, *141*, 78–87. [CrossRef]
5. Schafer, R.M. *The Soundscape: Our Sonic Environment and the Tuning of the World*; Destiny Books: New York, NY, USA, 1977.
6. Kang, J.; Aletta, F.; Gjestland, T.T.; Brown, L.A.; Botteldooren, D.; Schulte-Fortkamp, B.; Lercher, P.; Van Kamp, I.; Genuit, K.; Fiebig, A.E.; et al. Ten questions on the soundscapes of the built environment. *Build. Environ.* **2016**, *108*, 284–294. [CrossRef]
7. International Organization for Stadardization. *Acoustics—Soundscape—Part 1: Definition and Conceptual Framework*; ISO 12913-1; ISO: Geneva, Switzerland, 2014.
8. Herranz-Pascual, K.; Aspuru, I.; García, I. Proposed conceptual model of environmental experience as framework to study the soundscape. In Proceedings of the 39th International Congress and Exposition on Noise Control Engineering (INTERNOISE), Lisbon, Portugal, 13–16 June 2010; pp. 1–9.
9. Zwicker, E.; Fastl, H. *Psychoacoustics: Facts and Models*; Springer Science & Business Media: Berlin, Germany, 2013; Volume 22.

10. Brown, A.L. A review of progress in soundscapes and an approach to soundscape planning. *Int. J. Acoust. Vib.* **2012**, *17*, 73–81. [CrossRef]

11. Aletta, F.; Kang, J.; Axelsson, Ö. Soundscape descriptors and a conceptual framework for developing predictive soundscape models. *Landsc. Urban Plan.* **2016**, *149*, 65–74. [CrossRef]

12. Liu, F.; Kang, J. A grounded theory approach to the subjective understanding of urban soundscape in Sheffield. *Cities* **2016**, *50*, 28–39. [CrossRef]

13. Schulte-Fortkamp, B.; Fiebig, A. Soundscape analysis in a residential area: An evaluation of noise and people's mind. *Acta Acust. United Acust.* **2006**, *92*, 875–880.

14. Yang, W.; Kang, J. Soundscape and sound preferences in urban squares: A case study in sheffield. *J. Urban Des.* **2005**, *10*, 61–80. [CrossRef]

15. De Coensel, B.; Bockstael, A.; Dekoninck, L.; Botteldooren, D.; Schulte-Fortkamp, B.; Kang, J.; Nilsson, M.E. The soundscape approach for early stage urban planning: A case study. Pcoceedings of the Internoise, Lisbon, Portugal, 13–16 June 2010; pp. 1–10.

16. Guastavino, C.; Katz, B.F.G.; Polack, J.-D.; Levitin, D.J.; Dubois, D. Ecological Validity of soundscape reproduction. *Acta Acust. United Acust.* **2005**, *91*, 333–341.

17. International Organization for Stadardization. *Acoustics—Soundscape Part 2: Data Collection and Reporting Requirements*; ISO 123913-2; ISO: Geneva, Switzerland, 2017.

18. Vorländer, M. *Auralization*; Springer-Verlag Berlin Heidelberg: Berlin, Germany, 2008.

19. Hong, J.Y.; Jeon, J.Y. Designing sound and visual components for enhancement of urban soundscapes. *J. Acoust. Soc. Am.* **2013**, *134*, 2026–2036. [CrossRef] [PubMed]

20. Kang, J.; Zhang, M. Semantic differential analysis of the soundscape in urban open public spaces. *Build. Environ.* **2010**, *45*, 150–157. [CrossRef]

21. Davies, W.J.; Bruce, N.S.; Murphy, J.E.; Bruce, N.S.; Murphy, J.E. Soundscape reproduction and synthesis. *Acta Acust. United Acust.* **2014**, *10*, 285–292. [CrossRef]

22. Sudarsono, A.S.; Lam, Y.W.; Davies, W.J. The effect of sound level on perception of reproduced soundscapes. *Appl. Acoust.* **2016**, *110*, 53–60. [CrossRef]

23. Ruotolo, F.; Maffei, L.; Di Gabriele, M.; Iachini, T.; Masullo, M.; Ruggiero, G.; Senese, V.P. Immersive virtual reality and environmental noise assessment: An innovative audio-visual approach. *Environ. Impact Assess. Rev.* **2013**, *41*, 10–20. [CrossRef]

24. Maffei, L.; Masullo, M.; Pascale, A.; Ruggiero, G.; Puyana Romero, V. On the validity of immersive virtual reality as tool for multisensory evaluation of urban spaces. *Energy Procedia* **2015**, *78*, 471–476.

25. Maffei, L.; Masullo, M.; Pascale, A.; Ruggiero, G.; Romero, V.P. Immersive virtual reality in community planning: Acoustic and visual congruence of simulated vs real world. *Sustain. Cities Soc.* **2016**, *27*, 338–345. [CrossRef]

26. Lacey, J. Site-specific soundscape design for the creation of sonic architectures and the emergent voices of buildings. *Buildings* **2014**, *4*, 1–24. [CrossRef]

27. Ranjan, R.; Gan, W. Natural listening over headphones in augmented reality using adaptive filtering techniques. *IEEE/ACM Trans. Audio Speech Lang. Process.* **2015**, *23*, 1988–2002. [CrossRef]

28. Boren, B.; Musick, M.; Grossman, J.; Roginska, A. I hear NY4D: Hybrid acoustic and augmented auditory display for urban soundscapes. In Proceedings of the 20th International Conference on Auditory Display, New York, NY, USA, 22–25 June 2014.

29. Eargle, J. *The Microphone Book: From Mono to Stereo to Surround—A Guide to Microphone Design and Application*; CRC Press: Boca Raton, FL, USA, 2012.

30. Bartlett, B.; Bartlett, J. *Practical Recording Techniques: The Step-by-Step Approach to Professional Audio Recording*; CRC Press: Boca Raton, FL, USA, 2016.

31. Huber, D.M.; Runstein, R.E. *Modern Recording Techniques*; CRC Press: Boca Raton, FL, USA, 2013.

32. Fletcher, H. Symposium on wire transmission of symphonic music and its reproduction in auditory perspective basic requirements. *Bell Syst. Tech. J.* **1934**, *13*, 239–244. [CrossRef]

33. Dunn, F.; Hartmann, W.M.; Campbell, D.M.; Fletcher, N.H.; Rossing, T.D. *Springer Handbook of Acoustics*; Rossing, T.D., Ed.; Springer: Berlin, Germany, 2015.

34. Benesty, J.; Chen, J.; Huang, Y. *Microphone Array Signal Processing*; Springer Science & Business Media: Berlin, Germany, 2008; Volume 1.

35. Rafaely, B. *Fundamentals of Spherical Array Processing*; Springer: Berlin, Germany, 2015; Volume 8.

36. Brüel & Kjær. Sound and Vibration Advanced Options for Holography and Beamforming. Available online: http://www.bksv.com/Products/analysis-software/acoustics/noise-source-identification/spherical-beamforming-8606?tab=descriptions (accessed on 5 December 2016).

37. Nokia OZO. Virtual Reality Camera with 360-Degree Audio and Video Capture. Available online: https://ozo.nokia.com/ (accessed on 5 December 2016).

38. Algazi, V.R.; Duda, R.O.; Thompson, D.M. Motion-tracked binaural sound. *J. Audio Eng. Soc.* **2004**, *52*, 1142–1156.

39. mh Acoustics Eigenmike®Microphone. Available online: https://mhacoustics.com/products (accessed on 28 December 2016).

40. Visisonics Products Visisonics. Available online: http://visisonics.com/products-2/#3daudio (accessed on 3 February 2017).

41. Mathworks Phased Array Gallery—MATLAB & Simulink Example. Available online: https://www.mathworks.com/help/phased/examples/phased-array-gallery.html (accessed on 5 December 2016).

42. Paquette, D.; McCartney, A. Soundwalking and the Bodily Exploration of Places. *Can. J. Commun.* **2012**, *37*, 135–145. [CrossRef]

43. Brüel & Kjær. Sound & Vibration Type 4101-B-Binaural in-Ear Microphone Set. Available online: http://www.bksv.com/Products/transducers/acoustic/binaural-headsets/4101?tab=overview (accessed on 15 December 2016).

44. Vorländer, M. Past, present and future of dummy heads. In Proceedings of the Acustica, Guimarães, Portugal, 13–17 September 2004; pp. 1–6.

45. G.R.A.S. Head & Torso Simulators. Available online: http://www.gras.dk/products/head-torso-simulators-kemar.html (accessed on 13 June 2017).

46. Brüel & Kjær. Sound & Vibration TYPE 4128-C HATS. Available online: https://www.bksv.com/en/products/transducers/ear-simulators/head-and-torso/hats-type-4128c (accessed on 22 December 2016).

47. HEAD Acoustics Binaural Recording Systems—Artificial Head System HMS IV—Product Description. Available online: https://www.head-acoustics.de/eng/nvh_hms_IV.htm (accessed on 18 December 2016).

48. Georg Neumann GmbH Products/Current Microphones/KU 100/Description. Available online: https://www.neumann.com/?lang=en&id=current_microphones&cid=ku100_description (accessed on 25 December 2016).

49. Gerzon, M.A. Width-height sound reproduction. *J. Audio Eng. Soc.* **1973**, *21*, 2–10.

50. Sennheiser. Sennheiser AMBEO VR MIC—Microphone 3D AUDIO Capture. Available online: https://en-us.sennheiser.com/microphone-3d-audio-ambeo-vr-mic (accessed on 13 June 2017).

51. Core Sound TeraMic. Available online: http://www.core-sound.com/TetraMic/1.php (accessed on 28 December 2016).

52. SoundField. SoundField SPS200 Software Controlled Microphone | Microphones and Processors with Unique Surround Sound Capabilities. Available online: http://www.soundfield.com/products/sps200 (accessed on 28 December 2016).

53. Bates, E.; Dooney, S.; Gorzel, M.; O'Dwyer, H.; Ferguson, L.; Boland, F.M. Comparing ambisonic microphones: Part 1. In *Audio Engineering Society Conference on Sound Field Control*; Audio Engineering Society: Guildford, UK, 2016.

54. Axelsson, Ö.; Nilsson, M.E.; Berglund, B. A principal components model of soundscape perception. *J. Acoust. Soc. Am.* **2010**, *128*, 2836–2846. [CrossRef] [PubMed]

55. Hong, J.Y.; Jeon, J.Y. The effects of audio-visual factors on perceptions of environmental noise barrier performance. *Landsc. Urban Plan.* **2014**, *125*, 28–37. [CrossRef]

56. Pheasant, R.; Horoshenkov, K.; Watts, G.; Barrett, B. The acoustic and visual factors influencing the construction of tranquil space in urban and rural environments tranquil spaces-quiet places? *J. Acoust. Soc. Am.* **2008**, *123*, 1446–1457. [CrossRef] [PubMed]

57. Truax, B. Genres and techniques of soundscape composition as developed at Simon Fraser University. *Organ. Sound* **2002**, *7*, 5–14. [CrossRef]

58. Boren, B.; Andreopoulou, A.; Musick, M.; Mohanraj, H.; Roginska, A. I hear NY3D: Ambisonic capture and reproduction of an urban sound environment. In *135th Audio Engineering Society Convention (2013)*; Audio Engineering Society Convention: New York, NY, USA, 2004; p. 5.

59. He, J. *Spatial Audio Reproduction with Primary Ambient Extraction*; SpringerBriefs in Electrical and Computer Engineering; Springer Singapore: Singapore, 2017.
60. Blumlein, A.D. British Patent Specification 394,325. *J. Audio Eng. Soc. April* **1958**, *6*, 91–98.
61. Cooper, D.H.; Bauck, J.L. Prospects for transaural recording. *J. Audio Eng. Soc.* **1989**, *37*, 3–19.
62. Auria, D.D.; Mauro, D.D.; Calandra, D.M.; Cutugno, F. A 3D audio augmented reality system for a cultural heritage management and fruition. *J. Digit. Inf. Manag.* **2015**, *13*, 203–209.
63. Vogel, C.; Maffiolo, V.; Polack, J.-D.; Castellengo, M. Validation subjective de la prise de son en extérieur. In Proceedings of the Congrès Français D'acoustique 97, Marseille, France, 14–18 April 1997; pp. 307–310.
64. Payne, S.R. The production of a perceived restorativeness soundscape scale. *Appl. Acoust.* **2013**, *74*, 255–263. [CrossRef]
65. *Methods for the Subjective Assessment of Small Impairments in Audio Systems Including Multichannel Sound Systems*; International Telecommunication Union: Geneva, Switzerland, 1997.
66. Guastavino, C.; Katz, B.F.G. Perceptual evaluation of multi-dimensional spatial audio reproduction. *J. Acoust. Soc. Am.* **2004**, *116*, 1105–1115. [CrossRef] [PubMed]
67. Fellgett, P. Ambisonics part one: General system description. *Media* **1975**, *17*, 20–22.
68. Bruce, N.S.; Davies, W.J.; Adams, M.D. Development of a soundscape simulator tool. In Proceedings of the 38th International Congress and Exposition on Noise Control Engineering 2009, INTER-NOISE, Ottawa, ON, Canada, 23–26 August 2009.
69. Omnitone. Available online: http://googlechrome.github.io/omnitone/#home (accessed on 13 June 2017).
70. BBC R & D Surround Sound with Height—BBC R & D. Available online: http://www.bbc.co.uk/rd/projects/surround-sound-with-height (accessed on 13 June 2017).
71. Facebook 360. Available online: https://facebook360.fb.com/spatial-workstation/ (accessed on 12 June 2017).
72. Blauert, J. *An Introduction to Binaural Technology*; Gilkey, R.H., Anderson, T.R., Eds.; Springer: Berlin, Germany, 1997.
73. Ranjan, R. 3D Audio Reproduction: Natural Augmented Reality Headset and Next Generation Entertainment System Using Wave Field Synthesis. Ph.D. Thesis, School of Electrical and Electronic Engineering, Singapore, 2016.
74. Madole, D.; Begault, D. 3-D sound for virtual reality and multimedia. *Comput. Music J.* **1995**, *19*, 99. [CrossRef]
75. Sunder, K.; Tan, E.L.; Gan, W.S. Individualization of binaural synthesis using frontal projection headphones. *AES J. Audio Eng. Soc.* **2013**, *61*, 989–1000.
76. He, J.; Ranjan, R.; Gan, W.S. Fast continuous HRTF acquisition with unconstrained movements of human subjects. In Proceedings of the 2016 IEEE International Conference on Acoustics, Speech and Signal Processing (ICASSP), Shanghai, China, 20–25 March 2016; pp. 321–325.
77. Ranjan, R.; He, J.; Gan, W.-S. Fast continuous acquisition of HRTF for human subjects with unconstrained random head movements in azimuth and elevation. In Proceedings of the Audio Engineering Society Conference: 2016 AES International Conference on Headphone Technology, Aalborg, Denmark, 24 August 2016; pp. 1–8.
78. Gardner, W.G. *Transaural 3D Audio*; MIT Media Laboratory: Cambridge, MA, USA, 1995.
79. Guastavino, C.; Larcher, V.; Catusseau, G.; Boussard, P. Spatial audio quality evaluation: Comparing transaural, ambisonics and stereo. In Proceedings of the 13th International Conference on Auditory Display, Montréal, QC, Canada, 26–29 June 2007; pp. 53–58.
80. SIGGRAPH. "Sounds Good to Me" SIGGRAPH 2002 Course Notes. Available online: https://www.cs.princeton.edu/~funk/course02.pdf (accessed on 12 June 2017).
81. Vorländer, M. *Auralization: Fundamentals of Acoustics, Modelling, Simulation*; Springer: Berlin, Germany, 2008.
82. Pätynen, J.; Lokki, T. Evaluation of concert hall auralization with virtual symphony orchestra. In Proceedings of the International Symposium on Room Acoustics, Melbourne, Australia, 29–31 August 2010; pp. 1–9.
83. Rindel, J.H. Modelling in auditorium acoustics–from ripple tank and scale models to computer simulations. In Proceedings of the Forum Acusticum, Sevilla, Spain, 19 September 2002; pp. 1–8.
84. De Coensel, B.; de Muer, T.; Yperman, I.; Botteldooren, D. The influence of traffic flow dynamics on urban soundscapes. *Appl. Acoust.* **2005**, *66*, 175–194. [CrossRef]

85. Kang, J. Numerical modeling of the sound fields in urban squares. *J. Acoust. Soc. Am.* **2005**, *117*, 3695–3706. [CrossRef] [PubMed]
86. Hothersall, D.C.; Horoshenkov, K.V.; Mercy, S.E. Numerical modelling of the sound field near a tall building with balconies near a road. *J. Sound Vib.* **1996**, *198*, 507–515. [CrossRef]
87. Pellegrini, R.S. Quality assessment of auditory virtual environments. In Proceedings of the 2001 International Conference on Auditory Display, Espoo, Finland, 29 July–1 August 2001; pp. 161–168.
88. Begault, D.R. *3-D Sound for Virtual Reality and Multimedia*; Academic Press Professional, Inc.: San Diego, CA, USA, 2000.
89. Sunder, K.; He, J.; Tan, E.L.; Gan, W.S. Natural sound rendering for headphones: Integration of signal processing techniques. *IEEE Signal Process. Mag.* **2015**, *32*, 100–113. [CrossRef]
90. Alves, S.; Estévez-Mauriz, L.; Aletta, F.; Echevarria-Sanchez, G.M.; Puyana Romero, V. Towards the integration of urban sound planning in urban development processes: The study of four test sites within the SONORUS project. *Noise Mapp.* **2015**, *2*, 57–85.
91. Maffei, L.; Masullo, M.; Aletta, F.; Di Gabriele, M. The influence of visual characteristics of barriers on railway noise perception. *Sci. Total Environ.* **2013**, *445–446*, 41–47. [CrossRef] [PubMed]
92. Maffei, L.; Iachini, T.; Masullo, M.; Aletta, F.; Sorrentino, F.; Senese, V.P.; Ruotolo, F. The effects of vision-related aspects on noise perception of wind turbines in quiet areas. *Int. J. Environ. Res. Public Health* **2013**, *10*, 1681–1697. [CrossRef] [PubMed]
93. Meta 2. Available online: https://www.metavision.com/ (accessed on 22 March 2017).
94. Microsoft. Microsoft HoloLens. Available online: https://www.microsoft.com/microsoft-hololens/en-us (accessed on 12 June 2017).
95. Härmä, J.; Jakka, M.; Tikander, M.; Karjalainen; Lokki, T.; Hiipakka, J.; Gaëtan, L. Augmented reality audio for mobile and wearable appliances. *J. Audio Eng. Soc.* **2004**, *52*, 618–639.
96. Tikander, M.; Karjalainen, M.; Riikonen, V. An augmented reality audio headset. In Proceedings of the International Conference on Digital Audio Effects (DAFx-08), Espoo, Finland, 1–4 September 2008; pp. 181–184.
97. Gupta, R.; Lam, B.; Hong, J.; Ong, Z.; Gan, W. 3D audio VR capture and reproduction setup for auralization of soundscapes. In Proceedings of the 24th International Congress on Sound and Vibration, London, UK, 23–27 July 2017.
98. Lindau, A.; Weinzierl, S. Assessing the plausibility of virtual acoustic environments. *Acta Acust. United Acust.* **2012**, *98*, 804–810. [CrossRef]
99. Hao, Y.; Kang, J.; Wörtche, H. Assessment of the masking effects of birdsong on the road traffic noise environment. *J. Acoust. Soc. Am.* **2016**, *140*, 978–987. [CrossRef] [PubMed]
100. Thomas, P.; Wei, W.; Van Renterghem, T.; Botteldooren, D. Measurement-based auralization methodology for the assessment of noise mitigation measures. *J. Sound Vib.* **2016**, *379*, 232–244. [CrossRef]
101. Galbrun, L.; Ali, T.T. Acoustical and perceptual assessment of water sounds and their use over road traffic noise. *J. Acoust. Soc. Am.* **2013**, *133*, 227–237. [CrossRef] [PubMed]
102. Jeon, J.Y.; Lee, P.J.; You, J.; Kang, J. Perceptual assessment of quality of urban soundscapes with combined noise sources and water sounds. *J. Acoust. Soc. Am.* **2010**, *127*, 1357–1366. [CrossRef] [PubMed]
103. Eckel, G. Immersive audio-augmented environments the LISTEN project. In Proceedings of the Fifth International Conference on Information Visualisation, London, UK, 25–27 July 2001; pp. 571–573.
104. Geier, M.; Ahrens, J.; Spors, S. The soundscape renderer: A unified spatial audio reproduction framework for arbitrary rendering methods. In Proceedings of the Audio Engineering Society—124th Audio Engineering Society Convention, Amsterdam, The Netherlands, 17 May 2008; pp. 179–184.
105. Baskind, A.; Carpentier, T.; Noisternig, M.; Warusfel, O.; Lyzwa, J. Binaural and transaural spatialization techniques in multichannel 5.1 production. In Proceedings of the 27th Tonmeistertagung, VDT International Convention, Cologne, Germany, 22–25 November 2012.
106. Contin, A.; Paolini, P.; Salerno, R. *Sensory Aspects of Simulation and Representation in Landscape and Environmental Planning: A Soundscape Perspective*; Springer International Publishing AG: Cham, Switzerland, 2008; Volume 10, pp. 93–106.
107. OpenPSTD Homepage. Available online: http://www.openpstd.org/ (accessed on 22 March 2017).
108. Freeman, J.; DiSalvo, C.; Nitsche, M.; Garrett, S. Soundscape composition and field recording as a platform for collaborative creativity. *Organ. Sound* **2011**, *16*, 272–281. [CrossRef]

109. CATT Acoustic. CATT Acoustic Software v9.1. Available online: http://www.catt.se/CATT-Acoustic.htm (accessed on 22 March 2017).

110. Ahnert, W.; Feistel, R. EARS Auralization Software. *J. Audio Eng. Soc.* **1993**, *41*, 894–904.

111. Kang, J.; Meng, Y.; Brown, G.J. Sound propagation in micro-scale urban areas: Simulation and animation. *Acta Acust.* **2003**, *89*, 1–6. [CrossRef]

112. Google. Google VR Spatial Audio Homepage. Available online: https://developers.google.com/vr//spatial-audio (accessed on 22 March 2017).

113. Slab3D Software Homepage. Available online: http://slab3d.sourceforge.net/ (accessed on 22 March 2017).

applied
sciences

MDPI

Article

Stereophonic Microphone Array for the Recording of the Direct Sound Field in a Reverberant Environment

Jonathan Albert Gößwein *, Julian Grosse and Steven van de Par

Acoustics Group, Cluster of Excellence "Hearing4All", Carl von Ossietzky University,
26111 Oldenburg, Germany; julian.grosse@uni-oldenburg.de (J.G.);
steven.van.de.par@uni-oldenburg.de (S.v.d.P.)
* Correspondence: jonathan.goesswein@uni-oldenburg.de; Tel.: +49-441-798-3248

Academic Editors: Woon-Seng Gan and Jung-Woo Choi
Received: 15 March 2017; Accepted: 17 May 2017; Published: 24 May 2017

Abstract: State-of-the-art stereo recording techniques using two microphones have two main disadvantages: first, a limited reduction of the reverberation in the direct sound component, and second, compression or expansion of the angular position of sound sources. To address these disadvantages, the aim of this study is the development of a true stereo recording microphone array that aims to record the direct and reverberant sound field separately. This array can be used within the recording and playback configuration developed in Grosse and van de Par, 2015. Instead of using only two microphones, the proposed method combines two logarithmically-spaced microphone arrays, whose directivity patterns are optimized with a superdirective beamforming algorithm. The optimization allows us to have a better control of the overall beam pattern and of interchannel level differences. A comparison between the newly-proposed system and existing microphone techniques shows a lower percentage of the recorded reverberance within the sound field.

Keywords: intensity-based stereo-recording; convex optimization; superdirective beamformer; white noise gain; logarithmic array design; spatial audio

---

1. Introduction

Sound reproduction systems play an important role in our everyday life. They allow us to listen to recordings from a different place and a past time. Many different methods for the recording and playback of sound exist, utilizing different combinations of microphone and loudspeaker setups. The most common one is a simple stereo reproduction, but there are more complex reproduction techniques, such as wave field synthesis [1] or ambisonics [2]. Even though the state-of-the-art methods achieve a very good accuracy in reproducing sound fields, they do not consider the interaction between the acoustics of the recording and playback environment. In particular, extra reverberation is created by the playback environment, and in addition, there is no control over the spatial distribution of the reverberant sound field, which may influence the apparent source width and perceived listener envelopment. For this reason, ongoing investigations aim to improve the performance of these methods.

In particular, Grosse and van de Par proposed a new way of recording and playing back sound fields [3]. The main idea behind their research was to record the direct and reverberant sound field separately in order to be able to render it in a playback room while optimizing certain perceptually-motivated criteria for the authentic audio reproduction. These criteria aim for recreating the reverberant sound field in the playback environment as faithfully as possible by optimizing the amount and spectral shape of the reverberation, as well as the interaural cross-correlation created by the reproduced reverberant sound field, such as it is created in the reproduction room, including its added reverberant effect. In their paper, Grosse and van de Par assumed that optimizing these perceptual criteria is sufficient for an authentic reproduction of the sound field present in the recording room, which is created by a single source. This claim was supported by subjective evaluations. The playback and recording configuration can be seen in Figure 1. In addition to the two basic stereo loudspeakers, the proposed approach used two dipole loudspeakers to excite and equalize the reverberant sound field. For the optimized rendering, the system relies on the presence of a relatively dry direct signal to be rendered on the frontal loudspeakers and a reverberant signal to be optimized and rendered on the dipole loudspeakers. To record the direct sound, a microphone (C) was positioned close to the sound source. This also avoided early reflections, which could cause a change in coloration [4,5]. For recording the reverberant sound field, two microphones (B^l, B^r) were placed at two distant positions in the diffuse field.

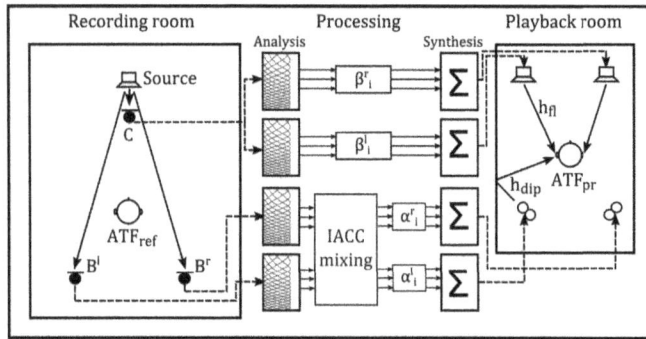

Figure 1. Recording and playback configuration with a processing stage in between to maintain the acoustical perception of a recording room. The microphone (C) records the direct sound, which is played later by two conventional loudspeakers, whereas the two microphones (B^l) and (B^r) record the reverberant sound field, which is played later by two dipole loudspeakers. Figure reproduced with permission from [3], Copyright IEEE, 2015.

Since the method of Grosse and van de Par [3] until now is limited to a single source and only records the direct sound field with one microphone, an extension is needed to also represent the spatial distribution of sources within the direct sound field signals as perceived at the listener position. Although this could in principle be achieved by using multiple close microphones and an appropriate mixing scheme, in this contribution, we want to provide a method with only a single 'true-stereo' microphone setup that is placed at the intended listener position within the recording room. Particular attention has to be paid to reduce the reverberant sound field in the direct sound field signals to be able to separately optimize the rendering of the direct and reverberant sound fields according to perceptual criteria within the playback room [3].

Although the specific design criteria for the proposed microphone array are envisioned to be used in the audio reproduction system of Grosse and van de Par [3], it can also be considered to use the proposed microphone array to record a relatively dry spatial image of the sound sources on stage to be combined with a reverberant track that can be mixed at a level that the recording engineer deems suitable. In this case, however, it will not necessarily fulfill the optimization criteria as formulated in Grosse and van de Par [3] that create a faithful audio reproduction.

The state-of-the-art true stereo systems combine two microphones with a characteristic directivity pattern, placed at different distances and under different angles relative to one another. Depending on these parameters, a deviating spatial rendering of the distributed sources can be observed [6]. Despite this, for use in the method proposed by Grosse and van de Par [3], these systems have some disadvantages that make them unsuitable to be implemented in this specific sound reproduction system because there is a high percentage of recorded reverberant sound, which should be avoided in the system of [3].

We overcome these disadvantages with the development of a new method of a true stereo microphone array, using a superdirective beamforming algorithm that is applied on two logarithmically-spaced microphone arrays. Correct, frequency-dependent interchannel level differences are captured by optimizing the shape of the two main lobes of the arrays. Together, they create the proper interchannel level difference required for an accurate spatial reproduction of the sound field while ensuring that no interchannel phase differences occur that can result in unintended changes in the perceived location of sound sources. Additionally, an optimal side lobe suppression is applied to reduce the influence of the reverberant sound field on the recording of the direct sound. This proposed stereo microphone array is compared to the state-of-the-art stereo microphone configurations mentioned earlier that shows a clearly reduced level of the reverberant sound field.

2. Methods

The following section is divided into five parts. The first Section 2.1 gives a brief introduction to the most relevant theory on beamforming needed for our proposed method. Section 2.2 focuses on the issue of the robustness of beamforming algorithms. The desired directivity pattern is specified in Section 2.3, which is based on a stereo intensity-panning rule related to the auditory processing of the interaural level differences. Section 2.4 introduces an optimal array design to suppress side lobes and, in this way, reduce the influence of the reverberant sound field on the recording of the direct sound. Further, a specific filter design is proposed in Section 2.5, which will be used and evaluated throughout this study. The design is based on a superdirective beamforming algorithm and describes how the directivity pattern that is specified in Section 2.3 can be used for the optimization.

2.1. Beamforming

Beamforming describes the process of forming the directivity pattern of several microphones, which are arranged into an array, with signal processing techniques to obtain a specific, frequency-dependent directivity pattern. The directivity pattern $b(f, \phi)$ of a linear discrete microphone array, consisting of N microphones, is calculated as follows [7]:

$$b(f, \phi) = \sum_{n=-\frac{N-1}{2}}^{\frac{N-1}{2}} w_n(f) G_n(f, \phi) \tag{1}$$

where ϕ denotes the angle ranging from $-\pi$ to π, f the frequency, $w_n(f)$ the frequency-dependent complex weighting filtering applied to microphone n and $G_n(f, \phi)$ the steering vector denoting the direction and frequency-dependent transfer function from the sound source to microphone n. Such a microphone array is illustrated in Figure 2.

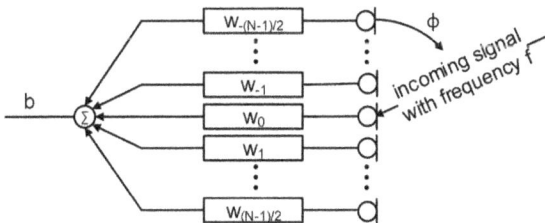

Figure 2. Microphone array receiving a signal with frequency f and angle of incidence ϕ. The incoming wavefront is captured with a microphone n, modified with the respective filter w_n and, at the end, summed up to form the directivity pattern $b(f, \phi)$.

Assuming a far field condition with the microphones that have an omnidirectional directivity pattern, the transfer function states:

$$G_n(f, \phi) = e^{-i\frac{2\pi f}{c} x_n \cos(\phi)} \tag{2}$$

where c is the speed of sound and x_n represents the distance of the n-th microphone to the center of the array [7].

The influence on the directivity patterns of the microphones in the array can be taken into account by changing the transfer function G_n. The filter optimization used to match the directivity pattern of the array with a desired one is called beamforming. The look direction of the microphone array is defined as the angle of the main lobe of the desired directivity pattern, which is also called the steering angle.

There are several beamforming algorithms based on an analytic solution for the optimal filter $w_n(f)$ and some others on a numerical approximation. Analytic solutions allow us to set N constraints on the directivity pattern for a finite number of frequencies, as for example described in [8]. Since we have a higher number of constraints in our problem, we will use numerical methods that allow accommodating a higher number of constraints to control the directivity pattern.

Equation (1) will be solved numerically, and for this purpose, the frequency range is discretized into P frequencies $f_p, p = 0, \ldots, P - 1$ and the angular range into M angles $\phi_m, m = 0, \ldots, M - 1$:

$$b(f_p, \phi_m) = \sum_{n=-\frac{N-1}{2}}^{\frac{N-1}{2}} w_n(f_p) G_n(f_p, \phi_m) \tag{3}$$

Equation (3) is reformulated in matrix notation as:

$$\mathbf{b}_m(f_p) = \mathbf{G}_{mn}(f_p) \mathbf{w}_n(f_p) \tag{4}$$

where the directivity pattern is an $M \times 1$ vector $\mathbf{b}_m^T(f_p) = [b(f_p, \phi_0), b(f_p, \phi_1), \ldots, b(f_p, \phi_{M-1})]$, the transfer function an $M \times N$ matrix $[\mathbf{G}(f_p)]_{mn} = e^{-i\frac{2\pi f_p}{c} x_n \cos(\phi_m)}$ and the filter a $N \times 1$ vector $\mathbf{w}_n(f_p) = [w_{-\frac{N-1}{2}}(f_p), w_{-\frac{N-3}{2}}(f_p), \ldots, w_{\frac{N-1}{2}}(f_p)]^T$ [7]. All bold variables are either vectors or matrices in the remainder of this manuscript.

2.2. Robustness and White Noise Gain

One of the problems that beamforming algorithms often have is their lack of robustness. This property is related to a resistance to the presence of spatially white noise and can be impaired by deviations from the specified microphone characteristics and microphone position errors. These imperfections affect the beamformer in a manner similar to a recorded spatially white noise that is amplified. Hence, the White Noise Gain (WNG) is a measure commonly used for quantifying the robustness of a beamformer design. The WNG shows the ability of a beamformer to suppress spatial white noise, because it expresses the gain of the beamformer in the desired look direction relative to the amplification of spatially white noise.

The WNG $A(f_p)$ is defined as follows:

$$A(f_p) = \frac{|b_{steer}(f_p)|^2}{\mathbf{w}_n^H(f_p) \mathbf{w}_n(f_p)} \tag{5}$$

where $b_{steer}(f_p)$ denotes the value of the directivity pattern in steer direction [7]. A high value of the WNG $A(f_p) > 1$ corresponds to a robust beamforming design, whereas a small value $A(f_p) < 1$

effectively corresponds to an amplification of spatial white noise [7]. The maximum possible value of the WNG is equal to the number of microphones used:

$$\max(A(f_p)) = N \tag{6}$$

which corresponds to a uniform filter [7]:

$$|w_n(f_p)| = \frac{1}{N} \tag{7}$$

2.3. Desired Directivity Pattern

The playback of the recorded signals should be in a stereophonic configuration, as mentioned in Section 1 and illustrated in Figure 3a.

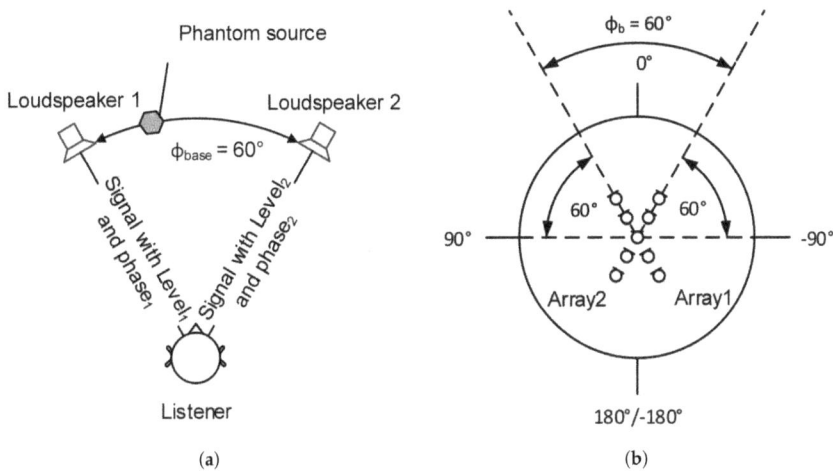

Figure 3. The stereophonic recording configuration is based on the playback one. Recorded level and phase differences with the two end-fire microphone arrays generate a phantom source between the two loudspeakers in the playback configuration. The signal emitted from Loudspeaker 1 has the level $Level_1$ and the phase $phase_1$. The signal emitted from Loudspeaker 2 has the level $Level_2$ and the phase $phase_2$. (**a**) Typical stereophonic playback configuration [9]; (**b**) proposed stereophonic recording configuration with sketched microphone positions. The absolute microphone positions are shown in Section 3.

The playback approach proposed by Grosse and van de Par [3] uses two loudspeakers for the direct sound reproduction with a typical base angle of $\phi_{base} = 60°$ relative to the listener's position [9]. There are several approaches to shift a phantom source from one loudspeaker to the other, utilizing phase differences $\Delta phase = phase_1 - phase_2$ and/or level differences (amplitude panning) $\Delta Level = Level_1 - Level_2$ applied on the two loudspeaker signals.

Based on this playback configuration, the recording configuration presented in this paper consists of two crossed end-fire microphone arrays with a 60° opening angle, sharing one center microphone and using omnidirectional microphones, illustrated in Figure 3b. The microphone positions in this figure can only be considered as a sketch, the absolute positions can be found in Section 3. The phantom-source shifting approaches of the playback configuration can be used to formulate either the correct phase and/or level differences between the two arrays. In this way, the perceived location of the sound source in the playback situation is identical to the one of the recording provided that the distribution of recorded sound sources does not span more than 60° of angle. Although not evaluated here, in principle, a different opening angle could be used for the microphone arrays, thus effectively compressing or expanding the reproduced sound stage. We restrict our proposed method to have only level differences, and for this reason, the desired directivity pattern $\hat{b}$ is purely real valued. With this desired directivity pattern, the phase of the directivity pattern is mainly controlled by the array design, which will be explained in Section 2.4.

In this paper, the phantom source shifting approach of amplitude panning is used for formulating the desired directivity pattern of Array 1 $\hat{b}_{array1}$ and Array 2 $\hat{b}_{array2}$ [9]:

$$
\begin{aligned}
\hat{b}_{array1}(\phi_\delta) &= \sqrt{\left(1 + \left(\frac{\tan(\phi_\delta) - \tan(\phi_b/2)}{\tan(\phi_\delta) + \tan(\phi_b/2)}\right)^2\right)^{-1}} \\[2mm]
\hat{b}_{array2}(\phi_\delta) &= \sqrt{\left(1 + \left(\frac{\tan(\phi_\delta) + \tan(\phi_b/2)}{\tan(\phi_\delta) - \tan(\phi_b/2)}\right)^2\right)^{-1}}
\end{aligned}
\tag{8}
$$

The angle area ϕ_δ between both arrays is defined by:

$$
\phi_\delta = \{\phi_m| - \phi_b/2 \leq \phi_m \leq \phi_b/2\}
\tag{9}
$$

with the constant $\phi_b = \phi_{base} = 60°$. The derivation of the desired directivity patterns according to [9] gives two possible recording room assumptions: an anechoic chamber or a real room. The latter one is chosen for Equation (8) since the microphone array configuration will be used in real rooms, such as concert halls.

The desired directivity pattern of the one array is the mirror-flipped version of the other array. This symmetry of the recording configuration makes it possible to formulate one desired directivity pattern, which is the same for both arrays. The following parts of the desired directivity pattern, the first $\hat{b}_{beam}$ valid for the beam area and the second $\hat{b}_{steer}$ valid for the steering angle, consider a microphone array aligned on the 0° axis corresponding to the steering angle $\phi_{steer} = 0°$:

$$
\hat{b}_{beam} =
\begin{cases}
\sqrt{\left(1 + \left(\frac{\tan(\phi+\phi_b/2)-\tan(\phi_b/2)}{\tan(\phi+\phi_b/2)+\tan(\phi_b/2)}\right)^2\right)^{-1}} & \text{for } -\phi_b \leq \phi < 0° \\[3mm]
\sqrt{\left(1 + \left(\frac{\tan(\phi-\phi_b/2)+\tan(\phi_b/2)}{\tan(\phi-\phi_b/2)-\tan(\phi_b/2)}\right)^2\right)^{-1}} & \text{for } 0° < \phi \leq \phi_b
\end{cases}
\tag{10}
$$

$$
\hat{b}_{steer}(\phi_{steer} = 0°) = 1
\tag{11}
$$

In the following subsections, an optimal array design in terms of optimal microphone positions and an optimal filter design is proposed to achieve the desired directivity pattern.

2.4. Array Design

The positions of the microphones have an influence both on the filter $w_n(f_p)$ and the transfer function $G_{mn}(f_p)$, and thus, on the directivity pattern itself. The optimal microphone positions selected for this paper maximize the spatial aliasing frequency and, at the same time, minimize the frequency from which beamforming is effectively possible. The spatial aliasing frequency describes the lowest frequency f_{al} for which aliasing effects occur, which is caused by a spatial undersampling of the array for sound waves at high frequencies. The aliasing leads to side lobes with the same amplitude as the main lobe. The spatial aliasing frequency of an array with linear microphone spacing is usually given in the literature as:

$$f_{al} = \frac{c}{2\triangle x} \tag{12}$$

with $\triangle x$ as the space between the microphones [10].

A small microphone spacing sets an upper limit to the spatial aliasing frequency. In contrast, a large microphone spacing sets a lower limit to the frequency from which beamforming is effectively possible. In order to have good directional properties of the microphone array across a wide frequency range, an irregularly-spaced microphone array is used in which both kinds of spacing can occur. A linear-shaped, logarithmically-spaced, to the reference microphone ($n = 0$), symmetrical array is used in this paper. Consequential, the number of the used microphones N has to be uneven ($N \in \mathbb{N}_U$). The symmetry around one central microphone ensures a purely real directivity. The microphone positions are calculated as follows [11].

$$\begin{aligned}(x_{n+1} - x_n) &= (x_n - x_{n-1})\xi \quad \text{if } n > 0 \\ (x_{n-1} - x_n) &= (x_n - x_{n+1})\xi \quad \text{if } n < 0\end{aligned} \tag{13}$$

with:

$$x_0 = 0$$

$$\xi = \left(l_{spread}\right)^{\frac{2}{N-3}}$$

$$(x_1 - x_0) = (x_0 - x_{-1}) = \frac{Length}{2\sum_{n=1}^{\frac{N-1}{2}}\xi^{n-1}}$$

where *Length* is the total length of the array. The array parameter $l_{spread} \in \mathbb{R}^{>0}$ is a free variable describing the ratio between the spacing of the microphones at the extremities of the array and the spacing of the microphones at the center of the array. Linear microphone spacings are archived with $l_{spread} = 1$. If $l_{spread} < 1$, the spacing of the microphones at the extremities of the array is smaller than the one at the center of the array. In the case of $l_{spread} > 1$, it is the opposite.

2.5. Filter Design

In this section, an optimal filter design is proposed to fit the directivity pattern of the array, whose design was specified in Section 2.4, to the desired directivity pattern specified in Section 2.3. The following filter design is based on numerical convex optimization and has the advantage that only one global minimum exists. In general, this end-fire design can also be used with different desired directivity patterns and array designs. In Section 3, we indicate the ideal values of the constants for the desired directivity pattern and array design proposed in this study.

The aim of this algorithm is to minimize the quadratic error $\mathbf{error}_m$ between the directivity pattern obtained by a microphone array $\mathbf{b}_m(f_p)$ and a desired frequency independent directivity pattern $\hat{\mathbf{b}}_m$ [7]:

$$\begin{aligned}\mathbf{error}_m &= \mathbf{G}_{mn}(f_p)\mathbf{w}_n(f_p) - \hat{\mathbf{b}}_m = \mathbf{b}_m(f_p) - \hat{\mathbf{b}}_m \\ \min_{\mathbf{w}_n(f_p)} &\|\mathbf{error}_m\|_2^2\end{aligned} \tag{14}$$

This minimization task will be subjected to additional constraints, and therefore, the beamformer will be termed the Constrained Least-Squares Beamformer (CLSB).

In the following subsections, the main minimization task and the used constraints will be explained paying particularly attention to the WNG and different spatial areas. These areas are shown in Figure 4.

Additionally, this optimization process is placed within an optimization loop in order to optimize several important constants. This optimization procedure will be explained in the last subsection of this section.

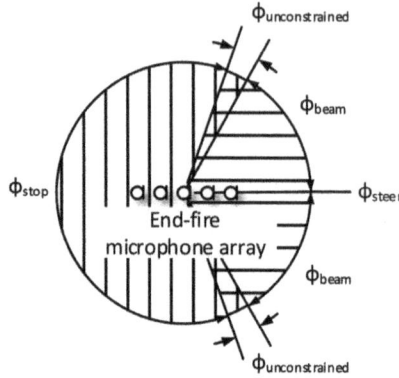

Figure 4. Different spatial areas in the directivity pattern optimization problem. The steering angle ϕ_{steer}, the beam area ϕ_{beam} (indicated by horizontal hash lines), an area without any constraints $\phi_{unconstrained}$ (indicated by crossed hash lines) and the stop area ϕ_{stop} (indicated by vertical hash lines).

2.5.1. White Noise Gain

Such a convex optimization procedure allows including a frequency-dependent lower bound $\gamma(f_p)$ for the WNG when optimizing the filters $w_n(f_p)$ [7]:

$$A(f_p) = \frac{|b_{steer}(f_p)|^2}{\mathbf{w}_n^H(f_p)\mathbf{w}_n(f_p)} \geq \gamma(f_p)$$

$$\text{with } \gamma(f_p) \in \mathbb{R}^{\geq 0}$$

$$(15)$$

This constraint has a direct influence on the robustness and on how well the desired directivity pattern can be achieved. A high value for the lower bound reduces the accuracy of forming the directivity pattern because the filter is too restricted by this constraint, whereas a low value leads to a not robust filter. In Section 3, an optimal value for this lower bound will be discussed.

2.5.2. Steering Angle

In the direction of the steering angle ϕ_{steer}, representing the direction of the main lobe of the microphone array, the directivity pattern obtained by the array is constrained to the value of the desired directivity pattern [7]:

$$G_{steer,n}(f_p)\mathbf{w}_n(f_p) = b_{steer}(f_p) \overset{!}{=} \hat{b}_{steer} \tag{16}$$

In this way, the directivity pattern is normalized to $\hat{b}_{steer}$. The steering angle is limited to the array-axis, since the goal is an end-fire array.

2.5.3. Beam Area

The area around the steering angle is the beam area, which defines the main lobe of the directivity pattern:

$$\phi_{beam} = \{\phi_m | \phi_{steer} - \phi_b \leq \phi_m \leq \phi_{steer-1} \land \phi_{steer+1} \leq \phi_m \leq \phi_{steer} + \phi_b\}$$

$$\text{with } \phi_b \in \mathbb{R}^{\geq 0}$$

$$(17)$$

$\phi_{steer-1}$ and $\phi_{steer+1}$ indicate one discrete angle before and after the steering angle, respectively. The constant ϕ_b can be chosen freely and defines the width of the beam area. Fitting the directivity pattern to the desired one, an angle-dependent upper bound ϵ_{beam} is set to the error (cf. Equation (14)) in this area:

$$\text{abs}(\textbf{error}_{beam}) \leq \epsilon_{beam}$$
$$\text{with } \epsilon_{beam} \in \mathbb{R}^{\geq 0} \tag{18}$$

where abs() denotes the absolute value of every entry of the vector argument. In this case, ϵ_{beam} is a column vector with as many entries as the directivity pattern in the beam area.

2.5.4. Unconstrained Area

An angle area without any constraints is defined to avoid an effective discontinuity in the intermediate zone between the beam and the stop area, which would have a negative impact on the optimized solution that would be obtained:

$$\phi_{unconstrained} = \{\phi_m | \phi_{steer} - \phi_b - \phi_u \leq \phi_m < \phi_{steer} - \phi_b \wedge \phi_{steer} + \phi_b < \phi_m \leq \phi_{steer} + \phi_b + \phi_u\}$$
$$\text{with } \phi_u \in \mathbb{R}^{\geq 0} \tag{19}$$

The constant ϕ_u can be chosen freely and defines the width of the unconstrained area.

2.5.5. Stop Area

The remaining area is called the stop area:

$$\phi_{stop} = \{\phi_m | \phi_{steer} + \phi_b + \phi_u < \phi_m < \phi_{steer} - \phi_b - \phi_u\} \tag{20}$$

The main optimization task is applied to this area. In the context of this work, the sound from this direction can be assumed to be mainly reverberant sound that does not belong to the direct sound and is therefore undesired. For this reason, the desired directivity pattern in this area is set to zero to suppress sound coming from this area as much as possible [7]:

$$\min_{\textbf{w}_n(f_p)} \|\textbf{error}_{stop}\|_2^2$$
$$\text{with } \hat{\textbf{b}}_{stop} = 0 \tag{21}$$

In addition to this optimization, an upper bound ϵ_{stop} is set to the uniform norm of the directivity pattern:

$$\|\textbf{error}_{stop}\|_\infty \leq \epsilon_{stop}$$
$$\text{with } \epsilon_{stop} \in \mathbb{R}^{\geq 0} \tag{22}$$

This upper bound is not angle-dependent, but restricted to the stop area because of the uniform norm and will play an important role in the following loop design.

2.5.6. Loop Design

Choosing the correct upper bound for the beam area is difficult: on the one hand, a low upper bound for the beam area leads to a good fit in this area (low **error**$_{beam}$ values), but to undesired side lobes in the stop area (high **error**$_{stop}$ values). Consequential, the direct sound will be recorded correctly, but is mixed with the undesired reverberant sound field, which should be ideally suppressed. On the other hand, a high upper bound for the beam area leads to the opposite, a bad fit in the beam area (high **error**$_{beam}$ values), but low undesired side lobes (low **error**$_{stop}$ values). The following loop design finds a frequency-dependent optimal upper bound for the beam area, which is a compromise between a good fit in the beam area and only small side-lobes in the stop area.

As a first step in the loop design, the upper bound of the beam area is initialized in matrix notation:

$$
\epsilon_{beam}^{k} = \begin{array}{c} \phi_{steer} - \phi_b \\ \vdots \\ \phi_{steer} + \phi_b \end{array}
\begin{pmatrix}
\begin{array}{cccc}
k=1 & k=2 & \dots & k=K \\
0 & \alpha & \dots & \hat{b}_{steer} - \hat{b}(\phi_{steer} - \phi_b) \\
\vdots & \vdots & \ddots & \vdots \\
0 & \alpha & \dots & \hat{b}_{steer} - \hat{b}(\phi_{steer} + \phi_b)
\end{array}
\end{pmatrix}
\tag{23}
$$

$$
\text{with } K \in \mathbb{N}^{>1}, \ k \in \mathbb{N}^{\le K} \text{ and } \alpha = [\hat{b}_{steer} - \hat{b}(\phi_{steer} \pm \phi_b)]/K
$$

The rows cover the beam area, whereas the columns cover the different iterations of the following loops with k as the counter, where $k = K$ indicates the last iteration. The upper bound starts in the first iteration with $\epsilon_{beam}^{k=1} = 0$ and continues linearly spaced with step size α. The step size is designed in such a way that the maximum value of the upper bound of the beam area $\hat{b}_{steer} - \hat{b}(\phi_{steer} \pm \phi_b)$ is reached in overall K steps. Either $\hat{b}(\phi_{steer} - \phi_b)$ or $\hat{b}(\phi_{steer} + \phi_b)$ can be chosen to calculate α, since they are equal according to the symmetry of the desired directivity pattern. The upper bound then ends with the difference between $\hat{b}_{steer}$ and $\hat{b}_{beam}$ at the row specific angle. If this difference is reached before the last iteration ($k < K$), this value will stay till this iteration is reached. This will be the case for every row, except the first and the last one. This procedure ensures that $\hat{b}_{steer}$ stays the maximum value of the directivity pattern.

In contrast to the upper bound of the beam area, the bound of the stop area is initialized as a vector, since there is no angle dependency:

$$
\epsilon_{stop}^{l} = \begin{pmatrix}
\begin{array}{ccc}
l=1 & \dots & l=L \\
\hat{b}_{steer} \cdot b_{stop}^{first} & \dots & \hat{b}_{steer}
\end{array}
\end{pmatrix}
\tag{24}
$$

$$
\text{with } L \in \mathbb{N}^{>1}, \ l \in \mathbb{N}^{\le L} \text{ and } b_{stop}^{first} \in \mathbb{R}^{\ge 0, \le 1}
$$

The entries with the counter l, where $l = L$ indicates the last iteration, correspond to the iterations of the following loops and are linearly spaced. The constant b_{stop}^{first} controls the maximum allowed value of the directivity pattern in the stop area for the first iteration.

The loop design itself can be seen in Figure 5 and is repeated for every frequency f_p, where the constants K_{temp} and K_{step} can be chosen freely so that $K/K_{temp} \in \mathbb{N}$ and $K_{temp}/K_{step} \in \mathbb{N}$, respectively. These two constants regulate the part of the upper bound of the beam area, which is used in the looped optimization process.

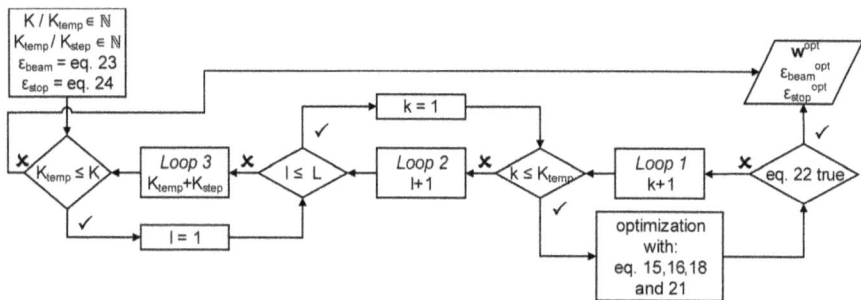

Figure 5. Loop design to determine the optimal filter, as well as the optimal upper bound for the beam and the stop area.

The first loop repeats the optimization with the first part of the upper bound of the beam area (from $\epsilon_{beam}^{k=1}$ to $\epsilon_{beam}^{k=K_{temp}\leq K}$) till Equation (22) with ϵ_{stop}^1 is true. A result of the optimization, fulfilling Equation (22), is denoted as valid. If this is not the case, Loop 2 repeats Loop 1 with different upper bounds of the stop area (from ϵ_{stop}^2 to ϵ_{stop}^L). If still no valid result is found, Loop 3 increases K_{temp} with the step width of K_{step}. The upper bounds, for which the loop design finds a valid solution, are denoted as optimal ϵ_{beam}^{opt} and ϵ_{stop}^{opt}. The filter $\mathbf{w}$, which corresponds to these upper bounds, is also denoted as optimal $\mathbf{w}^{opt}$. For the case that K_{step} increases K_{temp} over K ($K_{temp} + K_{step} > K$), the last $k = K$ calculated result of the optimization is taken as a valid solution.

3. Setup

The following setup is used for the numerical simulations, whose results are described in Sections 4 and 5. The angular range is discretized into $M = 360$ linearly-spaced angles $\{\phi_0 = 0°,$ $\phi_1 = 1°, \ldots, \phi_{359} = 360°\}$. The frequency range covers the range of $f_{p=0} = 0\,\text{Hz}$ to $f_{p=256} = 24\,\text{kHz}$ generated at a sampling rate of $f_s = 48\,\text{kHz}$ using a filter length of 512 samples. This results in $P = 257$ linear spaced frequency bins. This frequency range covers the spectral content of music [12] that is to be recorded by these microphone arrays. To obtain impulse responses of the filters, the complex spectrum was mirrored, conjugated and transformed towards the time domain via an ifft.

The microphone array consists of $N = 9$ omnidirectional microphones and has a total length of $Length = 1\,\text{m}$. The array design is done with $l_{spread} \approx 35$, so that the smallest microphone spacing (s) in the center of the array is $s = 0.01\,\text{m}$. Following that, the spatial aliasing frequency can be maximized to a frequency of $f_{al} \approx 17{,}000\,\text{Hz}$. For practical reasons, the limitation is set to $s = 0.01\,\text{m}$ to ensure enough space for the microphones. The absolute microphone positions are set as follows (displayed in millimeter precision): $x_{n=-4} = -0.500\,\text{m}$, $x_{n=-3} = -0.150\,\text{m}$, $x_{n=-2} = -0.043\,\text{m}$, $x_{n=-1} = -0.010\,\text{m}$, $x_{n=0} = 0\,\text{m}$, $x_{n=1} = 0.010\,\text{m}$, $x_{n=2} = 0.043\,\text{m}$, $x_{n=3} = 0.150\,\text{m}$, $x_{n=4} = 0.500\,\text{m}$.

After having specified the microphone positions, the convex functions of the CLSB, shown in Section 2.5, are solved utilizing CVX, a package for specifying and solving convex programs [13,14]. Parts of these convex functions are the WNG constraint and the loop design.

For the WNG constraint, the lower bound γ for the WNG $A(f_p)$ is set up as follows:

$$\gamma(f_p) = \begin{cases} 5 & \text{for } f_p = 0\,\text{Hz} \\ \text{CSI} & \text{for } 0\,\text{Hz} < f_p < 187.5\,\text{Hz} \\ 1 & \text{for } 187.5\,\text{Hz} \leq f_p \leq f_s/2\,\text{Hz}) \end{cases} \tag{25}$$

The lower bound starts with $\gamma(f_p = 0\,\text{Hz}) = 5$ and ends with $\gamma(f_p \geq 187.5\,\text{Hz}) = 1$. In the intermediate zone, a Cubic Spline Interpolation (CSI) connects both points. The CSI in the intermediate zone avoids rapid changes of the directivity pattern across frequency below ($f_p < 187.5\,\text{Hz}$). In the high frequency range ($f_p \geq 187.5\,\text{Hz}$), a lower bound of $\gamma = 1$ ensures a robust beamforming design.

For the loop design, the constants are set up as follows:

$$K = 100,\ K_{temp} = K_{step} = 10,\ \alpha = 0.01 \text{ cf. Equation (23)}$$
$$L = 9,\ b_{stop}^{first} = 0.2 \tag{26}$$
$$\phi_u = 10°$$

The constants ϕ_b and ϕ_{steer}, as well as the parts of the desired directivity pattern $\hat{b}_{beam}$ and $\hat{b}_{steer}$ are set up according to Section 2.3.

The values of the constants K, K_{temp} and K_{step} are chosen in such a way that Loop 1 scans the beam area from $\epsilon_{beam}^{k=1} = 0$ in steps of $\alpha = 0.01$ till $\epsilon_{beam}^{k=K_{temp}=10} = K_{temp} \cdot \alpha = 0.1$. If necessary, Loop 3 increases the value of the upper bound of the beam area according to the value of the constant K_{step} (cf. Section 2.5).

An increase of the value of the constant K leads to an improvement in the beam area (lower $error_{beam}$ values), because the step size α is smaller. The validity (cf. Section 2.5) of more possible directivity patterns with small $error_{beam}$ values is checked by the loop design. In fact, to find a valid solution, Loop 2 has to increase ϵ_{stop} further than before, which leads to a worsening in the stop area (higher $error_{stop}$ values). A decrease of the value of the constant K leads consequently to the opposite effect.

An increase of the values of the constants K_{temp} and K_{step} leads to a worsening in the beam area (higher $error_{beam}$ values), because the first end point of Loop 1 $\epsilon_{beam}^{k=K_{temp}}$, as well as all of the other ones $\epsilon_{beam}^{k=K_{temp}+K_{step}+K_{step}+\cdots}$ is now higher. More possible directivity patterns with high $error_{beam}$ values are checked by the loop design: Loop 2 does not have to increase ϵ_{stop} so much than before, because these directivity patterns are in general more likely to be valid. This leads then to an improvement in the stop area (lower $error_{stop}$ values). A decrease of the values of the constants K_{temp} and K_{step} leads consequently to the opposite effect.

The values of the constants L and b_{stop}^{first} are chosen in such a way that Loop 2 scans the stop area from $\epsilon_{stop}^{l=1} = 0.2$ in steps of $(\hat{b}_{steer} - b_{stop}^{first} \cdot \hat{b}_{steer})/(L-1) = 0.1$ till $\epsilon_{stop}^{l=L} = \hat{b}_{steer} = 1$.

An increase of the value of the constant b_{stop}^{first} and at the same time a decrease of the value of the constant L, preserving the step width of 0.1 as mentioned earlier, lead to a worsening in the stop area. The start point of Loop 2 is now higher, allowing higher $error_{stop}$ values from the beginning. It is now easier for Loop 1 to find a valid solution, which leads to an improvement in the beam area. A decrease of the value of the constant b_{stop}^{first} and a coherent increase of the value of the constant L lead to the opposite effect.

Overall, it can be said that a variation of the values of the constants K, K_{temp}, K_{step}, L and b_{stop}^{first} leads to a changed balance, fulfilling the constraints between the beam and the stop area. For every desired directivity pattern and intended purpose of the microphone array has to be found separately optimal values.

A variation of the value of the constant ϕ_u does not significantly change the results in terms of the error in the beam and the stop area. Nevertheless, the value should not be chosen too big to avoid undesired results (very big differences between the obtained and the desired directivity pattern), since there is no control over the directivity pattern in the unconstrained area. The maximum value of ϕ_u till there are no undesired results depends in a complex manner on the number of used microphones and the desired directivity pattern.

With the setup shown in Equation (26), we achieved best results in fitting the directivity pattern to the desired one. Different initializations of the constants are also possible, as mentioned before (a detailed analysis of the effect on the results regarding the variation of the constants' values given in Equation (26) is beyond the scope of this article). Our results are, however, discussed in the following Sections 4 and 5.

4. Objective Evaluation

The following section is divided into four parts. In Section 4.1, two array designs are compared to each other to show the improvement of the spatial aliasing of a logarithmically-spaced array over a linearly-spaced one. In the second Section 4.2, the new stereo system proposed in this study is compared to the state-of-the-art ones, which utilize two microphones. In the third Section 4.3, the WNG constraint and the frequency response are analyzed. Finally, in the last Section 4.4, the angular constraints, as well as the phase of the directivity pattern are investigated.

4.1. Directivity Index Comparison

The directivity pattern of the logarithmically-spaced array ($l_{spread} \approx 35, s = 0.01\,\text{m}$) is more directive for high frequencies than the one of a linearly-spaced array ($l_{spread} = 1, s = 0.125\,\text{m}$) having the same total length of $Length = 1\,\text{m}$. Less reverberant sound is recorded by the first type of array than by the latter one. As a measure, we choose the directivity index DI, which is the logarithm of the directivity D [15]:

$$D(f_p) = \frac{\sum_{m=0}^{M-1} \max_{\phi_m} (|b(f_p, \phi_m)|^2)}{\sum_{m=0}^{M-1} |b(f_p, \phi_m)|^2}$$

$$DI(f_p) = 10 \log_{10}(D(f_p))$$

(27)

In fact, Figure 6 shows that the linearly-spaced array has lower DI values for high frequencies $(f_p > 1200\,\text{Hz})$ than the logarithmically-spaced one. This is caused by aliasing effects, as the aliasing frequency for the linearly-spaced array is $f_{al} \approx 1460\,\text{Hz}$. There is a big drop of the DI values $(DI < 7\,\text{dB})$ for the logarithmically-spaced array for very high frequencies $(f_p > 10,500\,\text{Hz})$, which is also caused by aliasing effects. The lowest values of the DI for the logarithmically-spaced array are located around the aliasing frequency $f_{al}(\Delta x = s) \approx 17,000\,\text{Hz}$.

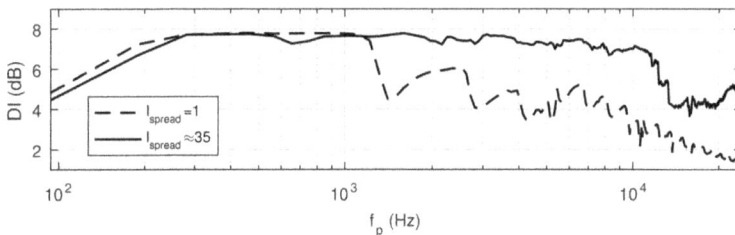

Figure 6. Directivity index $DI(f_p)$ of a linearly-spaced array ($l_{spread} = 1, s = 0.125\,\text{m}$) (dashed line) and the logarithmically-spaced one ($l_{spread} \approx 35, s = 0.01\,\text{m}$) (solid line) with the same total length of $Length = 1\,\text{m}$.

4.2. Comparison Stereo Systems

The necessary phase and/or level differences for a stereophonic recording as mentioned in Section 2.3 can also be obtained by only two microphones. Different angles and distances between these two microphones, as well as different microphone directivity patterns are possible, as described, for example, by the A-B or the X-Y technique [12]. A unified theory of these two-microphone systems for stereophonic sound recording can be found in [6].

Assuming no phase differences, this theory states that a level difference of $\Delta Level = \pm 15\,\text{dB}$ determines the left or right lateral shift towards the loudspeakers of a phantom sound source in the playback situation. This level difference is achieved in the recording situation with different angles between two microphones with specific directivity patterns. The angle covering this level difference is called recording angle ϕ_{rec}. If $\phi_{rec} > \phi_{base}$, the recorded sound scene is compressed in the playback configuration, whereas $\phi_{rec} < \phi_{base}$, the recorded sound scene is expanded [6]. Therefore, we can assume that if we have $\phi_{rec} = \phi_{base}$, the recorded spatial properties are the same after playback. Table 1 shows the possible microphone directivities and base angles between the microphone pairs.

The microphone array stereo system described in this study records less reverberant sound than these state-of-the-art two-microphone stereo systems. As a measure, we choose a modified definition of the directivity index DI_{mod}, which is the logarithm of a modified directivity D_{mod}, mentioned in Section 4.1:

$$D_{mod} = \frac{\sum_{m=0}^{M-1} 2\max_{\phi_m} (b_{mic1}(\phi_m)^2)}{\sum_{m=0}^{M-1} b_{mic1}(\phi_m)^2 + b_{mic2}(\phi_m)^2}$$

$$DI_{mod} = 10 \log_{10}(D_{mod})$$

(28)

where $b_{mic1}(\phi_m)$ and $b_{mic2}(\phi_m)$ are the directivity patterns of the first and the second microphone, respectively. The modified directivity index includes the sum of the directivity patterns of the two microphones. The modified directivity index considers the angle between these two directivity patterns, which determines the percentage of recorded reverberant sound in addition to the directivity pattern itself. As shown in Table 1, the proposed microphone array stereo system is, in fact, more directive than the two-microphone stereo ones, taking also into account the angle between the two microphone arrays.

Table 1. The modified directivity index DI_{mod} of the state-of-the-art two-microphone stereo systems and the microphone array stereo system described in this study. For the latter one, the desired directivity patterns are used. Only stereo systems with $\phi_{rec} = \phi_{base}$ are displayed. This angle constraint avoids angular compression or angular expansion in the playback situation.

Two-Microphone Stereo Systems		
Microphone Directivity	Angle between the Microphones (°)	DI_{mod}
Figure of Eight	101	5.95
Hypercardioid (back attenuation = −6 dB)	136	8.29
Hypercardioid (back attenuation = −10 dB)	156	8.7
Microphone Array Stereo System		
$DI_{mod} = 11.29$ with $b_{mic1}(\phi_m) = \hat{b}_{array1}(\phi_m)$ and $b_{mic2}(\phi_m) = \hat{b}_{array2}(\phi_m)$		

4.3. WNG and Frequency Response

The algorithm successfully fits the WNG $A(f_p)$ to the lower bound $\gamma(f_p)$ specified in Section 3, as shown in Figure 7a.

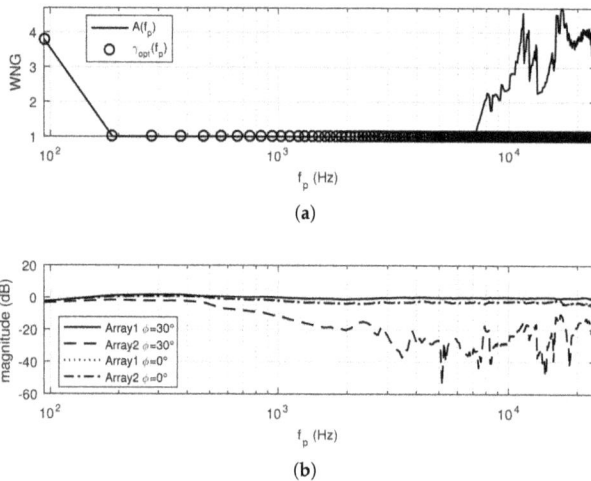

(a)

(b)

Figure 7. (**a**) White Noise Gain (WNG) $A(f_p)$, as well as the lower bound for the WNG $\gamma(f_p)$ across frequency; (**b**) shown are frequency responses of both arrays for two sound sources emanating from $\phi = 30°$ and $\phi = 0°$ according to the configuration illustrated in Figure 3.

This ensures a robust beamforming design. For high frequencies $f_p \geq 7031$ Hz, the algorithm finds even higher WNG values than the lower bound.

Figure 7b shows the frequency response of both arrays according to the configuration that is shown in Figure 3. The responses for both arrays were calculated for a sound source emanating from $\phi = 30°$ (resulting in a sound source perceived at the location of the left loudspeaker, solid and dashed line) according to Figure 3 and $\phi = 0°$ (resulting in a phantom source between both speakers, dotted and dash-dotted line). It can be seen that for $\phi = 0°$, the responses of both arrays show a high similarity in terms of level differences and have only minor fluctuations of approximately $\pm 2\,dB$ above 1000 Hz. Below 1000 Hz, it can be observed that there is a boost of approximately 3 dB, which might be attributed to a violation of a constraint at low frequencies. When the sound source is emanating from $\phi = 30°$, a flat frequency response can be observed for Array 1 (on axis) with minor fluctuations of approximately 1 dB across frequency. Array 2 shows a considerably lower level, but larger fluctuations. It can be assumed that these fluctuations will not be perceivable because the location of the sound source will be determined by Array 1.

4.4. Beam and Stop Area Constraints

The results of the loop design mentioned in Section 2.5 are shown in Figure 8. This loop design finds a compromise between a good fit in the beam area and low directivity pattern values in the stop area.

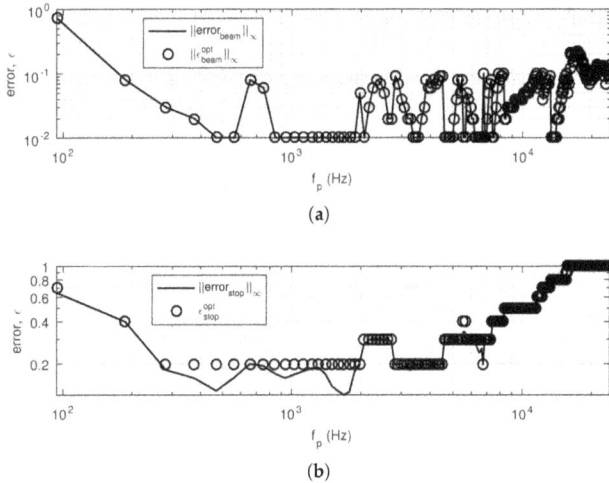

(a)

(b)

Figure 8. The difference between the simulated directivity pattern and the desired one (*error*) in the beam (**a**) and the stop (**b**) area, as well as the corresponding upper bounds of both areas as function of the frequency.

For low frequencies $f_p < 187.5\,Hz$, the directivity pattern is quite omnidirectional ($\|\mathbf{error}_{stop}\|_\infty > 0.2$ and $\|\mathbf{error}_{beam}\|_\infty > 0.1$), so that Loop 3 has to increase ϵ_{beam} to $\|\epsilon_{beam}^{opt}\|_\infty > 0.1$. For higher frequencies $f_p \geq 187.5\,Hz$, there is a good fit in the beam area $\|\mathbf{error}_{beam}\|_\infty \leq 0.1$ so that Loop 1 and Loop 2 find the ideal upper bound for the beam and the stop area. Overall, it can be said that the best result is found in the frequency range of $281.3\,Hz \leq f_p \leq 1969\,Hz$: a good fit in the beam area combined with low directivity pattern values in the stop area $\|\mathbf{error}_{stop}\|_\infty \leq 0.2$. At high frequencies ($f_p \geq 16,690\,Hz$), Figure 8b shows aliasing effects ($\|\mathbf{error}_{stop}\|_\infty = 1$), which are expected, since the aliasing frequency of the logarithmically-spaced array is $f_{al}(\Delta x = s) \approx 17,000\,Hz$.

Figure 9 shows the polar plot of the desired directivity pattern in addition to the absolute value of the directivity patterns of the frequencies $f_p = 250\,Hz$, $f_p = 1000\,Hz$, $f_p = 4000\,Hz$ and $f_p = 8000\,Hz$. For all frequencies, there is a good fit (a small difference between desired and obtained directivity pattern) in the beam area, as already quantified by Figure 8a. Comparing the side-lobe-levels of the different frequencies, the following can be stated: the side-lobe-level decreases from $f_p = 250\,Hz$ to 1000 Hz; there is no big difference in side-lobe-level between $f_p = 1000\,Hz$ and $f_p = 4000\,Hz$;

the side-lobe-level increases from $f_p = 4000\,\text{Hz}$ to $f_p = 8000\,\text{Hz}$. This analysis is described in a quantified matter in Figure 8b.

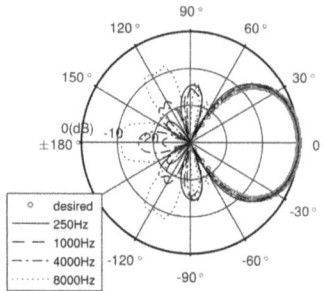

Figure 9. Polar plot of the desired directivity pattern (grey markers) and the absolute value of the obtained directivity patterns of the frequencies $f_p = 250\,\text{Hz}$ (solid line), $f_p = 1000\,\text{Hz}$ (dashed line), $f_p = 4000\,\text{Hz}$ (dashed-dotted line) and $f_p = 8000\,\text{Hz}$ (dotted line).

Figure 10a allows for a more detailed analysis, as it shows the absolute value of the difference between the directivity pattern and the desired one in the whole angular range $|error(\phi_m, f_p)|$. The omnidirectional behavior of the directivity pattern up to $f_p = 187.5\,\text{Hz}$ can be also seen there. For higher frequencies, side lobes appear at $\phi_m = \pm180\,°$ and move with increasing frequency into the direction of the beam $-60\,° \le \phi_m \le 60\,°$. Aliasing effects can be seen in Figure 10a, like in Figure 8b.

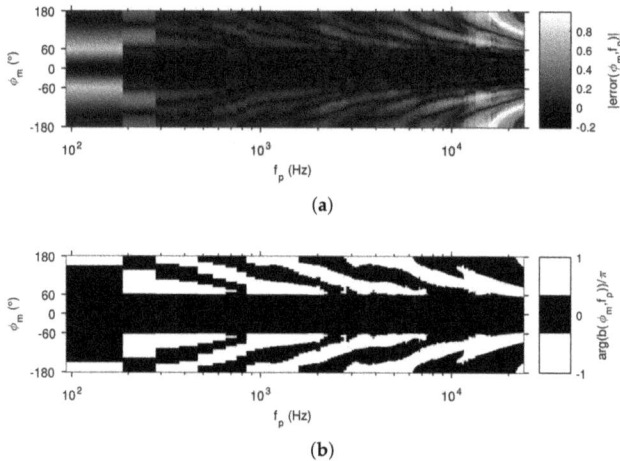

(a)

(b)

Figure 10. The difference between the directivity pattern and the desired one $|error(f_p, \phi_m)|$ (**a**), as well as the phase of the directivity pattern $\arg(b(f_p, \phi_m))$ (**b**).

In addition to the absolute value of the directivity pattern, the phase $\arg(b(f_p, \phi_m))$ is represented in Figure 10b.

The directivity pattern is purely real: the phase shows only three possible values $\arg(b) = \{-\pi, 0, \pi\}$ as mentioned in Section 2.3. In the beam area, the phase has, in fact, only values $\arg(b) = 0$, which leads to no phase differences between the two arrays in the recording configuration mentioned in Section 2.3.

5. Subjective Evaluation

In this section, the proposed microphone array is subjectively evaluated. For this purpose, a listening experiment was performed, whose results are shown.

5.1. Subjective Evaluation: Localization Accuracy

In order to evaluate the proposed stereophonic-microphone array in terms of localization accuracy when simulating spatially-distributed sound sources, subjective data were obtained in a localization experiment within a real room from listeners. The loudspeaker signals were generated using a single sound source and by simulating the delays between the microphones and the sound source. The optimized filters w^{opt} were applied on each microphone signal to obtain the output signal for the left and right array, which was then played back via the two loudspeakers during the listening experiment. The loudspeaker and array configurations are shown in Figure 3.

The sound sources were placed on virtual locations between $-30°$ and $+30°$ in a five degree resolution, resulting in a phantom source stereo image based on intensity-panning between the left and the right loudspeakers. The evaluation took place in a reverberant room with the dimensions $(7.5, 7.1, 2.97)$ m with a reverberation time of $T_{60} = 0.45$ s. The distance between the loudspeakers was 3 m, and the listeners were seated at the position that created a $60°$ stereo triangle with the loudspeakers (cf. Figure 3). As a source signal, three short pink noise bursts with a total length of 1.1 s were presented to the listeners. The noise covered a frequency rang from 100 Hz to $f_s/2$ covering the spectral content of musical signals. Data were obtained from seven listeners, and the 13 source position angles were presented in random order. For each subject, the experiment covered one training session and three measurement sessions. The task of the participants was to indicate the perceived direction between the loudspeaker using indicators placed between the loudspeakers in five degree steps.

5.2. Subjective Evaluation: Results

Figure 11 shows the perceived directions of the subjective evaluation. The dotted line indicates perfect correspondence between the true source location and the perceived location. Circles show the average perceived location in dependence of the simulated source location. As can be seen, there is a rather linear behavior on localization, indicating a mostly precise representation of the presented directions. Exceptions can be observed around ±20 degrees at which the presented source is perceived more lateral than the simulated source location. The maximum localization error of ≈6 degrees that can be observed can probably be attributed to the target functions that were used to optimize the directivity pattern, which may cause too high level differences when both arrays are used in combination.

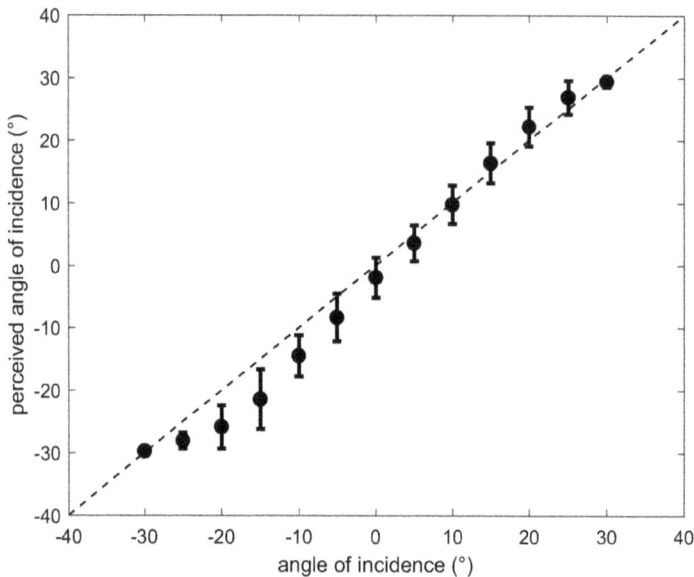

Figure 11. Illustrated are the mean-values of the perceived angle of incidence with the standard deviation across seven participants' means. The *x*-axis represents the simulated angle of incidence ϕ of the presented noise sources. The dotted line indicates a perfect match between simulated and perceived localization.

6. Discussion and Conclusions

In this study, a new approach for intensity stereophonic recording has been investigated. Guided by the playback situation and its auditory requirements, we decided to postulate a setup consisting of two crossed end-fire microphone arrays and a fitting desired directivity pattern. The difference between the directivity pattern obtained and the one desired was minimized by a superdirective beamforming algorithm. It was based on convex numeric optimization and also contains a frequency-dependent WNG constraint to ensure a robust beamforming design.

In addition to designing the filters of the microphones via beamforming algorithms, we found an ideal array design. This design maximizes the spatial aliasing frequency and also takes practical issues into account, which will appear in an actualization of the arrays. The extent of the microphones demands a particular spacing, also to avoid interferences between them.

A comparison between the new stereo system and the state-of-the-art ones, which use two microphones, has shown that the former has the advantage of less recorded reverberant sound, as it is more directive in the look direction than the latter are. This matches the requirements posed by the recording method proposed in Grosse and van de Par [3], which requires separate dry and reverberated representations of the audio signal. The reverberated sound field can be taken from single microphone signals.

Future research could develop a method to optimize the directivity pattern of both arrays as one system rather than handling them separately. Furthermore, two additional beams pointing into the diffuse field could be introduced for optimization to replace the two microphones placed in that field and to use only the array system.

A final assessment of the proposed recording and playback system needs to run listening tests and investigate the perception of the recording and playback room.

Appl. Sci. **2017**, *7*, 541

Acknowledgments: We would like to thank the Deutsche Forschungsgemeinschaft for supporting this work as part of the Forschergruppe Individualisierte Hoerakustik (FOR-1732). We also would like to thank the reviewers for their helpful and insightful comments.

Author Contributions: Steven van de Par and Julian Grosse formulated the constraints for the true stereo microphone array. Jonathan Albert Gößwein developed and evaluated the methods for optimizing the true stereo microphone array. Julian Grosse planed and performed the localization experiment.

Conflicts of Interest: The authors declare no conflict of interest

Abbreviations

The following abbreviations are used in this manuscript:

WNG	White Noise Gain
CLSB	Constrained Least-Squares Beamformer
CSI	Cubic Spline Interpolation

References

1. Berkhout, A.J. A holographic approach to acoustic control. *J. Audio Eng. Soc* **1988**, *36*, 977–995.
2. Gerzon, M.A. Periphony: With-Height sound reproduction. *J. Audio Eng. Soc* **1973**, *21*, 2–10.
3. Grosse, J.; van de Par, S. Perceptually accurate reproduction of recorded sound fields in a reverberant room using spatially distributed loudspeakers. *IEEE J. Sel. Top. Signal Process.* **2015**, *9*, 867–880.
4. Schroeder, M.R. Statistical parameters of the frequency response curves of large rooms. *J. Audio Eng. Soc* **1987**, *35*, 299–306.
5. Haeussler, A.; van de Par, S. Theoretischer und subjektiver Einfluss des Aufnahmeraumes auf den Wiedergaberaum. In Proceedings of the 40th DAGA'14 Jahrestagung fuer Akustik, Oldenburg, Germany, 10–13 March 2014.
6. Williams, M. Unified theory of microphone systems for stereophonic sound recording. In Proceedings of the 82th Audio Engineering Society Convention, London, UK, 10–13 March 1987.
7. Mabande, E.; Schad, A.; Kellermann, W. Design of robust superdirective beamformers as a convex optimization problem. In Proceedings of the 2009 IEEE International Conference on Acoustics, Speech and Signal Processing, Taipei, Taiwan, 19–24 April 2009; pp. 77–80.
8. Frost, O.L. An algorithm for linearly constrained adaptive array processing. *Proc. IEEE* **1972**, *60*, 926–935.
9. Pulkki, V. Compensating displacement of amplitude-panned virtual sources. In Proceedings of the Audio Engineering Society Conference: 22nd International Conference: Virtual, Synthetic, and Entertainment Audio, Espoo, Finland, 15–17 June 2002.
10. McCowan, I.A. Robust Speech Recognition using Microphone Arrays. Ph.D. Thesis, Queensland University of Technology, Brisbane City, QLD, Australia, 2001.
11. Corteel, E. On the use of irregularly spaced loudspeaker arrays for wave field synthesis, potential impact in spatial aliasing frequency. In Proceedings of the 9th international converence on Digital Audio Effects (DAFx'06), Montreal, QC, Canada, 18–20 September 2006; pp. 209–214.
12. Dickreiter, M.; Dittel, V.; Hoeg, W.; Woehr, M. *Handbuch der Tonstudiotechnik*, 7th ed.; K. G. Sauer Verlag: München, Germany, 2008; Volume 1.
13. Grant, M.; Boyd, S. CVX: Matlab Software for Disciplined Convex Programming, version 2.1. 2014. Available online: http://cvxr.com/cvx (accessed on 18 May 2017).
14. Grant, M.; Boyd, S. Graph implementations for nonsmooth convex programs. In *Recent Advances in Learning and Control: Lecture Notes in Control and Information Sciences*; Blondel, V., Boyd, S., Kimura, H., Eds.; Springer: New York, NY, USA, 2008; pp. 95–110. Available online: http://stanford.edu/~boyd/graph_dcp.html (accessed on 18 May 2017).
15. Kinsler, L.; Frey, A.; Coppens, A.; Sanders, J. *Fundamentals of Acoustics*; John Wiley and Sons, Inc.: New York, NY, USA, 2000.

applied
sciences

|MDPI|

Article

The Reduction of Vertical Interchannel Crosstalk: The Analysis of Localisation Thresholds for Natural Sound Sources

Rory Wallis and Hyunkook Lee *

Applied Psychoacoustics Lab, University of Huddersfield, Huddersfield HD1 3DH, UK; rory.wallis@hud.ac.uk
* Correspondence: h.lee@hud.ac.uk; Tel.: +44-1484-471893

Academic Editors: Woon-Seng Gan and Jung-Woo Choi
Received: 8 February 2017; Accepted: 10 March 2017; Published: 14 March 2017

Abstract: In subjective listening tests, natural sound sources were presented to subjects as vertically-oriented phantom images from two layers of loudspeakers, 'height' and 'main'. Subjects were required to reduce the amplitude of the height layer until the position of the resultant sound source matched that of the same source presented from the main layer only (the localisation threshold). Delays of 0, 1 and 10 ms were applied to the height layer with respect to the main, with vertical stereophonic and quadraphonic conditions being tested. The results of the study showed that the localisation thresholds obtained were not significantly affected by sound source or presentation method. Instead, the only variable whose effect was significant was interchannel time difference (ICTD). For ICTD of 0 ms, the median threshold was −9.5 dB, which was significantly lower than the −7 dB found for both 1 and 10 ms. The results of the study have implications both for the recording of sound sources for three-dimensional (3D) audio reproduction formats and also for the rendering of 3D images.

Keywords: vertical interchannel crosstalk; 3D; psychoacoustics; microphone technique; audio; reproduction; image rendering

1. Introduction

In audio reproduction systems for three-dimensional (3D) sound, such as Auro 3D [1] and Dolby Atmos [2], the loudspeakers can generally be divided into two layers: the lower (main) layer and the upper (height) layer. In the context of sound recording made using a microphone array in an acoustic space, the frontal loudspeakers of the main layer, which are located on the horizontal plane, are predominantly used for sound source positioning. Conversely, the height layer, which is typically elevated by between 30° and 45°, primarily aims to enhance the perceived listener envelopment (LEV) by presenting ambient signals, although it can also be used to reproduce elevated sound sources. When recording for such formats, it is necessary to pay close attention to the amount of direct sound present in the height layer signal. The reason for this is as follows. Should there be excessive direct sound in the height layer then, at the reproduction stage, sound sources may be perceived as vertically-oriented phantom images at intermediate positions between the main and height loudspeaker layers. Additional spatial and timbral effects may also be perceived, depending on the time and level relationships between the direct sounds in the respective layers. Collectively, these properties comprise an interference effect referred to as 'vertical interchannel crosstalk'.

To date, the few studies that have considered vertical interchannel crosstalk have primarily been concerned with preventing the direct sound present in the height layer from affecting the perceived location of the main channel signal. Although this has received little attention in the literature, suggestions as to the nature of the effect can be garnered from studies undertaken within the context of vertical amplitude panning. Within such studies, the literature generally agrees that increases in interchannel level difference (ICLD) between vertically arranged stereophonic loudspeakers will cause the resultant phantom image to be localised in a position biased towards the loudspeaker of greater amplitude [3–7]. Despite this, such studies do not necessarily indicate that sufficient ICLD alone will prevent the signal in the height layer from affecting the perceived location of the main channel signal. For example, whilst Barbour [5] demonstrated that ICLDs between 6 and 9 dB resulted in the perceived phantom image position matching the physical position of the lower loudspeaker for pink noise and speech sources, Somerville et al. [3] and Kimura and Ando [7] found that the phantom images remained somewhat elevated for ICLDs up to and including 15 dB when the test stimuli were musical sources, pink noise and speech. It should be noted that differences in the experimental setup and, in particular, the physical position of the loudspeakers might have contributed to these differences in results.

With respect to more direct experiments into the effects of vertical interchannel crosstalk, Lee [8] conducted an analysis into the 'localisation threshold', which was defined as the minimum amount of attenuation of direct sound necessary in the height layer for the main channel signal to be localised at the position of the main layer. It is important to note that the localisation threshold is not a complete masking of the direct sound in the height layer. Instead, although the perceived location of the main channel signal would be unaffected, the aforementioned spatial and timbral effects of vertical interchannel crosstalk would remain somewhat audible. In [8], cello and bongo sources were presented from vertically-arranged stereophonic loudspeakers located directly in front of the listening position. With respect to the listening position, the lower (main) loudspeaker was not elevated, whilst the upper (height) loudspeaker was elevated by 30°. Delays ranging from 0 to 50 ms were applied to the height loudspeaker with respect to the main. A subsequent study conducted by Stenzl et al. [9] was generally similar, although for that study phantom images were formed between diagonally-arranged loudspeakers (e.g., the left loudspeaker in the main layer and right loudspeaker in the height layer). In addition, their test stimuli also included male speech alongside the cello and bongo sources. The results of both studies revealed the following with respect to localisation thresholds. Firstly, for delays in the range of 0–10 ms, they were not significantly affected by interchannel time difference (ICTD). In addition to this, the effect of sound source was not significant. The thresholds reported by each study then, for ICTDs up to 10 ms, were in the range of −6 to −7 dB, which shows good agreement with the amplitude panning experiments of Barbour [5] and Wendt et al. [6].

From a practical standpoint, the results presented in [8] and [9] can be used to influence techniques both for the rendering of 3D images and for the design of microphone configurations for recording in 3D audio formats. In either case, the results are informative as to the maximum levels of direct sound that can be present in the height layer without the perceived location of the main channel signal being affected. For example, with respect to microphone techniques, it can be seen that the direct sound in the height layer must be attenuated by a minimum of 6 dB when the spacing between the main and height layers of microphones is less than 3.4 m (corresponding to an ICTD of 10 ms). This can be achieved through the use of cardioid microphones in the height layer. In the case that a vertically-coincident configuration is used, angling the height microphones at least 90° away from the sound source should provide the necessary attenuation, as was suggested in [8]. The microphones in the main layer should be positioned on axis with respect to the sound source.

In a more recent study conducted by the authors [10], it was demonstrated that localisation thresholds have a frequency dependency. Octave bands of pink noise, with centre frequencies ranging from 125 Hz to 8 kHz, as well as broadband pink noise, were presented to subjects from vertically-arranged stereophonic loudspeakers in an anechoic chamber. Delays ranging from 0 to 10 ms were applied to the height layer with respect to the main. The results of the study showed that the localisation thresholds were not significantly affected by ICTD, which agreed with the results reported in [8,9]. The thresholds for the 125 and 250 Hz bands were in the range of −3 to −5 dB, which was significantly higher than the −9 to −11 dB thresholds found for the 1, 2 and 8 kHz bands. In addition, the threshold for the broadband source was the lowest of all stimuli tested, being −11.5 dB. These results seem to provide an implication for the analysis of localisation thresholds for natural sound sources with different spectral balances; it might be suggested that the threshold for a high frequency dominant source would be lower than that for a low frequency dominant source. As mentioned above, previous studies [8,9] reported that the localisation threshold was not source dependent. However, since the sources used in those studies were somewhat limited (cello and bongo in [8], cello, bongo and speech in [9]), a wider range of sources would need to be tested in order to confirm the source dependency of the localisation threshold.

Of further interest in the present study is how localisation thresholds are affected by the way in which the test stimuli are presented to subjects (the presentation method). In two recent localisation experiments conducted by the authors [11,12], continuous broadband pink noise was presented to subjects from loudspeakers arranged in two layers. The loudspeakers positioned on the main layer were not elevated with respect to the listening position, whilst those in the height layer were elevated by 30°. In the first experiment [11], each layer consisted of a single loudspeaker positioned with 0° azimuth. Under such conditions, localisation judgments for the pink noise sources were accurate. Conversely, in the second experiment [12] each layer consisted of stereophonic loudspeakers with a base angle of 60° (±30°). The results of this study showed that the pink noise was perceived as being elevated with respect to the physical position of each layer. This difference in results is indicative of the phantom image elevation effect, in which stimuli are perceived as being more elevated when presented as stereophonic phantom images compared to single source only presentation [13,14]. This has notable implications for the reduction of vertical interchannel crosstalk. If main channel images are elevated with respect to the physical position of the main channel layer as a result of the phantom image elevation effect then it could be argued that the location-based effects of vertical interchannel crosstalk would be less distracting. Consequently, the localisation threshold might be much lower under such circumstances or, alternatively, might not be necessary at all. It is therefore of interest to determine how the localisation thresholds would vary when stimuli are presented as vertically-arranged quadraphonic phantom images compared to for vertical stereophonic presentation.

From the above background the following research questions were derived:

- Does there exist a sound source dependency for localisation thresholds?
- How do localisation thresholds vary for vertical quadraphonic stimulus presentation compared to vertical stereophonic?

The present paper is organised as follows. An experiment is first described in which the effects of sound source, presentation method and ICTD on localisation threshold were analysed. Following this, a second experiment is presented in which the thresholds obtained in the first experiment were applied to sound sources and verified in localisation tests. The paper concludes with discussions pertaining to the results of each experiment, as well as the implications for image rendering and microphone techniques. This also includes suggestions for future work.

2. Experiment One: Localisation Thresholds for Natural Sound Sources

2.1. Materials and Methods

2.1.1. Physical Setup

Figure 1 shows the physical setup used for the experiment, which was conducted in the ITU-R BS.1116-compliant listening room [15] at the University of Huddersfield. The experiments utilised six Genelec 8040A loudspeakers, which were arranged in two layers, 'height' and 'main'. The main layer consisted of centre (C), left (L) and right (R) loudspeakers, which were each positioned 1.2 m above the ground and 2 m from the listening position. With respect to the listening position, the centre loudspeaker was located at 0° azimuth, with the left and right loudspeakers at ±30°. The height layer comprised the three remaining loudspeakers each positioned 1.15 m directly above a loudspeaker in the main layer: Height Left (HL), Height Right (HR) and Height Centre (HC). With respect to the listening position, the main layer was not elevated, whilst the height layer was elevated by 30°. Appropriate time and level alignment was applied to the main layer with respect to the height layer to accommodate for the difference in distance between the loudspeakers in each layer and the listening position. An acoustically-transparent curtain was positioned between the listening position and the loudspeakers in order to obscure the nature of the test setup from subjects. The ear height of subjects was aligned to the centre point between the woofer and tweeter on the main layer of loudspeakers using a height-adjustable chair.

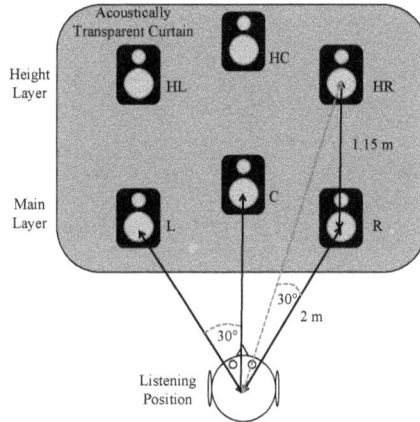

Figure 1. The physical setup used for Experiment One. C: centre loudspeaker; L: left loudspeaker; R: right loudspeaker; HC: height centre loudspeaker; HL: height left loudspeaker; HR: height right loudspeaker.

2.1.2. Test Stimuli

The test stimuli used for the experiment were anechoically-recorded guitar, speech, conga, quartet and oboe excerpts (Figure 2). These stimuli were chosen primarily due to their variations in spectral content. The predominant energy of the oboe source, for example, was in the range of 600 Hz–2 kHz, with notable peaks around 700 Hz and 1.5 kHz, whilst that for the conga ranged from 150 to 500 Hz. Given that in [10] it was reported that the localisation thresholds for low frequency octave bands were significantly higher compared to those for the mid-high frequency bands, it was thought that these two sources in particular would be beneficial in analysing the source dependency of localisation thresholds for natural sound sources. The speech source was chosen due to its broadband nature. It was reasoned that if a source dependency could be identified based on frequency then the inclusion of a wideband source would potentially make it possible to identify the frequency region that is more dominant when determining the localisation threshold for broadband sources. The guitar and quartet sources were chosen due to their varying balance between low and high frequency content, which was greater for the quartet compared to the guitar (although both were more dominant in the region below 1 kHz). This again was due to the aim of analysing whether or not the frequency dependency of localisation thresholds could translate to complex sound sources. It should also be noted that the sources contained a varied blend of continuous and transient characteristics, as can be seen in Figure 2, although it has not yet been reported in the literature how this would affect the localisation threshold.

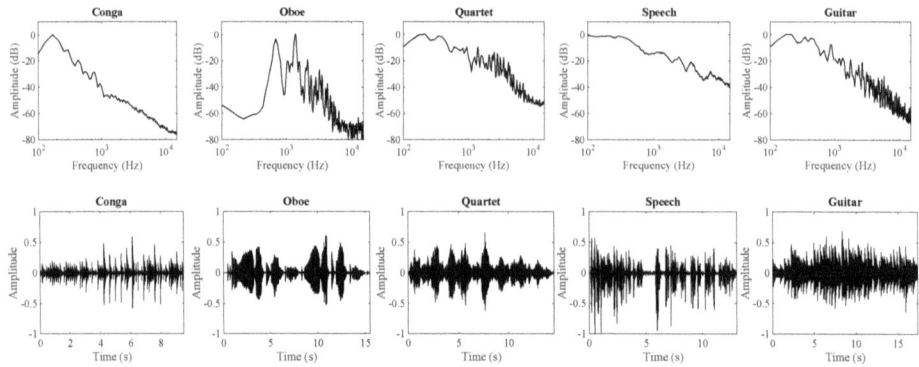

Figure 2. Long-term average spectra and waveforms of test stimuli used for Experiment One.

The test stimuli were presented to subjects as vertically oriented phantom images using the following two conditions (Figure 3):

1. Vertical stereophonic: stimulus presentation from the C (main layer) and HC (height layer) loudspeakers.
2. Vertical quadraphonic: stimulus presentation from the L, R (main layer), HL and HR (height layer) loudspeakers.

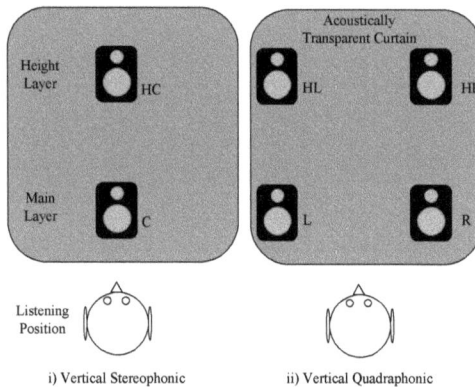

Figure 3. Presentation methods for test stimuli.

For each condition, the resultant phantom image was formed directly in front of the subject (i.e., on the median plane). The height layer was delayed with respect to the main layer for both conditions by 0, 1 and 10 ms. The delay times were chosen to emulate different spacings between the main and height microphone layers in the context of concert hall recording, with 10 ms being a likely maximum spacing (3.4 m path difference between the direct sound arriving at the main and height layers, respectively). In total, there were 30 stimuli (five sources, three delay times and two presentation methods). The amplitude of each stimulus at the listening position when presented from the main layer only (either C or L and R) was 70 dB LAeq. The amplitude of the stimulus when presented as a phantom image was dependent on the amplitude of the height layer relative to the main, which was to be varied by the subject as described in Section 2.1.4.

2.1.3. Subjects

Ten subjects, comprising staff and both postgraduate and final year undergraduate students from the University of Huddersfield's Music Technology courses, participated in the listening tests. These subjects were chosen due to their critical listening experience in spatial audio, making them better suited than more naïve subjects to determine the subtle localisation differences caused by vertical interchannel crosstalk. They all reported normal hearing.

2.1.4. Test Method

For each stimulus, subjects were presented with a 'test' and 'reference' sound. The 'reference' was the stimulus presented from the main layer only. The 'test' sound was the stimulus presented as a vertically-oriented phantom image with one of the three test ICTDs applied to the height layer. For each 'test' sound, subjects were required to reduce the amplitude of the height layer until they perceived the location of the resultant phantom image to be matching that of the 'reference'. To ensure the localisation threshold was found in each case, they were asked to set the amplitude of the height layer to the highest possible point at which this condition was met.

The threshold detection method used in the current study was based on the method of adjustment (MOA). This is an indirect scaling method that requires subjects to reduce the amplitude of a stimulus until it is equivalent to that of a reference [16]. Cardozo [17] asserted that the principal application of MOA is in situations whereby stimuli differ from one another by more than one attribute. This was applicable to the present study, as, although subjects were tasked with identifying localisation shifts, there would inevitably be some timbral changes due to the use of ICTD. However, despite such benefits, the MOA is limited in that it presents subjects with a large range from which to find the threshold, making it difficult for answers to be precise [18]. This limitation was addressed for the present experiment by requiring subjects to complete a three-stage MOA for each stimulus. Each stage was designed to be a more refined version of the previous stage as follows:

- Stage 1: The amplitude of the height layer could be adjusted from 0 to −25 dB in 5-dB steps. The localisation threshold was therefore found to within 5 dB.
- Stage 2: The amplitude of the height layer could be adjusted in the 5-dB range determined by the previous stage. The step size was 1 dB.
- Stage 3: The amplitude of the height layer could be adjusted in the 1-dB range determined by the previous stage. The step size was 0.25 dB.

This method can be considered as combining the standard MOA with adaptive testing. Adaptive threshold detection methods, which include Parameter Estimation by Sequential Testing (PEST) [19] and Up-Down [20], present stimuli to subjects at amplitudes determined by the history of the test run. This allows testing to be made at levels closer to the threshold, increasing efficiency [20]. It was however decided against using an adaptive method outright based on research conducted by Hesse [21], which indicated that the subsequent duration of the test is between three and five times longer compared to when MOA is used. As the method used for the present experiment used elements of both MOA and adaptive testing, it was named as an 'adaptive method of adjustment' (AMOA). It was considered that this fusion of methods would improve the accuracy of the test, whilst still making it relatively quick and easy for subjects to complete. The graphical user interface for the AMOA task was created using the Max7 software (Cycling'74, Walnut, CA, USA).

During the test, subjects were strictly instructed to face forwards, keeping their head still and using only their eyes to look at the test interface. The heads of subjects were not fixed, however head movements were monitored using a motion tracker device [22]. The tracker instructed subjects if their head position had deviated from an acceptable range of natural motion (10 mm in any direction). Additionally, a guide point for the ear height and distance was placed on the right hand side of the subject to help maintain the correct listening position throughout the test. Prior to the start of each test, all subjects sat a supervised practice, which utilised a speech source, in order to ensure that the instructions were understood. The test was completed in two sittings, each of which contained 15 stimuli and lasted around 20 min. The order of the tests, as well as the stimulus order, was randomised for each subject.

2.2. Results

2.2.1. The Effect of Presentation Method

Figure 4 shows the median localisation thresholds for each stimulus at each ICTD for both presentation methods. The medians have been plotted with notch edges, which is a method suggested by McGill et al. [23]. In general, there is considerable overlap between the notch edges for each presentation method, which, according to [23], indicates that pairs of stimuli are not significantly different from one another with 95% confidence. However, it is clear that in some cases the overlap between notches is minimal (e.g., conga at 1 ms, quartet at 10 ms). In order to analyse this further, the results for vertical stereophonic and quadraphonic presentation were compared for each stimulus using Wilcoxon tests. The critical p value was 0.05. According to this analysis, the effect of stimulus presentation was only significant for the Oboe with ICTD of 0 ms ($p = 0.036$). However, it is clear from Figure 4 that there is a large overlap between the notch edges for this stimulus. In addition to this, the effect size r calculated based on Cohen [24] was 0.49, which is not considered as being a large effect [24]. It could be argued then that the significant effect identified in the Wilcoxon test was a type-I error, being a false positive when in fact there is no true effect [25]. It can therefore be concluded that the effect of presentation method on the localisation thresholds obtained was not significant.

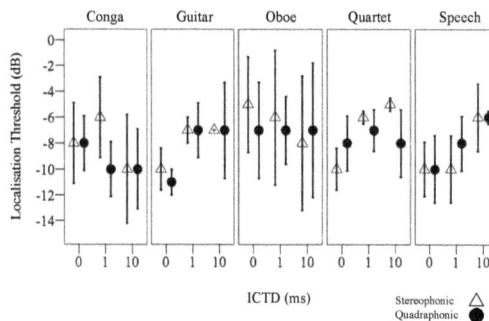

Figure 4. Medians and associated notch edges for each experimental condition. ICTD: interchannel time difference.

2.2.2. The Effect of ICTD

As it was identified that the effect of stimulus presentation on the localisation threshold was not significant, the results for stereophonic and quadraphonic presentation were combined. Figure 5 shows the effect of ICTD on the localisation thresholds obtained, with combined results for stimulus presentation. As before, the median localisation thresholds have been plotted with notch edges. Consideration of the notch edges suggests that the effect of ICTD on the localisation threshold was significant for at least some of the sound sources. The median threshold for the guitar, for example, looks significantly lower for 0 ms (−10 dB) than for 1 and 10 ms (−7.5 dB). Equally, the quartet at 0 ms (−9.5 dB) looks significantly lower than for 1 ms (−6 dB). Friedman tests (critical p value = 0.05) showed that the effect of ICTD was significant for the guitar ($p = 0.002$), speech ($p = 0.021$) and quartet ($p = 0.005$). The effect was not significant for the oboe ($p = 0.418$) and conga ($p = 0.788$). A Wilcoxon test was subsequently conducted for the guitar, speech and quartet sources, with the Bonferroni correction being applied to reduce type-I errors [26]. The results showed the following: For the guitar, significant differences were identified between the 0-ms ICTD and both the 1-ms ($p = 0.015$) and 10-ms ($p = 0.027$) ICTDs. For speech, there were significant differences between the 10 ms ICTD and both the 0 ms ($p = 0.012$) and 1 ms ($p = 0.024$) ICTDs. For the quartet, the 0-ms and 10-ms ($p = 0.015$) ICTDs were significantly different from one another. These results generally agree with the notch edges shown in Figure 5, although there are some small differences. It can therefore be concluded that the effect of ICTD on localisation threshold was significant.

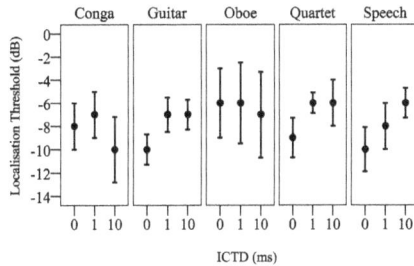

Figure 5. Medians and associated notch edges with the results for both presentation methods combined.

2.2.3. The Effect of Sound Source

Figure 6 shows the median localisation thresholds for each stimulus at each ICTD. The medians have been plotted with notch edges. The notch edges alone suggest that the effect of sound source on the localisation threshold was not significant. However, it should be noted that there are a number of notch edges that have minimal overlap (e.g., the guitar and oboe at 0 ms). A Friedman test conducted on the data indicated that the effect of sound source was significant for the 0 ($p = 0.001$) and 10 ms ($p = 0.039$) ICTDs. A Wilcoxon test was subsequently conducted to identify which pairs of stimuli were significantly different from one another, again with the Bonferroni correction being applied. The results of this analysis showed no significantly different pairs for the 10 ms ICTD. This suggests that sound source had no significant effect on the localisation threshold for this ICTD, which agrees with the overlap of notch edges in Figure 6. It should also be noted that although the overlap between conga and speech is notably minimal, the effect size indicated a small effect ($r = 0.28$). For the 0 ms ICTD, significant differences were identified between the oboe and both the guitar ($p = 0.01$) and speech ($p = 0.05$). However, the effect size was not large in either case ($r = 0.49$ between the oboe and guitar and $r = 0.42$ between the oboe and speech). In addition, there is overlap between all notch edges. Further, the effect size (Kendall's W), which was calculated during the Friedman test, was low (0.262). Based on this analysis, it can therefore be concluded that the effect of sound source on localisation threshold was not significant. This would suggest that the same localisation thresholds could be applied to all sources tested in the present study.

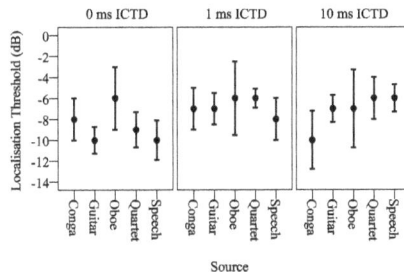

Figure 6. Medians and associated notch edges with the results for both presentation methods combined, arranged to compare the localisation thresholds for each sound source at each ICTD.

2.2.4. Localisation Thresholds for Combined Sources

As it was shown that the effect of sound source on the localisation threshold was not significant, the results for each sound source were combined. This is shown in Figure 7. The median threshold for sources with 0 ms ICTD was −9.5 dB. Based on the notch edges, the threshold for this ICTD appears to be significantly lower than the −7 dB median threshold found for the 1 and 10 ms ICTDs. This significance was confirmed with the results of both Friedman ($p = 0.000$), and Wilcoxon ($p = 0.01$ between 0 ms and 1 ms, $p = 0.00$ between 0 ms and 10 ms) tests. It can therefore be concluded that the only variable whose effect was significant on the localisation thresholds obtained in the present study was ICTD. The effects of sound source and presentation method were not significant.

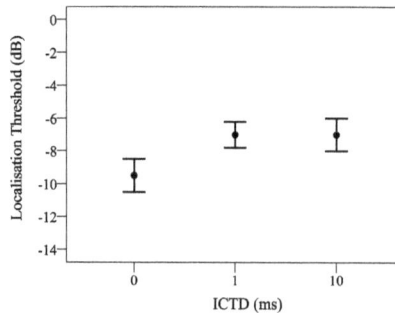

Figure 7. Localisation thresholds for combined sources.

3. Experiment Two: Verification of the Localisation Thresholds

3.1. Materials and Methods

3.1.1. Physical Setup

The physical setup for the verification test is shown in Figure 8. The experiment was conducted in the same room as was used in Experiment One and used an almost identical setup. However, as Experiment One demonstrated that the localisation thresholds were not affected by presentation method, only the L, R, HL and HR loudspeakers were used (i.e., the vertical quadraphonic condition); the C and HC loudspeakers were removed. The vertical quadraphonic condition was favoured to the vertical stereophonic condition as existing 3D audio systems, such as Auro 3D [1], tend to make use of elevated L and R loudspeakers, however they do not always use an elevated centre loudspeaker. It was therefore considered that the vertical quadraphonic condition would be more relevant to practical situations. A light-emitting diode (LED) strip was positioned directly in front of the listening position. This was located behind the acoustically-transparent curtain and was to be used by subjects to make localisation judgments.

3.1.2. Test Stimuli

The stimuli used for the experiment were the same sources used in Experiment One. The test stimuli were presented to subjects in the following conditions: (1) main layer only; (2) height layer only; (3) vertically oriented phantom image with 0 dB interchannel level difference (ICLD) and; (4) vertically oriented phantom image with the localisation threshold applied to the height layer. The ICTDs applied to the height layer for the phantom image conditions were 0 and 1 ms. The 10 ms condition was not tested for the following reasons. Firstly, as there was no significant difference between the localisation thresholds obtained for 1 and 10 ms it was deemed unnecessary to test both conditions. Furthermore, as discussed earlier, 10 ms represents a condition whereby the path difference between the direct sound arriving the main and height layers respectively is around 3.4 m, which is fairly large in practice. It was therefore decided that the 1-ms condition would be more representative of a practical configuration, with the resultant path difference being only around 0.34 m.

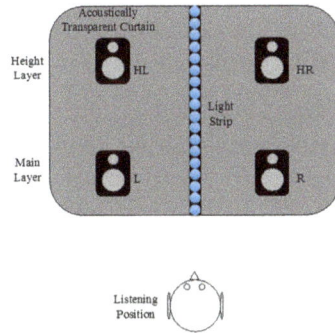

Figure 8. Physical setup for the localisation threshold verification test.

Although not necessarily integral to the verification test, the 0-dB ICLD and height layer only conditions were included in the experiment in order to reduce any expectation biases. During preliminary tests, in which only the main layer only and localisation threshold conditions were considered, subjects reported that hearing all stimuli originate from the same position in a localisation test was confusing. Furthermore, some were led to believe that the stimuli could not all be coming from the same location and this forced them to provide different answers to what was actually being perceived. The height layer only and 0-dB conditions were therefore included in order to introduce stimuli that were in a position away from the main layer only condition. This was found to prevent the issue.

All stimuli were presented at 70 dB LAeq at the listening position when presented from the main layer only. The increase in amplitude when the stimuli were presented as vertically arranged quadraphonic phantom images was dependent on the localisation threshold applied to the height layer. In the case of the 0-ms condition, the height layer was attenuated by 9.5 dB with respect to the main layer, whilst for 1 ms the attenuation was 7 dB, which was based on the results from Experiment One. In total, there were 30 stimuli, being the main and height layer-only conditions (10), the localisation threshold conditions (10—five sources, two ICTDs) and the 0-dB ICLD conditions (10—five sources, two ICTDs).

3.1.3. Test Method

The test was completed by the same 10 subjects who participated in Experiment One. Localisation judgments were made using the LED strip located directly in front of the listening position. For each test, subjects were provided with a handheld knob, which controlled which LED on the strip was turned on. Subjects were required to adjust the knob until the position of the active LED matched the perceived location of the focal point of each stimulus. This method was chosen following research conducted by Lee et al. [27], who found that it was faster and produced results with greater accuracy and consistency compared to the numbered scale method, which had been used in a number of previous vertical localisation studies [11,28,29]. The position of the LED selected for each stimulus was converted into an elevation angle. The heads of subjects were not fixed, however they were instructed to sit up and face forwards at all times, using only their eyes to look at the light strip. To help maintain the correct seating position, a small headrest was positioned behind the head of each subject. The test was completed four times by each subject, with each sitting containing all 30 stimuli and taking around 10 min to complete. The presentation order of stimuli was randomised for each test.

3.2. Results

Levene and Shapiro–Wilk tests were first conducted, using the SPSS Statistics 22 software (IBM, New York, NY, USA), in order to determine the suitability of the collected data for parametric statistical analysis. The Shapiro–Wilk test showed that not all scores in each condition featured normal distribution, although the results of the Levene test showed homogeneity of variance for all sound sources. For these reasons, non-parametric tests were chosen for the statistical analysis.

Figure 9 shows the median perceived elevation of each of the test stimuli, plotted with notch edges. Consideration of the data reveals the following. Firstly, the localisation thresholds derived in Experiment One resulted in perceived elevation judgments similar to those for the same source presented from the main layer only. This was the case for all sources, with the median difference in perceived elevation between the main layer only and localisation threshold conditions ranging between $-4.0°$ and $-0.8°$ for the 0-ms ICTD and between $3.9°$ and $0.0°$ for the 1-ms ICTD. In addition to this, the notch edges for all the localisation threshold conditions overlap with those for main layer-only presentation. It is also interesting to note that, for the 0-ms ICTD, the median perceived elevation for the stimuli with the localisation threshold applied was slightly lower than the main layer only condition for all sources.

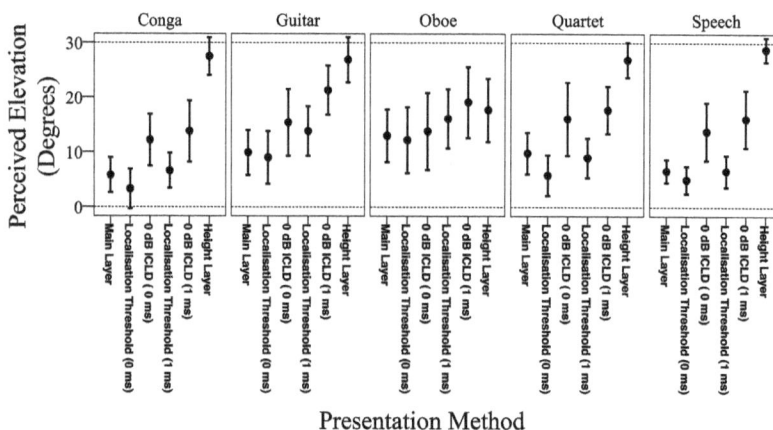

Figure 9. Medians and associated notch edges for the results of the verification test showing the perceived elevation of each of the test stimuli. The dotted lines at 0 and 30° represent the physical positions of the main and height layers respectively. ICLD: interchannel level difference.

In order to further determine whether or not the localisation thresholds derived from Experiment One were successful in preventing vertical interchannel crosstalk from affecting the perceived location of the main channel signal, Wilcoxon tests were conducted. The results suggested that, generally, there were no significant differences between the elevation judgments for the localisation threshold and main layer-only conditions. However, the data did suggest that the difference was significant for the quartet at 0 ms ($p = 0.041$) and for both the guitar ($p = 0.002$) and speech ($p = 0.025$) at 1 ms. Despite this, there is a clear overlap between the notch edges for each of these stimuli. In addition, the Pearson's correlation coefficient did not show a large effect in any case ($r = 0.25$ for quartet at 0 ms, $r = 0.39$ for guitar at 1 ms, $r = 0.28$ for speech at 1 ms). It can therefore be suggested that the difference in median perceived elevation between the localisation threshold and main layer only conditions was not significant. Consequently, the localisation thresholds derived in the present study are appropriate in preventing vertical interchannel crosstalk from affecting the perceived location of the main channel signal.

With respect to the 0-dB ICLD conditions, a series of interesting results can be seen. Firstly, for the oboe it is clear that perceived elevation was not significantly affected by changes in how the stimulus was presented to subjects. In all cases, the perceived elevation was similar, which would indicate that this source was less affected by the migration of the main channel signal from the main layer as a result of vertical interchannel crosstalk. This result might suggest that the application of localisation thresholds would not always be necessary. Furthermore, for the other stimuli it is clear that the median perceived elevation was greater for the 0-dB ICLD condition compared to the main layer only condition. For 0 ms, the difference in median perceived elevation ranged from 5.5° to 7.2°, whilst for 1 ms the difference ranged from 7.9° to 11.4°. This result indicates that the perceived location of the main channel signal would be more affected by vertical interchannel crosstalk when the height layer is delayed with respect to the main. However, despite this result it is clear that the difference was not always significant, with there being a notable overlap between notch edges between the 0-dB ICLD conditions and the main layer-only conditions. This is particularly noticeable for the guitar and quartet sources at 0 ms. Nevertheless, is clear that vertical interchannel crosstalk at the very least resulted in an increase in the median perceived elevation of the main channel signal, which was notably reduced when the localisation thresholds derived from Experiment One were applied.

A further result of note can be seen with respect to the main and height layer-only conditions. Firstly, for the latter condition it would appear that perceived elevation judgments were generally accurate for all sources, excluding the oboe, with respect to the physical position of the height layer. Conversely, for the main layer-only condition the judgments were less accurate, with perceived source elevation being in the range of 5.8°–13.0° with respect to the main layer's physical location. This elevation of the sound source with respect to the main layer was also maintained for the conditions whereby a localisation threshold was applied to the height layer. The results of a Wilcoxon signed rank test, which compared the results for the main layer-only condition to the physical position of the main layer (0°), showed that each source was perceived to be significantly higher than the physical height from which the source was presented ($p = 0.000$ for all sources).

4. Discussion

The experimental data obtained in Experiment One showed that localisation thresholds for natural sound sources are not significantly affected either by sound source or presentation method. Instead, the only variable whose effect was significant was ICTD. When the ICTD was 0 ms, the threshold was found to be -9.5 dB, which was significantly lower than the -7 dB found for ICTDs of both 1 and 10 ms. In verification tests (Experiment Two), these thresholds were found to be effective at preventing vertical interchannel crosstalk from affecting the perceived location of the main channel signal. This section discusses potential physical causes for the subjective results and suggests practical implications of the results.

4.1. Sound Source Dependency

One of the key aims of the present study was to determine whether or not there existed a sound source dependency of localisation thresholds. The results of Experiment One showed that, although the median localisation threshold varied for different sound sources, these differences were not significant, which agrees with the results reported in [8,9]. Additionally, in Experiment Two the non-significant effect of sound source on the localisation threshold was demonstrated further. In this regard, the results showed that, when the same threshold was applied to each source, the resultant phantom image position was not significantly different from that for main layer only presentation. In order to explain the reasons for these results, considerations were given to three different aspects of vertical localisation: (1) the spectral energy distribution of the ear input signal; (2) the so-called 'pitch-height' effect [29]; and (3) the effect of vertical image spread (VIS) on vertical localisation.

Since the primary cue for elevation perception is known to be the spectral filtering of the pinnae in the 4–10 kHz range [30], it was first considered how the ear input spectra are affected by the presence of the height layer. In Figure 10, the difference in spectral energy between main and height layer only presentation has been shown for both presentation methods as delta spectra. The head-related impulse responses (HRIRs) used for this exercise were taken from the Massachusetts Institute of Technology (MIT)'s HRIR database created using the KEMAR dummy head microphone [31]. Any points where the line falls below 0 dB represents dominance of the main layer over the height layer and vice versa. From both delta spectra, it can be seen that the predominant difference in energy between the two layers is a peak (height layer dominant) in the range of 7–9 kHz, a region that is associated with localisation above the subject [30,32]. It is therefore reasonable to suggest that the increased energy in this region is a key reason that stimuli presented from the height layer only were perceived as being more elevated than those presented from the main layer only in Experiment Two. Since the resultant spectrum for phantom image presentation will depend on the relative strengths of each layer, the following can be suggested. When the ICLD is small, the contribution of each layer to the resultant spectrum will be similar. As a consequence of this, a given source presented as a vertically oriented phantom image will feature more energy in the 7–9 kHz region compared to the same source presented from the main layer only. This will manifest in differences in perceived elevation between the two conditions, as was demonstrated in [33]. However, as the ICLD increases, the main layer becomes more dominant in determining the resultant ear input spectrum, which means that the differences between the phantom image and main layer only conditions in the 7–9 kHz region would decrease. At the localisation threshold then, this difference is not sufficient enough to be interpreted as an elevation difference and therefore the main layer only and phantom image conditions are perceived as being in the same location. Based on this hypothesis, it can be argued that the primary mechanism used for the subject's localisation threshold judgment might be the relative spectral energy weighting between the main and height layers in head-related transfer function (HRTF), rather than being related to the fine spectral details of the sound source, which agrees with a previous study conducted by the authors [10]. It should be noted, however, that the ear input spectra are dependent on the subject and that it is therefore difficult to generalise HRTF characteristics. Based on this, it is apparent that further study is necessary, using measured HRTFs of different subjects, before the importance of the 7–9 kHz region in particular in determining the localisation threshold can be confirmed.

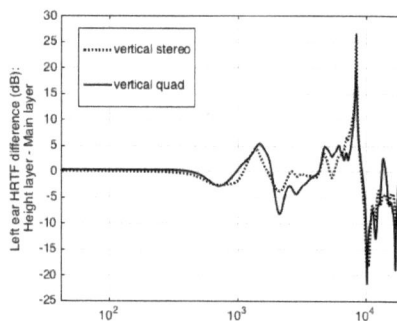

Figure 10. The differences in spectral energy between the main and height layer only conditions for both vertical stereophonic and vertical quadraphonic source presentation.

An interesting point of note with respect to the non-significant effect of sound source relates to the results of Experiment Two, which showed that localisation judgments for the oboe source were generally consistent regardless of how the source was presented to subjects. A potential explanation as to why this result was obtained is offered thus. According to the literature, narrowband stimuli incident from the median plane are localised on the basis of frequency, with increases in frequency corresponding to increases in perceived elevation [11,30,34]. This phenomenon is known as the 'pitch-height effect' [29]. As can be seen from Figure 2, the spectrum for the oboe source was notably narrow, with a bandwidth ranging from around 500 Hz to 4 kHz and with its predominant energy focused around 1 to 2 kHz. According to the literature, band-limited stimuli in this frequency range are localised at a similar vertical position regardless of which loudspeaker layer presented the source for both vertical stereophonic [11,29] and vertical quadraphonic [12] loudspeaker arrangements. Therefore, it might be that localisation judgments for the oboe were determined by the pitch-height effect rather than the HRTF-based vertical localisation mechanism discussed above. This would indicate that the latter mechanism is predominantly applicable to broadband sources, with less relevance to sources that are both narrowband and absent in high frequency energy.

That ICLD was always necessary to reach the localisation threshold for the oboe source, despite there being no significant difference in perceived elevation between the 0 dB and main layer only conditions, might be explained by an alternative hypothesis proposed by the authors [10]. When sound source presentation shifts from main layer only to vertical phantom image, a key difference is an increase in perceived VIS. Such an increase contributes to an increase in localisation blur [35], which inevitably makes the sound source more difficult to localise in vertical space. Therefore, given that the test conditions required a direct comparison between the positions of stimuli presented using the main layer only and vertical phantom image conditions, it is possible that differences in perceived VIS were perceived as elevation differences. Further, these differences would have decreased as the amplitude of the height layer was reduced. At the localisation threshold then, the difference in VIS is sufficiently small for stimuli presented using the two conditions to be perceived as being in the same location. As with the hypothesis regarding the importance of the spectral energy weighting in determining the localisation threshold, this hypothesis is able to explain why the effect of sound source was not significant in Experiment One. Simply, the difference in perceived VIS between main layer only and vertical phantom image presentation is primarily a function of ICLD and is not affected by the sound source itself. Based on these discussions, it is clear that the exact mechanisms that determine whether or not the localisation threshold has been met requires further study.

4.2. The Effect of Presentation Method

A further aim of the present study was to analyse how the localisation thresholds would be affected by changing the presentation method from vertical stereophonic to vertical quadraphonic, with the results of Experiment One showing that the effect was not significant. In Section 4.1 it was discussed how a key mechanism in determining whether or not the localisation threshold had been met might be the balance of spectral cues provided by the main and height layers respectively. The results of Experiment One would seemingly indicate that this balance was not affected when source presentation changed from vertical stereophonic to vertical quadraphonic. This is apparent when Figure 10 is considered, which shows that the difference in spectral energy between the main and height layer-only conditions were somewhat similar for both presentation methods. In order to gain further objective insights into this result, the influence of the height layer on the frequency spectrum of the ear-input signal was analysed for each presentation method. For this, the spectral magnitude of the ear-input signal resulting from the main layer only was subtracted from that from both the main and height layers, using the MIT's KEMAR HRIR database [31]. Figure 11 plots the analysis results obtained when the ICLD was 0 dB (no height layer level reduction applied) and when the localisation threshold (-9.5 dB) was applied. From the plots, the following can be observed. Firstly, the spectral energy in the 7–9 kHz range for the 0-dB ICLD condition was dominant over that for the main layer-only condition, for both presentation methods, in a manner similar to that hypothesised in Section 4.1. In addition to this, when the localisation threshold was applied for each method, the difference in energy in this region was decreased between 8 and 10 dB, whereas that below this region was about 5 dB or less. This therefore supports the hypothesis that decreases in spectral energy in the 7–9 kHz region will result in the localisation threshold being met, with similar reductions for both presentation methods for a given ICLD likely being the reason for the non-significant effect of presentation method. There is also good agreement with the results of a previous study conducted by the present authors [10], which showed that the main layer only and two-layer conditions did not have to have equal energy in the frequency range in which the spectral cues for elevation lie for the localisation threshold to be met.

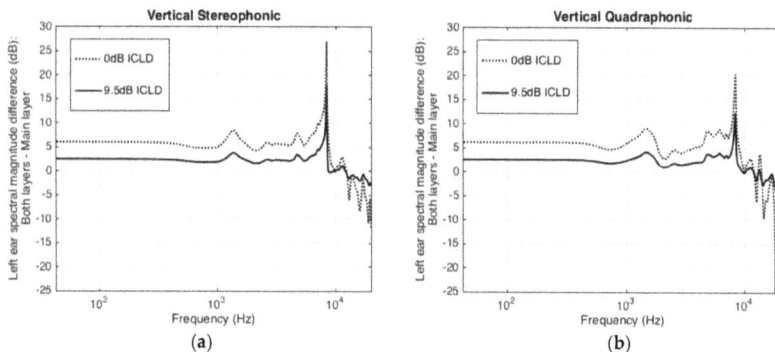

Figure 11. Difference in spectral energy between the main layer only and phantom image conditions for both presentation methods with 0-dB ICLD between the main and height layers and 9.5-dB ICLD (localisation threshold): (**a**) vertical stereophonic condition; (**b**) vertical quadraphonic condition.

The non-significant effect of presentation method is interesting when the results of previous localisation studies are considered. As shown in [11,28,29], for a single loudspeaker placed in front of the listener in the median plane, the perceived image of broadband noise tends to be localised accurately at the physical position of the loudspeaker. Conversely, the phantom centre image of the noise produced from stereophonic loudspeakers at the ear height (i.e., the main layer of the quadraphonic condition in the current study) would be elevated with respect to the physical position of the loudspeaker as reported in [12–14] (this was also observed for natural sound sources in Experiment Two of the present study). Furthermore, a similar degree of difference between real and phantom image conditions in perceived elevation would be observed also for elevated loudspeakers (i.e., the height layers of the current study), based on data presented in [11,12]. From the above, it can be inferred that, for 0-dB ICLD, sound sources presented using the vertical quadraphonic condition would be elevated with respect to those presented using the vertical stereophonic condition, with the difference in perceived elevation being similar to that for the same sources presented from the main layer only. This would therefore imply that the perceived differences in elevation between the main layer-only and phantom image conditions for a given ICLD would be similar for both presentation methods, as is demonstrated in Figure 12, which would further explain why the effect of presentation method was not significant.

4.3. The Localisation Dominance Effect

The results of the present experiment suggest that the only variable that had a significant effect on the localisation threshold was ICTD, with delays of both 1 and 10 ms requiring significantly less level reduction than did 0 ms. However, there was no condition whereby ICTD alone was sufficient for the localisation threshold; ICLD was always necessary. This result indicates that the precedence effect, in which an ICTD greater than 1.1 ms between coherent loudspeakers located on the horizontal plane will cause the resultant sound source to be localised at the exact position of the earlier loudspeaker [36], is not a feature of median plane localisation. This agrees with the conclusions reported in [8–11]. What is suggested, however, is somewhat of a localisation dominance effect, whereby the presence of an ICTD biases localisation towards the earlier loudspeaker. This can be considered as being similar to summing localisation [31] and has been shown to operate in numerous median plane localisation studies [37–39]. If it is the case that the earlier loudspeaker becomes dominant in determining perceived source location in the median plane, then this might explain why significantly less ICLD was necessary to meet the localisation threshold for the 1 and 10 ms ICTDs compared to for the 0 ms condition. It should be noted, however, that higher localisation thresholds as a result of a localisation dominance effect have not been reported in previous localisation threshold experiments, with both [8] and [9] reporting that there was no significant difference between the localisation thresholds in the range of 0–10 ms.

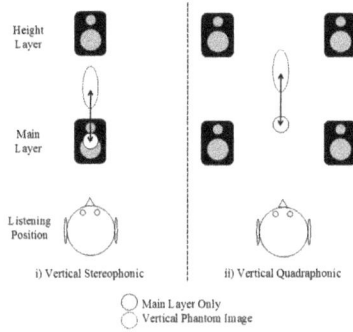

Figure 12. Illustration to show how presentation method would not affect localisation thresholds despite the presence of the phantom image elevation effect.

The data provided in Experiment Two enables further analysis as to whether or not the significant effect of ICTD was related to the operation of a localisation dominance effect. Figure 13 shows the experimental data for the main and height layer-only conditions alongside those for the 0-dB ICLD conditions (both 0- and 1-ms ICTD). The median perceived elevation for each has been plotted with notch edges. From the results, it is clear that there is no evidence to support the existence of a localisation dominance effect, with the median perceived elevation for all stimuli increasing in the presence of an ICTD. This result is somewhat similar to those of a previous study conducted by the authors [11], in which the perceived elevation of broadband pink noise presented from vertically-arranged stereophonic loudspeakers in anechoic conditions increased as the ICTD increased from 0 to 1 ms. It should also be noted in the present study that the differences between the 0- and 1-ms conditions were not significant. Based on these results, the hypothesis that the localisation thresholds are higher in the presence of an ICTD due to the operation of a localisation dominance effect can be rejected. As a consequence of this, further study would be required in order to adequately explain this result.

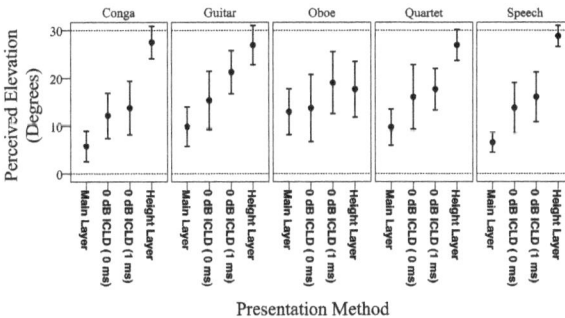

Figure 13. Medians and associated notch edges for the verification test results arranged for the analysis of the localisation dominance effect.

4.4. Practical Implications

A primary aim of the present study was to obtain localisation thresholds that could be used to influence both the placement of microphones and the rendering of 3D images in the context of 3D audio. The purpose of this was to prevent vertical interchannel crosstalk from affecting the perceived location of the main channel signal. With respect to 3D microphone configurations, a series of techniques have already been proposed in [8]. In that study, it was suggested that, since the localisation threshold needs to be applied to the height-layer microphone signals in order for the source image to be located at the position of the main layer, directional microphones should be used for the height layer rather than omni-directional ones. The results of the present study support this suggestion. In addition, it was proposed that, in case of using microphones with an 'ideal' cardioid polar pattern (i.e., −6 dB attenuation at 90°), the necessary ICLD could be achieved for both vertically coincident and spaced configurations by angling the height layer of microphones at least 90° away from the direct sound. However, the data reported in the present study indicates that a minimum angle of 105° would be necessary in the case that the main and height layers are spaced apart, whilst for a coincident configuration the angle should be 115°. This would provide attenuation of direct sounds in the height layer at the localisation thresholds for 0 ms and 1 ms found in the present study: around −7 and −9.5 dB, respectively.

The results of the current study are considered to be also useful for vertical image rendering in 3D sound mixing and upmixing applications. They indicate that direct sounds can be present in the height layer provided they are attenuated with respect to those in the main layer by either 9.5 dB (in the case of 0 ms ICTD) or 7 dB (in the case of 1–10 ms ICTD) without the perceived location of the main channel signal being affected. Such a technique could have potentially pleasing effects such as an increase in perceived VIS. However, it is currently not clear how the timbre of the main channel signal would be affected by such a technique and, further, if the end result would be pleasing. It can be seen from Figure 11, for example, that the resultant spectrum of the signal is different at the localisation threshold compared to main layer-only presentation, with a notable peak in the 7–9 kHz range. Alongside this, Halmrast [40] suggested that secondary vertical sources would result in orchestral music sounding 'boxy', whilst Barron and Marshall [41] indicated that timbral colouration as a result of vertical reflections are more audible than for lateral reflections. It would be necessary then to evaluate first of all what the perceptual differences are between the main layer only and vertical phantom image conditions with the localisation thresholds applied and further if the threshold conditions are considered as being preferable. Such a study would make it possible to determine whether the localisation threshold should be applied or, conversely, if the direct sound in the height layer should be either masked or absent entirely. This would provide further insights on both image rendering and microphone techniques in the context of 3D audio production.

It should also be noted that there are some, limited, applications with respect to the vertical panning of sound sources. It is indicated by the results that, depending on the ICTD, the threshold value for a source to be fully panned to the main loudspeaker layer is in the range of 7–9 dB, which agrees with the vertical localisation studies of both Barbour [5] and Wendt et al. [6]. However, further study would be needed to determine if this value is applicable to source localisation at the position of the height layer and, further, how changes in both ICLD and ICTD affect the perceived localisation of the resultant phantom image in between these extremes.

5. Conclusions

The present study carried out an analysis of localisation thresholds for natural sound sources. The study was divided into two experiments. In the first (Experiment One) the effects of sound source, ICTD and presentation method were examined. Anechoically recorded conga, quartet, speech, guitar and oboe sources were presented to subjects in a natural listening environment using two conditions: vertical stereophonic and vertical quadraphonic. For each condition, the loudspeakers were divided into two layers, being 'height' (30° elevation) and 'main' (0° elevation). Delays ranging from 0 to 10 ms were applied to the height layer with respect to the main. Subjects sat a listening test in which the minimum amount of attenuation necessary in the height layer for the resultant phantom image to match the position of the same source presented from the main layer alone was considered.

The results of the experiment showed that the localisation thresholds were affected only by ICTD. For delays of 0 ms the threshold was −9.5 dB, which was significantly lower than the −7 dB found for 1 and 10 ms. That less ICLD was necessary in the presence of a delay was initially interpreted based on the existence of a localisation dominance effect. In addition, attempts to explain the non-significant effect of sound source were made based on the hypothesis that the primary mechanism to determine whether or not the localisation threshold had been met was the balance of spectral energy provided by the main and height layer, particularly in the 7–9 kHz range, which is not related to the spectrum of the source itself. This hypothesis also explained the non-significant effect of presentation method, with it being demonstrated that the reduction in the difference in energy between the main layer only and phantom image conditions in the 7–9 kHz region was similar for both methods for a given ICLD.

In Experiment Two, the localisation thresholds obtained in Experiment One were applied to natural sound sources, with localisation tests being conducted in order to verify that they were effective at preventing vertical interchannel crosstalk from affecting the perceived location of the main channel signal. Stimuli were presented using the vertical quadraphonic condition, with the main and height layer-only, 0-dB ICLD and localisation threshold conditions all being tested. ICTDs of 0 and 1 ms were applied to the height layer with respect to the main. Subjects used a light strip, which was controlled by a handheld knob, in order to identify the perceived location of each stimulus. For all stimuli, there was no significant difference in perceived elevation between the main layer only and localisation threshold conditions.

A key result from Experiment Two was that no evidence was found to support the existence of a localisation dominance effect, with the perceived elevation of the sources with 1-ms ICTD being higher than those with 0-ms ICTD. It is therefore unclear why less level reduction was necessary in Experiment One in the case that an ICTD was present. The results also showed evidence of the phantom image elevation effect, which was used to suggest that the perceived difference in elevation between the main layer only and phantom image conditions would be similar for both presentation methods for a given ICLD. This therefore further explained why the effect of presentation method was not significant in Experiment One. In addition, the results implied that the oboe source was localised based on the pitch-height effect. This meant that the hypothesis regarding the balance of spectral cues provided by the main and height layers did not adequately explain the localisation thresholds obtained for this source. As a result of this, it was suggested that the results might be explained based on differences in perceived VIS between the main layer-only and phantom image conditions.

The practical implications of the results obtained in the study were also discussed. In particular, differences between suggestions made in previous studies and those indicated by the present results were considered. It was also stated that further study would need to be conducted into the spatial and timbral effects when the localisation thresholds are applied in order to determine whether or not it would be more appropriate for the direct sound in the height layer to be masked.

Acknowledgments: This work was supported by the Engineering and Physical Sciences Research Council (EPSRC), UK, Grant Ref. EP/L019906/1. The authors thank the staff members and students of the University of Huddersfield's music technology courses who participated in the listening tests.

Author Contributions: Rory Wallis conducted the experiment, analysed the data and wrote the paper. Hyunkook Lee supervised the project and contributed to the data analysis and discussion presented in the paper.

Conflicts of Interest: The authors declare no conflict of interest.

References

1. Listening Formats: Auro 3D. Available online: http://www.auro-3d.com/system/listening-formats (accessed on 13 October 2016).

2. Dolby Atmos. Available online: http://www.dolby.com/us/en/brands/dolby-atmos.html (accessed on 13 October 2016).

3. Somerville, T.; Gilford, C.L.S.; Spring, N.F.; Negus, R.D.M. *Recent Work on the Effects of Reflectors in Concert Halls and Music Studios*; British Broadcasting Corporation: London, UK, 1965.

4. Pulkki, V. Localization of Amplitude-Panned Virtual Sources II: Two- and Three-Dimensional Panning. *J. Audio Eng. Soc.* **2001**, *49*, 753–767.

5. Barbour, J. Elevation Perception: Phantom images in the vertical hemisphere. In Proceedings of the AES 24th International Conference on Multichannel Audio, Banff, AB, Canada, 26–28 June 2003.

6. Wendt, F.; Frank, M.; Zotter, F. Panning with height on 2, 3 and 4 loudspeakers. In Proceedings of the 2nd International Conference on Spatial Audio, Erlangen, Germany, 21–23 February 2014.

7. Kimura, T.; Ando, H. 3S Audio System Using Multiple Vertical Panning for Large-Screen Multiview 3D Video Display. *ITE Trans. MTA* **2014**, *2*, 33–45.

8. Lee, H. The relationship between interchannel time and level differences in vertical sound localisation and masking. In Proceedings of the Audio Engineering Society 131st Convention, New York, NY, USA, 20–23 October 2011. Preprint 8556.

9. Stenzl, H.; Scuda, U.; Lee, H. Localisation and masking thresholds of diagonally positioned sound sources and their relationship to interchannel time and level differences. In Proceedings of the International Conference on Spatial Audio, Erlangen, Germany, 21–23 February 2014.

10. Wallis, R.; Lee, H. Vertical Stereophonic Localisation in the Presence of Interchannel Crosstalk: The Analysis of Frequency-Dependent Localisation Thresholds. *J. Audio Eng. Soc.* **2016**, *64*, 762–770. [CrossRef]

11. Wallis, R.; Lee, H. The Effect of Interchannel Time Difference on Localisation in Vertical Stereophony. *J. Audio Eng. Soc.* **2015**, *63*, 767–776. [CrossRef]

12. Lee, H. Perceptual Band Allocation (PBA) for the Rendering of Vertical Image Spread with a Vertical 2D Loudspeaker Array. *J. Audio Eng. Soc.* **2016**, *64*, 1003–1013. [CrossRef]

13. De Boer, K. A Remarkable Phenomenon with Stereophonic Sound Reproduction. *Philips Tech. Rev.* **1947**, *9*, 8–13.

14. Lee, H. Investigation on the phantom image elevation effect. In Proceedings of the Audio Engineering Society 139th Convention, New York, NY, USA, 29 October–1 November 2015. Preprint 9441.

15. International Telecommunication Union. *Recommendation ITU-R BS.1116-1: Methods for the Subjective Assessment of Small Impairments in Audio Systems Including Multichannel Sound Systems*; International Telecommunications Union: Geneva, Switzerland, 1994.

16. Bech, S.; Zacharov, N. *Perceptual Audio Evaluation: Theory, Method and Application*; Wiley: Chester, UK, 2006.

17. Cardozo, B.L. Adjusting the Method of Adjustment: SD vs. DL. *J. Acoust. Soc. Am.* **1965**, *37*, 786–792. [CrossRef]

18. Lawless, H.T. *Quantitative Sensory Analysis: Psychophysics, Models and Intelligent Design*; Wiley-Blackwell: Chichester, UK, 2013.

19. Taylor, M.M.; Creelman, C.D. PEST: Efficient Estimates on Probability Functions. *J. Acoust. Soc. Am.* **1967**, *41*, 782–787. [CrossRef]

20. Levitt, H. Transformed Up-Down Methods in Psychoacoustics. *J. Acoust. Soc. Am.* **1970**, *49*, 467–477. [CrossRef]

21. Hesse, A. Comparison of Several Psychophysical Procedures with Respect to Threshold Estimates, Reproducibility and Efficiency. *Acta Acust. United Acust.* **1986**, *59*, 263–273.

22. Johnson, T.; Gibson, I.; Evans, B.; Wendl, M. An Investigation into Kinect and Middleware Error and Their Suitability for Academic Listening Tests. In Proceedings of the Audio Engineering Society 140th Convention, Paris, France, 4–7 June 2016. eBrief 273.

23. McGill, R.; Tukey, J.W.; Larsen, W.A. Variations of Box Plots. *Am. Stat.* **1978**, *32*, 12–16. [CrossRef]

24. Cohen, J. *Statistical Power Analysis for the Behavioral Sciences*; Lawrence Erlbaum Associates: New York, NY, USA, 1988.

25. Lieberman, M.D.; Cunningham, W.A. Type I and Type II Error Concerns in fMRI Research: Re-balancing the Scale. *Soc. Cog. Affect. Neurosci.* **2009**, *4*, 423–428. [CrossRef] [PubMed]

26. Simner, R. An Improved Bonferroni Procedure for Multiple Tests of Significance. *Biometrika* **1986**, *73*, 751–754.

27. Lee, H.; Johnson, D.; Mironovs, M. A new response method for auditory localisation and spread tests. In Proceedings of the Audio Engineering Society 140th Convention, Paris, France, 4–7 June 2016. e-Brief 240.

28. Roffler, S.K.; Butler, R.A. Factors that Influence the Localisation of Sound in the Vertical Plane. *J. Acoust. Soc. Am.* **1968**, *43*, 1255–1259. [CrossRef] [PubMed]
29. Cabrera, D.; Tiley, S. Vertical localisation and image size effects in loudspeaker reproduction. In Proceedings of the AES 24th International Conference on Multichannel Audio, Banff, AB, Canada, 26–28 June 2003.
30. Hebrank, J.; Wright, D. Spectral Cues used in the Localisation of Sound Sources on the Median Plane. *J. Acoust. Soc. Am.* **1974**, *56*, 1829–1834. [CrossRef] [PubMed]
31. Gardener, B.; Martin, K. HRTF Measurements of a KEMAR Dummy-Head Microphone. 2000. Available online: http://sound.media.mit.edu/resources/KEMAR.html (accessed on 17 November 2016).
32. Blauert, J. Sound Localisation in the Median Plane. *Acta Acust. United Acust.* **1969**, *22*, 205–213.
33. Chun, C.J.; Kim, H.K.; Choi, S.H.; Jang, S.; Lee, S. Sound Source Elevation Using Spectral Notch Filtering and Directional Band Boosting in Stereo Loudspeaker Reproduction. *IEEE Trans. Consum. Electron.* **2011**, *57*, 1915–1920. [CrossRef]
34. Pratt, C.C. The Spatial Character of High and Low Tones. *J. Exp. Psychol.* **1930**, *13*, 278–285. [CrossRef]
35. Blauert, J. *Spatial Hearing: The Psychophysics of Human Sound Localisation*; MIT Press: Cambridge, UK, 1997.
36. Wallach, H.; Newman, E.B.; Rosenzweig, M.R. The Precedence Effect in Sound Localisation. *Am. J. Psychol.* **1949**, *52*, 315–336. [CrossRef]
37. Blauert, J. Localisation and the Law of the First Wavefront in the Median Plane. *J. Acoust. Soc. Am.* **1971**, *50*, 466–470. [CrossRef] [PubMed]
38. Litovsky, R.Y.; Rakerd, B.; Tin, T.C.T.; Hartmann, W.M. Psychophysical and Physiological Evidence for a Precedence Effect in the Median Sagittal Plane. *J. Neurophysiol.* **1997**, *77*, 2223–2226. [PubMed]
39. Tregonning, A.; Martin, B. The Vertical Precedence Effect: Utilising Delay Panning for Height Channel Mixing in 3D Audio. In Proceedings of the Audio Engineering Society 139th convention, New York, NY, USA, 29 October–1 November 2015. Preprint 9469.
40. Halmrast, T. Orchestral Timbre: Comb-Filter Coloration from Reflections. *J. Sound Vib.* **2000**, *232*, 53–69. [CrossRef]
41. Barron, M.; Marshall, A.H. Spatial Impression due to Early Lateral Reflections in Concert Halls: The Derivation of a Physical Measure. *J. Sound Vib.* **1981**, *77*, 211–232. [CrossRef]

![applied sciences logo] *applied sciences*

MDPI

Article

A Measure Based on Beamforming Power for Evaluation of Sound Field Reproduction Performance

Ji-Ho Chang [1],* and Cheol-Ho Jeong [2]

[1] Center for Fluid Flow & Acoustics, Division of Physical Metrology,
    Korea Research Institute of Standards and Science, Daejeon 34113, Korea
[2] Acoustic Technology, Department of Electrical Engineering, Technical University of Denmark,
    DK-2800 Kongens Lyngby, Denmark; chj@elektro.dtu.dk
*   Correspondence: chang.jiho@gmail.com; Tel.: +82-42-868-5309

Academic Editors: Woon-Seng Gan and Jung-Woo Choi
Received: 16 January 2017; Accepted: 27 February 2017; Published: 3 March 2017

Abstract: This paper proposes a measure to evaluate sound field reproduction systems with an array of loudspeakers. The spatially-averaged squared error of the sound pressure between the desired and the reproduced field, namely the spatial error, has been widely used, which has considerable problems in two conditions. First, in non-anechoic conditions, room reflections substantially deteriorate the spatial error, although these room reflections affect human localization to a lesser degree. Second, for 2.5-dimensional reproduction of spherical waves, the spatial error increases consistently due to the difference in the amplitude decay rate, whereas the degradation of human localization performance is limited. The measure proposed in this study is based on the beamforming powers of the desired and the reproduced fields. Simulation and experimental results show that the proposed measure is less sensitive to room reflections and the amplitude decay than the spatial error, which is likely to agree better with the human perception of source localization.

Keywords: sound field reproduction; higher-order ambisonics (HOA); evaluation of reproduction performance; beamforming power; loudspeaker arrays

---

1. Introduction

Sound field reproduction methods, such as higher-order ambisonics (HOA) [1–8] and wave field synthesis (WFS) [9–11], attempt to reproduce a sound field that is similar both physically and perceptually to what is intended. Multi-zone sound field reproduction methods have the same purpose for each sound zone [12–15]. Sound field reproduction systems using those reproduction methods have been implemented in several places [16–27], and there have been a number of studies on physical validations [17,18,25,26,28–30] and perceptual evaluations [17–20,22,28,30–33] with the reproduction systems. There are spatial [17,18,25,26,28,30], temporal, and spectral features [29] in the physical validations, which are related to perceptual attributes, such as localization, source width, and sound quality. This paper is concerned with a physical evaluation for the spatial feature, which is related to human localization.

For the physical evaluation of the spatial feature, the spatially-averaged squared pressure error between the desired and the reproduced fields, called the spatial error in what follows, has been widely used [5–8,12,14,15,25,34], whilst listening tests are conducted for perceptual evaluation of human localization. However, the spatial error does not always correspond to human localization [29]. For example, if a sound source is reproduced in non-anechoic rooms, room reflections significantly increase the spatial error, although human localization is affected to a lesser degree. In addition, differences in the amplitude decay with the distance increase the spatial error consistently, but play a less important role from a perceptual view point [35–38]. The aim of this paper is to propose a physical measure to evaluate the spatial feature of the reproduced sound fields that potentially has better agreement with human localization than the spatial error does. Since it is nearly impossible that a physical measure perfectly reflects human localization as long as human perception of source localization is not fully understood, this paper attempts to reduce the effects of the room reflections and amplitude decay with distance while evaluating the directions of sound waves as effectively as the spatial error.

Room reflections and spatial amplitude decay are relevant issues to practical reproduction systems. Most reproduction systems are installed in highly-damped rooms, not in anechoic conditions, although most reproduction methods strictly assume a free-field condition. There are a few studies that attempt to make use of room reflections [39,40], but these methods are feasible only if the room geometry and absorption properties are known, which is not always possible in practice. Moreover, linear arrays and circular arrays are implemented in many cases, which is called a 2.5-dimensional (2.5D) reproduction. When a 2.5D reproduction is used to reproduce a monopole source, the reproduced monopole does not have the amplitude decay of $1/r$, where r is the distance to the monopole. Unwanted reflections and unsuccessful reproduction of the amplitude decay would consistently magnify the spatial error, of which the perceptual effect is found to be rather limited [35–38].

In order to overcome the limitations of the spatial error, the present study proposes a measure based on the beamforming power, or the beam-power. The beam-powers are calculated based on the sound field in a control region, whereas the other binaural measures, such as inter-aural time difference (ITD) or inter-aural level difference (ILD), are specific to the position and direction of the listener at each time. The main reason why beamforming is used for the measure is that it can concentrate the direct sound energy in a control region along its propagating direction, whereas the reflections are, more or less, uniformly distributed. Thus, the effect of the reflections can be suppressed by a directional filter that gives a higher weighting around the intended direction of a reproduced source. After applying the directional filter, the error of the beam-power between the desired and the reproduced fields is quantified and termed the beam-power error. The effect of the difference in the amplitude decay vanishes when the beam-power is normalized (See Section 3). A typical circular loudspeaker array with 2.5D HOA reproduction is used as an example.

The research problems of this paper are defined in Section 2. The proposed measure and its basic properties are investigated via simulations in Section 3. In Section 4, the beam-power error is compared with the spatial error in a simple pre-experiment and experiments with 2.5D HOA with a circular array of 16 loudspeakers in a damped room.

2. Problem Statements

2.1. Spatial Error in Sound Field Reproduction

Figure 1 illustrates a typical sound field reproduction system that consists of multiple loudspeakers in an array. The number of the loudspeakers is denoted as L, and the position of the lth loudspeaker is denoted as $\vec{r}_s^{(l)}$. The reproduced sound pressure at an arbitrary point in a room $\vec{r}$ can be described as:

$$P\left(\vec{r};\omega\right) = \sum_{l=1}^{L} H^{(l)}\left(\vec{r}\,\middle|\,\vec{r}_s^{(l)};\omega\right) Q^{(l)}(\omega)\, S(\omega), \tag{1}$$

where $H^{(l)}\left(\vec{r}\,\middle|\,\vec{r}_s^{(l)};\omega\right)$ is the transfer function characterizing sound propagation from the lth loudspeaker and sound pressure at the point in the room $\vec{r}$, $Q^{(l)}(\omega)$ is the filter for the lth loudspeaker, and $S(\omega)$ is the input signal. For simplicity, $S(\omega)$ is assumed to be 1, and the frequency ω is omitted in what follows.

A plane wave or a sound wave generated by a monopole in free field condition is defined as the desired field in many studies because any sound fields can be expressed as the superposition of plane waves or monopoles [5–7,9–11]. In this study, the desired field is defined by a monopole:

$$\tilde{P}_f\left(\vec{r}\right) = A\frac{e^{ikR_v}}{R_v},$$

(2)

where A is the monopole amplitude, R_v is the distance between the position of the monopole $\vec{r}_v$ and a point in the control region $\vec{r}$ ($R_v = \left|\vec{r} - \vec{r}_v\right|$).

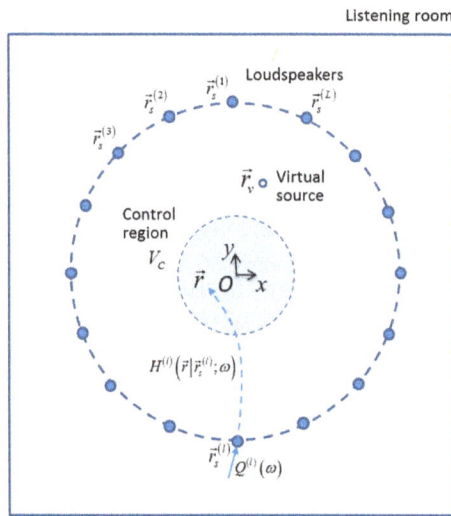

Figure 1. A typical sound field reproduction system in a listening room.

The filter $Q^{(l)}$ is obtained by the reproduction methods, such as WFS and HOA. These methods normally model the loudspeakers as monopoles or plane waves in free-field condition. Thus, the transfer function in a free field can be denoted as $H_f^{(l)}\left(\vec{r}\left|\vec{r}_s^{(l)}\right.\right)$. The reproduction methods attempt to make the reproduced field identical to the desired field in the control region V_c:

$$\tilde{P}_f\left(\vec{r}\right) \approx P_f\left(\vec{r}\right) = \sum_{l=1}^{L} H_f^{(l)}\left(\vec{r}\left|\vec{r}_s^{(l)}\right.\right)Q^{(l)}, \quad \vec{r} \in V_c$$

(3)

The spatially-averaged error $\bar{e}_f$ is defined as the error between the desired and the reproduced fields:

$$\bar{e}_f^{\,2} = \frac{\int_{V_C}\left|\tilde{P}_f\left(\vec{r}\right) - P_f\left(\vec{r}\right)\right|^2 dV}{\int_{V_C}\left|\tilde{P}_f\left(\vec{r}\right)\right|^2 dV}$$

(4)

This measure quantifies how similar the reproduced field is to the desired in the space domain in terms of the pressure distribution.

2.2. Effect of Room Reflections

Loudspeaker-based reproduction systems are installed in non-anechoic conditions in most practical cases, and consequently unintended room reflections are included in the transfer function of the loudspeakers as follows,

$$H^{(l)}\left(\vec{r}\,\middle|\,\vec{r}_s^{(l)}\right) = H_f^{(l)}\left(\vec{r}\,\middle|\,\vec{r}_s^{(l)}\right) + H_{ref}^{(l)}\left(\vec{r}\,\middle|\,\vec{r}_s^{(l)}\right),$$ (5)

where $H_{ref}^{(l)}\left(\vec{r}\,\middle|\,\vec{r}_s^{(l)}\right)$ is the transfer function for the reverberant sound field. The reproduced field also includes the room reflections $P_{ref}\left(\vec{r}\right)$, which increase the spatial error:

$$\bar{e}^2 = \frac{\int_{V_c}\left|\tilde{P}_f\left(\vec{r}\right) - P_f\left(\vec{r}\right) - P_{ref}\left(\vec{r}\right)\right|^2 dV}{\int_{V_c}\left|\tilde{P}_f\left(\vec{r}\right)\right|^2 dV}.$$ (6)

On the contrary, room reflections are known to affect human localization of the sound source to a lesser degree according to law of the first wave front [35]. Late reflections hardly affect human localization of the sound source, and early reflections up to 10 ms can smear or shift the image of the perceived source location [35], but do not significantly change the localized direction. By removing the late reflections with a time window [41,42], the effect of reflections on the spatial error can be suppressed to some extent. Yet, some early reflections cannot be perfectly removed due to overlap, and increase the spatial error.

Hence, in non-anechoic conditions, the spatial error gives little explanations on the reproduced field [29]. This means that the spatial error cannot show whether or not there are experimental errors, such as positioning errors of microphones and loudspeakers, or difference between loudspeaker models and actual loudspeaker responses.

2.3. Effect of Amplitude Decay with Distance

The amplitude of a monopole decays with distance by $1/r$, where r is the distance to the monopole position. If a linear array or a circular array of loudspeakers is used, e.g., a 2.5D reproduction, the amplitude decay will differ. Figure 2 shows the amplitude decay in a computer simulation where the reproduced field is generated with a circular array of 16 loudspeakers at 1 kHz. The radius of the circular array is 1.5 m, and the position of a desired monopole is at (10 m, 0). The desired and the reproduced decays are compared with a special focus in the controllable zone (9.6 m < x < 10.4 m). The magnitude is normalized to have 0 dB at the center of the controllable region (x = 10). The reproduced field differs clearly from the desired decay. The difference in the amplitude decay would significantly increase the spatial error. On the contrary, the amplitude decay hardly affects the human localization in terms of perception; directional perception is mainly due to the inter-aural time difference (ITD) and the inter-aural level difference (ILD), and the distance perception is more attributed to the loudness and coloration than the amplitude decay [35–38].

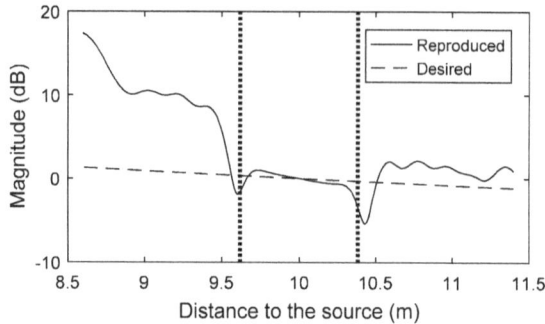

Figure 2. Amplitude decay of the reproduced and the desired field.

3. Proposed Method

3.1. Beam-Power and Directional Filter

Beamforming methods can distribute the acoustic energy in a sound field along the propagating directions. When the beamforming methods are applied to a successfully-reproduced field in a free-field condition, the beam-power should have the maximum value at the intended direction of the virtual monopole source. Even in non-anechoic conditions, the direction of the maximum value does not vary considerably because the beam-power of the direct sound is still dominant, while room reflections are distributed in many directions. The effect of the reflections can be reduced by using a directional filter that gives a greater weighting towards the monopole direction.

To obtain a consistent measure across the frequency, the main lobe needs to have a constant width. This width depends on the beamforming method and the aperture size of the sensor array. If HOA is used, the reproducible region is determined as $kr < N$ [5], where k is the wave number, N is the order of the HOA, and r is the radius of the region. This means that the aperture size decreases with the frequency. Delay-and-sum (DAS) beamforming is chosen [43], because DAS beamforming also has the narrower main lobe width at the higher frequency for a constant aperture size. Thus, HOA is used together with DAS beamforming. If the reproducible region is frequency-invariant, constant directivity beamforming methods need to be used.

The desired beamformer output with respect to the assumed azimuth angle ϕ_c for the DAS beamforming can be calculated in frequency domain as:

$$\widetilde{B}(\phi_c) = \int_0^{2\pi} \int_0^a \widetilde{P}_f(r, \phi) \exp(-ikR_c) 2\pi r dr d\phi, \tag{7}$$

where a is the radius of the controllable region of the N-th order HOA ($a = N/k$), and R_c is the distance between the measurement point (r, ϕ) and the assumed position of the source (r_c, ϕ_c), as shown in Figure 3. The intended position of the virtual monopole is known ($\overrightarrow{r}_v = (r_c, \phi_v)$) and the radius r_c is assumed to be identical to the distance of the monopole r_v. If r_v is assumed to be much greater than the radius of the reproducible region a ($r_v \gg a$) for simplicity, inserting Equation (2) into Equation (7) leads to:

$$\widetilde{B}(\phi_c) = \left| \frac{8\pi a^2 A}{r_v^2} \frac{J_1\left(2ka \sin\left(\frac{\phi_0 - \phi_v}{2}\right)\right)}{2ka \sin\left(\frac{\phi_0 - \phi_v}{2}\right)} \right|, \tag{8}$$

where J_1 is the first-order Bessel function of the first kind. See Appendix A for more details. This beam-former output is frequency-invariant as ka is constant as N. To make it independent of the distance, the normalized beam-power $\tilde{\beta}_N$ can be defined as:

$$\tilde{\beta}_N(\phi_c) = \frac{\tilde{B}(\phi_c)}{\max\{\tilde{B}(\phi_c)\}} = \left| \frac{J_1\left(2ka\sin\left(\frac{\phi_0 - \phi_v}{2}\right)\right)}{ka\sin\left(\frac{\phi_0 - \phi_v}{2}\right)} \right|, \tag{9}$$

Then, the effect of the amplitude decay with the distance vanishes.

The beam-former output of the reproduced field $B(\phi_c)$ can be obtained in the same way,

$$B(\phi_c) = \left| \int_0^{2\pi} \int_0^a P(r, \phi) \exp(-ikR_c) 2\pi r dr d\phi \right| \tag{10}$$

Unlike the ideal case (Equation (8)), the outcome of this integration is a complex number in general. The reproduced field contains the direct sound and reflections. The effect of these reflections can be reduced by a time window for excluding late reflections and a directional filter $W(\phi_c)$ for reducing the weighting at remote directions from ϕ_v. In this study, a directional filter that has the maximum at $\phi_c = \phi_v$ is defined as:

$$W(\phi_c) = 0.5[1 + \cos(\phi_c - \phi_v)] \tag{11}$$

The beam-power error is then defined as:

$$\bar{e}_{BF}^2 = \frac{\int_0^{2\pi} \left| W(\phi_c)\beta_N(\phi_c) - W(\phi_c)\tilde{\beta}_N(\phi_c) \right|^2 d\phi_c}{\int_0^{2\pi} \left| W(\phi_c)\tilde{\beta}_N(\phi_c) \right|^2 d\phi_c} \tag{12}$$

where $\beta_N(\phi_c)$ is the normalized beam-power of the reproduced field.

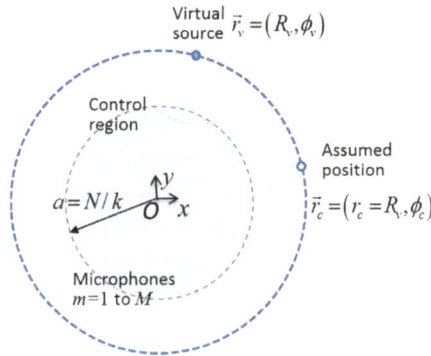

Figure 3. Setup for the beamforming method of a simple example.

3.2. Properties of the Beam-Power Error

In order to investigate the properties of the beam-power error, two sound fields are evaluated with the spatial error and the beam-power error via numerical simulations. A reference sound field has a monopole in a free field, corresponding to the desired field. To compare with the reference condition, two modifications are made: inclusion of room reflections and change in the monopole position, i.e., change in wave fronts. The position of the monopole $\vec{r}_v$ is (0, 0.9 m), and the order of HOA, N, is 7.

In the first modification, the reproduced field has room reverberation on top of the desired field. A small room (4.5 m × 4.4 m × 2.5 m) is assumed, and reflections are simulated by the image source method [44,45] with the pressure reflection coefficient, ρ, varying from 0.1 to 0.9. Late reflections that arrive after the mixing time ($t_{mix} = \sqrt{V} = 7.0$ ms) are excluded with a time window. The mixing time has been proposed as a criterion that separates early reflections from late reflections [46–48]. Figure 4 shows the beam-power distributions of the desired and the reproduced fields when the reflection coefficient is 0.9. The side lobes of the reproduced field have greater values than the desired field due to the strong reflections, but the main lobe has a small difference in these two curves, decreasing the beam-power error. Figure 5 compares the beam-power error with the spatial error with respect to the reflection coefficient. The beam-power error has at least 20 dB lower values than the spatial error, having lower values than −20 dB even for high reflection coefficients.

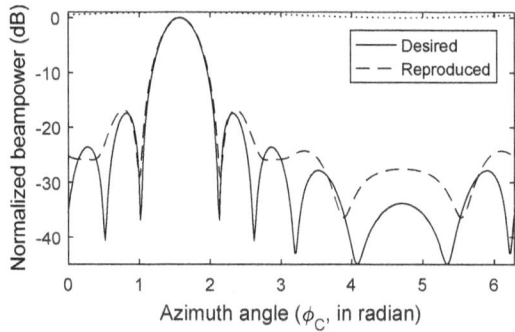

Figure 4. Normalized beam-power distributions of the desired and the reproduced fields with $\rho = 0.9$.

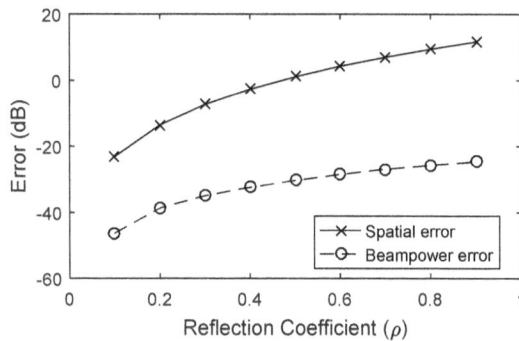

Figure 5. Spatial error and beam-power error with respect to the reflection coefficient.

In the second modification, a wrong monopole location is used for the reproduced field in an anechoic condition. This error should be effectively detected because it is induced by the error in the direction of sound waves, which can be easily perceived by human listeners. Figure 6 shows the spatial error and the beam-power error with respect to the difference of the monopole direction, $\Delta\phi$, where both errors differ by 1.7 dB on average and 2.0 dB at the most. This means that the beam-power error can detect the error due to the change in the direction of the monopole as effectively as the spatial error does.

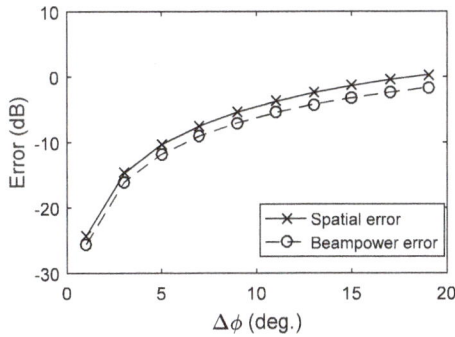

Figure 6. Spatial error and beam-power error with respect to $\Delta\phi$.

In summary, the beam-power error is much less sensitive to room reflections than the spatial error, but detects faults in the direction of arrival as effectively as the spatial error does.

4. Experimental Example

4.1. Pre-Experiment with One Source On

As a pre-experiment, the proposed measure was compared with the spatial error in a real reproduction room where only one loudspeaker was turned on. The reproduction room was an acoustically-damped room (4.5 m × 4.4 m × 2.5 m) [27,28,30] built at the Technical University of Denmark, where the reverberation time T_{30} was 0.16 s in the 125 Hz octave band, and below 0.1 s for higher frequencies. The loudspeaker was located at (0, 1.8 m). Figure 7 illustrates this setup, and the loudspeaker is indicated as #1.

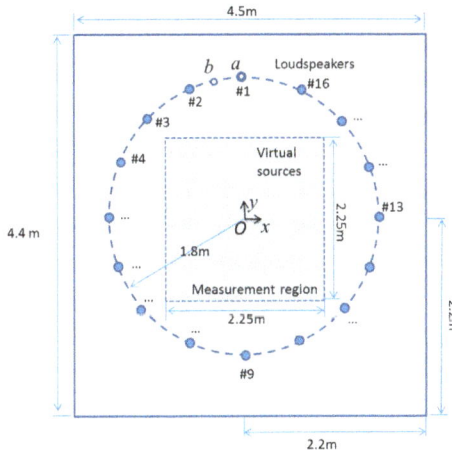

Figure 7. The listening room, loudspeaker array, and virtual sources.

The measurement was conducted with a planar array of 60 (6 × 10) microphones (B&K type 4957) at 15 positions (5 × 3). The measurement region was a square in the horizontal plane (2.25 m × 2.25 m) indicated in Figure 7, and the spacing of the measurement points was 7.5 cm. The total number of the measurement points was, thus, 900 (30 × 30) (Figure 7). The microphones were calibrated, and the phase difference among the microphones was ±5° up to 3 kHz. The microphone signals were recorded with a multi-channel analyzer (B&K frontend frame 3560D, modules 7537A and 3038B). To synchronize the measurement at each position, the signal fed into the loudspeaker was also measured. The background noise was lower than 10 dB SPL above 80 Hz, and the temperature and the humidity were 21 °C and 42%. The sampling frequency was 16,384 Hz.

The control region was assumed to be a circular region ($kr < N$) with the maximum order $N = 7$. The measurement points located in the control region were used for the beamforming and, thus, the number of the measurement points varied with the frequency. For frequencies lower than 340 Hz, the entire control region was not measured because some of the control region was outside the measurement region. The late reflections that arrived after the mixing time were excluded by the time window as used in the simulation.

Figure 8 shows the error measures with frequency. The spatial error varies from around −10 to 0 dB values at all frequencies, which is a considerable amount. Since only one loudspeaker was used, the sound field was close to that generated by a monopole at the loudspeaker position, except for room reflections, and human localization is expected to be stable because the room is acoustically damped. Informal listening tests also confirmed a stable localization performance. In contrast, the beam-power error has values lower than −20 dB above 200 Hz. The spatial error has periodic fluctuations approximately every 280 Hz. The period of this frequency corresponds to the time delay between the direct sound and the reflection from the floor. Hence, the floor reflection is the main cause of the periodic increased spatial error, whereas the beam-power error is hardly affected by such reflections. The beam-power error is mainly affected by the difference in the angle ϕ_c of the maximum beam-power between the desired and the reproduced field.

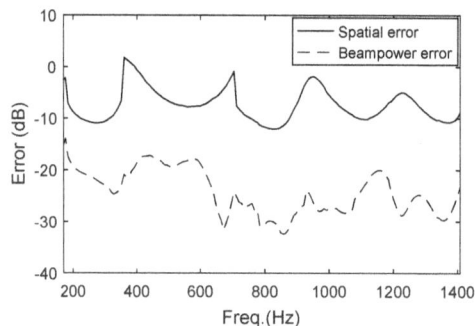

Figure 8. Spatial error and beam-power error in the pre-experiment.

4.2. Experiment with 2.5D HOA

The spatial error and the beam-power error were compared in the same room, yet with 2.5D HOA [6]. A circular array of 16 loudspeakers at equiangular positions was used, and the radius of the circle was 1.8 m (Figure 7). The loudspeakers had ±3 dB deviation from the mean value in the entire frequency range of interest. The details of the filter used were described in Appendix B. The loudspeakers were modeled as monopoles.

Two virtual source positions were aimed for reproduction as illustrated in Figure 7. A monopole is reproduced in each experiment as indicated by '*a*' and '*b*' in Figure 7. The position '*a*' is (1.8 m, 90 degrees), and the position '*b*' is (1.8 m, 101.25 degrees). The first position corresponds to that of the first loudspeaker, and the second one is at the middle point of the first and the second loudspeakers.

Figure 9 compares the spatial error and the beam-power error for the desired monopole located at *a* (left) and *b* (right). As the position '*a*' coincide with the loudspeaker #1, one might expect a better reproduction performance than that for the position '*b*'. The spatial error, however, shows similar values and fluctuations with the frequency for those two cases. The beam-power error for position 'a' turns out to be smaller than that for position 'b'.

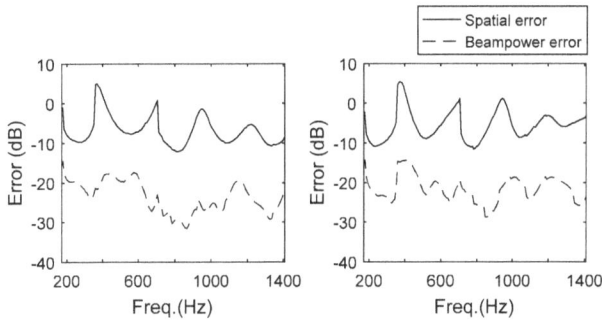

Figure 9. Spatial error and beam-power error in reproduction of a monopole at 'a' (**left**) and 'b' (**right**).

As explained in Figure 8, these peaks in the spatial error are ascribed to the floor reflection. On the other hand, the beam-power error is less affected by the floor reflection.

5. Discussion

A time window was used in the experiment to exclude the late reflections based on the concept of mixing time. In general, as the window becomes shorter, both the spatial and beam-power error becomes smaller because the direct sound becomes more predominant. Controlling such a window length can be useful to check whether or not the sound field is close to a free-field condition. Early reflections were included in this study, which occasionally shifted the peak in the beam-power. This might be related to the fact that early reflections of 5–10 ms can shift the perceived source position in human listening [35,49]. This relation needs to be further studied, but it is beyond the scope of the present study.

As shown in Figure 4, the beam-power distribution shows the difference between the desired and the reproduced fields with respect to the assumed direction. The shift of the propagating direction of the direct sound is shown in the main-lobe, and the effect of excessive reflections is shown mainly in the side lobes. This means that the beam-power distribution gives more information than the spatial distribution does.

In addition, as shown in Figure 6, if there are no reflections, it is expected that the beam-power error can detect the difference in the main lobe as effectively as the spatial error. Thus, for reproduced sound fields by simple panning methods, such as VBAP (vector-based amplitude panning) [50], the beam-power error is expected to be as large as the spatial error. On the other hand, because of the directional filter, the difference in the side lobes can be underestimated compared with the spatial error.

Recently, another measure called planarity has been proposed for evaluation of the reproduced fields [51]. The planarity quantifies the similarity to a plane wave sound field using the ratio between the intensity component in the direction of the plane wave and the total energy flux by the beamforming technique. Although the planarity and the proposed beam-power error use the beamforming method in common, the main difference is that the planarity does not take the shape of the main lobe into consideration because the planarity is calculated as the weighted sum of the beam-power. This makes the planarity insensitive to the difference in the wavefronts. Figure 10 shows the planarity for the case considered in Figure 6. The planarity decreases as the angle difference increases as expected, but the change is not significant. For example, the spatial error and the beam-power error increases by around 15 dB for the shift of five degrees, while the planarity decreases from 0.970 to 0.967. Consequently, it is concluded that the beam-power error is more effective to detect the difference in wave fronts.

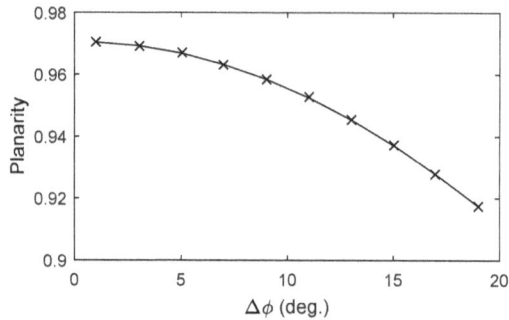

Figure 10. Planarity with respect to the $\Delta\phi$ in the same condition of Figure 6.

The experimental example used a specific reproduction method (2.5D HOA) in a specific listening room, and sound fields in the controllable region on the horizontal plane, which is frequency dependent ($kr < N$), were used to obtain the beam-power distribution. However, this measure can be used in other circumstances. For example, a frequency-independent spherical region that just includes the listener's head can be chosen, because the sound field in this region is directly related to the sound pressure that the listener would have at two ears. If 3D reproduction methods are used, and the elevation angle of the virtual source is also of concern, the beamformer output can be extended to the two-dimensional case that the beamformer output is expressed as $B(\theta_c, \phi_c)$.

The present study is limited to the physical evaluation of sound fields. Subjective tests are needed to prove that the proposed measure has good agreements with human localization in future works.

6. Conclusions

The present study proposed a measure that can quantify how well a sound field is reproduced based on a beamforming technique. Instead of directly comparing the reproduced field pressure with that of the desired field, as the spatial error does, the proposed measure compares the beam-power between the reproduced and desired sound field. The spatial error is overly affected by the amplitude decay with distance and early reflections in rooms and overestimated for a simple sound field generated by one loudspeaker in a highly-damped room. The proposed beam-power measure is less sensitive to the amplitude decay over distance and room reflections, which could better correspond with human perception. The beam-power error can pick up the difference in wavefronts of the direct sound in an equally effective manner as the spatial error. This measure can be useful particularly when a reproduction system is installed in non-anechoic rooms or a 2.5D reproduction is conducted with a linear or a circular array of loudspeakers.

Acknowledgments: This work was partially supported by KRISS (Grant No. 17011008).

Author Contributions: Ji-Ho Chang designed, performed the experiments, and analyzed the data; Ji-Ho Chang and Cheol-Ho wrote the paper.

Conflicts of Interest: The authors declare no conflict of interest.

Appendix A Beamforming Output of a Circular Region

The distance R_c in Equation (2) can be approximated as:

$$R_c = \sqrt{r_s^2 + r^2 - 2r_s r \cos(\phi_c - \phi)}$$
$$\cong r_s - r \cos(\phi_s - \phi), \quad if \; r_s \gg a. \tag{A1}$$

The desired field in free field $\widetilde{P}_f(r, \phi)$ can also be approximated as:

$$\begin{aligned}
\widetilde{P}_f(r, \phi) &= A\frac{\exp(ikR_c)}{R_c} \\
&\cong A\frac{\exp(ikr_s)}{r_s}\exp(-ikr\cos(\phi_0 - \phi)).
\end{aligned} \tag{A2}$$

Inserting Equations (A1) and (A2) into Equation (7):

$$\begin{aligned}
\widetilde{B}(\phi_c) &= \frac{A}{r_s^2}\int_0^{2\pi}\int_0^a \exp(ikr[\cos(\phi_c - \phi) - \cos(\phi_0 - \phi)])2\pi r\,dr\,d\phi \\
&= \frac{A}{r_s^2}\int_0^{2\pi}\int_0^a \exp\left(ikr\left[2\sin\left(\frac{\phi_c+\phi_0}{2} - \phi\right)\sin\left(\frac{\phi_0-\phi_c}{2}\right)\right]\right)2\pi r\,dr\,d\phi.
\end{aligned} \tag{A3}$$

ϕ can be substituted by $\phi' = \frac{\pi}{2} - \frac{\phi_s+\phi_0}{2} + \phi$:

$$\begin{aligned}
\widetilde{B}(\phi_c) &= \frac{A}{r_s^2}\int_0^a\int_0^{2\pi} \exp\left(i2kr\sin\left(\frac{\phi_0-\phi_c}{2}\right)\cos\phi'\right)d\phi'\,2\pi r\,dr \\
&= \frac{8\pi A}{r_s^2}\int_0^a\int_0^{\pi/2}\cos\left(2kr\sin\left(\frac{\phi_0-\phi_c}{2}\right)\cos\phi'\right)d\phi'\,r\,dr.
\end{aligned} \tag{A4}$$

Using the properties of the Bessel function J_n, this equation is reduced as:

$$\begin{aligned}
\widetilde{B}(\phi_c) &= \frac{8\pi A}{r_s^2}\int_0^a J_0\left(2kr\sin\left(\frac{\phi_0-\phi_s}{2}\right)\right)r\,dr \\
&= \frac{8\pi A}{r_s^2}\left[\frac{r^2 J_1\left(2kr\sin\left(\frac{\phi_0-\phi_s}{2}\right)\right)}{2kr\sin\left(\frac{\phi_0-\phi_s}{2}\right)}\right]_0^a \\
&= \frac{8\pi a^2 A}{r_s^2}\frac{J_1\left(2ka\sin\left(\frac{\phi_0-\phi_s}{2}\right)\right)}{2ka\sin\left(\frac{\phi_0-\phi_s}{2}\right)}
\end{aligned} \tag{A5}$$

Appendix B Near-Field Compensated Higher-Order Ambisonics

The reproduction method, 2.5D NFC-HOA, can generate a monopole source with a circular array of loudspeakers [30]. The distance and the angle of the monopole source can be controlled in the horizontal plane of the loudspeakers. The loudspeakers are modeled as monopoles.

The NFC-HOA filter can be written with complex cylindrical harmonics as follows. For each monopole source located at (R_v, ϕ_v) (in polar coordinates). Ambisonic signals were encoded up to an order N and decoded to an array of L loudspeakers located at $(r_s, \phi_s^{(l)})$. The filter for the lth loudspeaker $Q_{NFC}^{(l)}$ can be written as:

$$Q_{NFC}^{(l)} = \sum_{n=-N}^{N} W_n H_n^{NFC(R_s,\phi_s)}e^{in\phi_v}e^{-in\phi_s^{(l)}}. \tag{A6}$$

$H_n^{NFC(R_v,\phi_v)}$ is the NFC filter [6]:

$$H_n^{NFC(R_s,\phi_s)} = \frac{h_n^{(1)}(kR_v)}{h_0^{(1)}(kR_v)}\frac{h_0^{(1)}(kr_s)}{h_n^{(1)}(kr_s)}, \tag{A7}$$

where $h_n^{(1)}$ is the spherical Hankel function of the first kind, and W_n is the regularization function [26]:

$$W_n = \frac{2}{\left|H_n^{NFC(R_v,\phi_v)}\right|^2 + 1}. \tag{A8}$$

The number of the coefficients for a given N in Equation (A6) is $2N + 1$. It is well known that the number of loudspeakers should be greater than $2N + 1$ for an accurate reproduction [5]. In this experiment, 16 loudspeakers were used, which allow for a maximum order of $N = 7$.

References

1.   Fellgett, P.B. Ambisonic reproduction of directionality in surround-sound systems. *Nature* **1974**, *252*, 534–538. [CrossRef]
2.   Gerzon, M.A. Periphony: Width-height sound field reproduction. *J. Audio Eng. Soc.* **1973**, *21*, 2–10.
3.   Gerzon, M.A. Ambisonics in multichannel broadcasting and video. *J. Audio Eng. Soc.* **1985**, *33*, 859–871.
4.   Bamford, J.S.; Vanderkooy, J. Ambisonic sound for us. In Proceedings of the 99th Audio Engineering Society Convention, New York, NY, USA, 6–9 October 1995.
5.   Ward, D.B.; Abhayapala, T.D. Reproduction of a plane-wave sound field using an array of loudspeakers. *IEEE Trans. Speech Audio Proc.* **2001**, *9*, 697–707. [CrossRef]
6.   Daniel, J. Spatial sound encoding including near field effect: Introducing distance coding filters and a viable, new ambisonic format. In Proceedings of the 23th Audio Engineering Society International Conference, Helsingør, Denmark, 23–25 May 2003.
7.   Poletti, M.A. Three-dimensional surround sound systems based on spherical harmonics. *J. Audio Eng. Soc.* **2005**, *53*, 1004–1025.
8.   Wu, Y.J.; Abhayapala, T.D. Theory and design of soundfield reproduction using continuous loudspeaker concept. *IEEE Trans. Audio Speech Lang. Proc.* **2009**, *17*, 107–116. [CrossRef]
9.   Berkhout, A. A holographic approach to acoustic control. *J. Audio Eng. Soc.* **1988**, *36*, 977–995.
10. Berkhout, A.; de Vries, D.; Vogel, P. Acoustic control by wave field synthesis. *J. Acoust. Soc. Am.* **1993**, *93*, 2764–2778. [CrossRef]
11. Spors, S.; Rabenstein, R.; Ahrens, J. The theory of wave field synthesis revisited. In Proceedings of the 124th Convention Audio Engineering Society, Amsterdam, The Netherlands, 17–20 May 2008.
12. Wu, Y.J.; Abhayapala, T.D. Spatial multizone soundfield reproduction: Theory and design. *IEEE Trans. Audio Speech Lang. Process.* **2011**, *19*, 1711–1720. [CrossRef]
13. Poletti, M.A.; Fazi, F.M. An approach to generating two zones of silence with application to personal sound systems. *J. Acoust. Soc. Am.* **2015**, *137*, 598–605. [CrossRef] [PubMed]
14. Jin, W.; Kleijn, W.B. Theory and design of multizone soundfield reproduction using sparse methods. *IEEE/ACM Trans. Audio Speech Lang. Process.* **2015**, *23*, 2343–2355.
15. Zhang, W.; Abhayapala, T.D.; Betlehem, T.; Fazi, F.M. Analysis and control of multi-zone sound field reproduction using modal-domain approach. *J. Acoust. Soc. Am.* **2016**, *140*, 2134–2144. [CrossRef] [PubMed]
16. De Vries, D. *Wave Field Synthesis*; Audio Engineering Society Inc.: New York, NY, USA, 2009; pp. 44–83.
17. De Vries, D.; Vogel, P. Experience with a sound enhancement system based on wave field synthesis. In Proceedings of the 95th Convention Audio Engineering Society, New York, NY, USA, 7–10 October 1993.
18. Corteel, E.; Nguyen, K.-V.; Warusfel, O.; Caulkins, T.; Pellegrini, R. Objective and subjective comparison of electrodynamic and MAP loudspeakers for wave field synthesis. In Proceedings of the 30th Conference Audio Engineering Society, Saariselka, Finland, 15–17 March 2007.
19. Klehs, B.; Sporer, T. Wave field synthesis in the real world: Part 1—In the living room. In Proceedings of the 114th Convention Audio Engineering Society, Amsterdam, The Netherlands, 1 March 2003.
20. Sporer, T.; Klehs, B. Wave field synthesis in the real world: Part 2—In the movie theatre. In Proceedings of the 116th Convention Audio Engineering Society, Berlin, Germany, 8–11 May 2004.
21. Buchner, H.; Spors, S.; Kellermann, W. Full-Duplex systems for sound field recording and Auralization based on wave field synthesis. In Proceedings of the 116th Convention Audio Engineering Society, Berlin, Germany, 1 May 2004.
22. Geier, M.; Wierstorf, H.; Ahrens, J.; Wechsung, I.; Raake, A.; Spors, S. Perceptual evaluation of focused sources in wave field synthesis. In Proceedings of the 128th Convention Audio Engineering Society, London, UK, 1 May 2010.
23. Lopez, J.J.; Cobos, M.; Pueo, B.; Aguilera, E. Wave field synthesis for next generation videoconferencing. In Proceedings of the 4th International Symposium Communications, Control and Signal Processing, (ISCCSP 2010), Limassol, Cyprus, 3–5 March 2010; pp. 1–4.
24. Farina, A.; Capra, A.; Martignon, P.; Fontana, S.; Adriaensen, F.; Galaverna, P.; Malham, D. Three-dimensional acoustic displays in a museum employing WFS (wave field synthesis) and HOA (high order Ambisonics). In Proceedings of the 14th International Congress on Sound and Vibration (ICSV 14), Cairns, Australia, 9–12 July 2007.

25. Fazi, F.M.; Nelson, P.A.; Christensen, J.E.N.; Seo, J. Surround system based on three-dimensional sound field reconstruction. In Proceedings of the 125th Convention Audio Engineering Society, San Francisco, CA, USA, 1 October 2008.

26. Epain, N.; Guillon, P.; Kan, A.; Kosobrodov, R.; Sun, D.; Jin, C.; van Schaik, A. Objective evaluation of a three-dimensional sound field reproduction system. In Proceedings of the 20th International Congress on Acoustics (ICA 2010), Sydney, Australia, 23–27 August 2010.

27. Favrot, S.; Buchholz, J.M. LoRA: A loudspeaker-based room Auralization system. *Acta Acust. United Acust.* **2010**, *96*, 364–375. [CrossRef]

28. Kaesbach, J.; Favrot, S.; Buchholz, J. Evaluation of a Mixed-Order Planar and Periphonic Ambisonics Playback Implementation. Available online: http://johannes.kaesbach.de/Acoustics_files/PaperMOA_EAAtemplate2.pdf (accessed on 3 March 2017).

29. Solvang, A. Spectral impairment of two-dimensional higher order ambisonics. *J. Audio Eng. Soc.* **2008**, *56*, 267–279.

30. Favrot, S.; Buchholz, J. Reproduction of nearby sound sources using higher-order ambisonics with practical loudspeaker arrays. *Acta Acust. United Acust.* **2012**, *98*, 48–60. [CrossRef]

31. Guastavino, C.; Katz, B.F.G. Perceptual evaluation of multi-dimensional spatial audio reproduction. *J. Acoust. Soc. Am.* **2004**, *116*, 1105–1115. [CrossRef] [PubMed]

32. Frank, M.; Zotter, F.; Sontacchi, A. Localization experiments using different 2D ambisonics decoders. In Proceedings of the 25th Tonmeistertagung, Leipzig, Germany, 13–16 November 2008.

33. Bertet, S.; Daniel, J.; Parizet, E.; Gros, L.; Warusfel, O. Investigation of the perceived spatial resolution of higher order ambisonic sound fields: A subjective evaluation involving virtual and real 3D microphones. In Proceedings of the 30th Convention Audio Engineering Society, Saariselka, Finland, 15–17 March 2007.

34. Chang, J.H.; Choi, J.W.; Kim, Y.H. A plane wave generation method by wave number domain point focusing. *J. Acoust. Soc. Am.* **2010**, *128*, 2758–2767. [CrossRef] [PubMed]

35. Blauert, J. *Spatial Hearing: The Psychophysics of Human Sound Localization*; MIT Press: Cambridge, MA, USA, 1983; pp. 222–235.

36. Gerzon, M.A. The design of distance panpots. In Proceedings of the 92nd Convention Audio Engineering Society, Vienna, Austria, 1 March 1992.

37. Shinn-Cunningham, B.G.; Kopco, N.; Martin, T.J. Localizing nearby sound sources in a classroom: Binaural room impulse responses. *J. Acoust. Soc. Am.* **2005**, *117*, 3100–3115. [CrossRef] [PubMed]

38. Pellegrini, R.S.; Horbach, U. Perception-based design of virtual rooms for sound reproduction. In Proceedings of the 112nd Convention Audio Engineering Society, Munich, Germany, 10–13 May 2002.

39. Betlehem, T.; Abhayapala, T.D. Theory and design of sound field reproduction in reverberant rooms. *J. Acoust. Soc. Am.* **2005**, *117*, 2100–2111. [CrossRef] [PubMed]

40. Jin, W.; Kleijn, W.B. Multizone sound field reproduction in reverberant rooms using compressed sensing techniques. In Proceedings of the IEEE ICASSP, Florence, Italy, 4–9 May 2014; pp. 4728–4732.

41. Farina, A. Simultaneous measurement of impulse response and distortion with a swept-sine technique. In Proceedings of the 108th Convention Audio Engineering Society, Paris, France, 19–22 February 2000.

42. Muller, S.; Massarani, P. Transfer-function measurement with sweeps. *J. Audio Eng. Soc.* **2001**, *49*, 443–471.

43. Dudgeon, D.; Johnson, D. *Array Signal Processing: Concepts and Techniques*; Prentice Hall: Englewood Cliffs, NJ, USA, 1993; pp. 111–119.

44. Allen, J.B.; Berkley, D.A. Image method for efficiently simulating small-room acoustics. *J. Acoust. Soc. Am.* **1979**, *65*, 943–950. [CrossRef]

45. Lee, H.; Lee, B.-H. An efficient algorithm for the image model technique. *Appl. Acoust.* **1988**, *24*, 87–115. [CrossRef]

46. Polack, J.-D. Playing billiards in the concert hall: The mathematical foundations of geometrical room acoustics. *Appl. Acoust.* **1993**, *38*, 235–244. [CrossRef]

47. Defrance, G.; Polack, J.-D. Estimating the mixing time of concert halls using the eXtensible Fourier Transform. *Appl. Acoust.* **2010**, *71*, 777–792. [CrossRef]

48. Jeong, C.-H.; Brunskog, J.; Jacobsen, F. Room acoustic transition time based on reflection overlap (L). *J. Acoust. Soc. Am.* **2010**, *127*, 2733–2736. [CrossRef] [PubMed]

49. Barron, M.F.E. *Auditorium Acoustics and Architectural Design*; E & FN Spon: London, UK, 1993; Chapter 3.

50. Pulkki, V. Virtual sound source positioning using vector base amplitude panning. *J. Audio Eng. Soc.* **1997**, *45*, 456–466.

51. Jackson, P.J.; Jacobsen, F.; Coleman, P.; Pedersen, J.A. Sound field planarity characterized by superdirective beamforming. In Proceedings of the International Congress on Acoustics, Montreal, QC, Canada, 2–7 June 2013.

applied sciences

MDPI

Article

Objective Evaluation Techniques for Pairwise Panning-Based Stereo Upmix Algorithms for Spatial Audio

Martin Mieth and Udo Zölzer *

Department of Signal Processing and Communications, Helmut Schmidt University, Hamburg 22043, Germany; martin.mieth@hsu-hh.de
* Correspondence: zoelzer@hsu-hh.de; Tel.: +49-40-6541-2761

Academic Editors: Woon-Seng Gan and Jung-Woo Choi
Received: 16 January 2017; Accepted: 31 March 2017; Published: 10 April 2017

Abstract: Techniques for generating multichannel audio from stereo audio signals are supposed to enhance and extend the listening experience of the listener. To assess the quality of such upmix algorithms, subjective evaluations have been carried out. In this paper, we propose an objective evaluation test for stereo-to-multichannel upmix algorithms. Based on defined objective criteria and special test signals, an objective comparative evaluation is enabled in order to obtain a quantifiable measure for the quality of stereo-to-multichannel upmix algorithms. Therefore, the basic functional principle of the evaluation test is demonstrated, and it is illustrated how possible results can be visualized. In addition, the proposed issues are introduced for the optimization of upmix algorithms and also for the clarification and illustration of the impacts and influences of different modes and parameters.

Keywords: objective evaluation; stereo upmix; spatial audio; quality measure

1. Introduction

While multichannel loudspeaker systems for home and car entertainment are becoming increasingly popular these days, the number of available multichannel audio recordings is still limited (note that multichannel audio recordings and mixed multichannel audio is meant subsequently). In contrast to movies on DVD or Blu-ray, the majority of audio recordings are only obtainable in the two-channel stereo format. In addition, the main content of digital radio and television and also the increasing significant streaming services for music and movies are only obtainable in the two-channel stereo format, too. So, it can be noted that there is a low availability of multichannel audio records. That is why a system is worthwhile, which extends an original stereo audio signal for playback over a multichannel loudspeaker system. As a result, the spatial quality and the listening experience can be enhanced compared with the pure stereo playback. For this reason, there is wide scholarly interest in novel stereo-to-multichannel upmix algorithms, e.g., [1–5] but there are many more recent publications on this topic. In order to determine the quality of such upmix algorithms, subjective listening tests were typically used [6–8]. Usher [6] presented specific design criteria for stereo-to-multichannel upmix algorithms to enhance spatial sound quality. He used formal listening tests for subjective evaluation of the design criteria according to [9], where three general sound quality issues for the evaluation of multichannel audio systems were defined. Choisel and Wickelmaier [7] used eight selected spatial attributes for sound quality evaluation in order to compare upmix algorithms. They derived a set of objectives measures from sound field analysis to predict auditory attributes. Barry and Kearney [8] used subjective listening tests for the assessment of source separation-based upmixing algorithms.

In addition, they used objective testing to measure the errors which could theoretically occur in source separation algorithms.

Formal listening tests have been the only appreciable approach to assess the quality of stereo-to-multichannel upmix algorithms. More importantly, subjective quality assessments are always connected with expenditure and are both time-consuming and expensive. That is why an objective evaluation technique for stereo upmix algorithms for spatial audio is desirable.

2. Objective Evaluation Test

For the objective evaluation test, the following assumptions about stereo-to-multichannel upmix algorithms are made: stereo-to-multichannel upmix algorithms should enhance and extend the listening experience without adding artificial effects or contents and provide virtual sound sources true to original. The virtual sound sources in an original stereo configuration are placed between the two (front) loudspeakers. So, it is assumed that upmix algorithms are designed to have no virtual sound sources in the rear, only between the front loudspeakers. Therefore, the remaining amount of direct signal in the surround channels should be as low as possible. Furthermore, stereo-to-multichannel upmix algorithms should provide the listener with front channels that are louder than the surround channels under the condition that there is always an existing virtual sound source in the used stereo input signal. In addition, it is assumed that the effects of the correlation of the surround channels are perceived subjective, and that the surround channels should have a certain correlation to prevent uncomfortable perception. Finally, stereo-to-multichannel algorithms should create a high subjective perceived spatial quality with all loudspeakers in order to enhance the listening experience.

For the objective evaluation of stereo-to-multichannel upmix algorithms, the following tests were defined: 1. panning test, 2. direct signal test, 3. volume test, 4. phase test, 5. perception test. In every single test a special test signal is used as input signal for the tested upmix. The generated output signals are then analyzed and evaluated according to defined criteria (see Figure 1). Note that the evaluation test will measure how well the assumptions were met according to defined criteria.

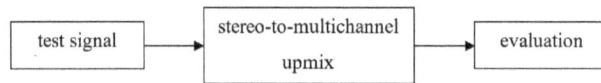

Figure 1. Schematic procedure of the evaluation tests.

The overall evaluation score $score_{upmix}$ of a tested upmix results from the weighted single-test evaluation scores $score_i$, and is given by

$$score_{upmix} = \frac{\sum_i g_i \cdot score_i}{\sum_i g_i} \tag{1}$$

The higher a score, the better the test result. Zero is the worst, one the best evaluation score. Appropriate results should be visualized here based on two upmix algorithms available on the market with two modes for music (a) and movies (b) in each case. Hereinafter, they are denoted as upmix 1(a), 1(b), 2(a) and 2(b).

2.1. Panning Test

Criterion: The direction of the virtual sound source in the stereo-to-multichannel upmix should correspond to the direction of the virtual sound source in the initial stereo configuration. This is accompanied with the result that the spatial representation of sound events is preserved true to original.

The panning test (see Figure 2) is conducted in two versions. The evaluation score $score_{PT}$ results from the evaluation scores of the time- and frequency-independent panning test $score_{PT1}$ and the time- and frequency-dependent panning test $score_{PT2}$, weighted with g_{PT1} and g_{PT2}, given by

$$score_{PT} = \frac{g_{PT1} \cdot score_{PT1} + g_{PT2} \cdot score_{PT2}}{g_{PT1} + g_{PT2}} \tag{2}$$

Initially, at an interval of $1°$ virtual test sound sources φ_i are defined with angles from $-30°$ to $30°$ according to the reference loudspeaker arrangement [10]. With the tangent law as the modified stereophonic law of sines, the two panning coefficients $a_{L,i}$ and $a_{R,i}$ can be calculated from φ_i [11]. A signal, weighted with the left panning coefficient $a_{L,i}$, represents the left part of a stereo signal $x_{L,i}$. A signal, weighted with the right panning coefficient $a_{R,i}$, represents the right part of a stereo signal $x_{R,i}$. For every angle φ_i a stereo test signal is generated as input signal for the tested upmix. This is done by multiplying the resulting panning coefficients $a_{L,i}$ and $a_{R,i}$ with white Gaussian noise x_{wgn} according to

$$\begin{aligned} x_{L,i}(n) &= a_{L,i} \cdot x_{wgn}(n) \\ x_{R,i}(n) &= a_{R,i} \cdot x_{wgn}(n) \end{aligned} \tag{3}$$

With the output signals of the upmix algorithm for the three front channels $x_{FL,i}$ (front left), $x_{C,i}$ (center) and $x_{FR,i}$ (front right), two panning coefficients $\hat{a}_{L,i}$ and $\hat{a}_{R,i}$ are determined, and from these panning coefficients the direction of the virtual sound source $\hat{\varphi}_i$ is calculated. Note that two-to-five upmix algorithms are tested, but only the three front channels are used for the panning test. The difference $\Delta\varphi_i = \varphi_i - \hat{\varphi}_i$, which is the deviation of the angle of the virtual sound source of the upmix from the defined test signal, serves as the basis for evaluation. The score of the time- and frequency-independent panning test $score_{PT1}$ is calculated from the mean of the normalized absolute deviation of all test cases with $\varphi = (\varphi_1, \ldots, \varphi_m) \in W (1 \le i \le m)$ and $W = \{n \in \mathbb{Z} | -30 \le n \le 30\}$ given by

$$score_{PT1} = 1 - \frac{1}{\max(|\varphi|)} \cdot \frac{1}{m} \sum_{i=1}^{m} |\Delta\varphi_i| \tag{4}$$

The division by $\max(|\varphi|)$ is needed to normalize the mean of the deviations so that the evaluation score assumes values ranging from 0 to 1.

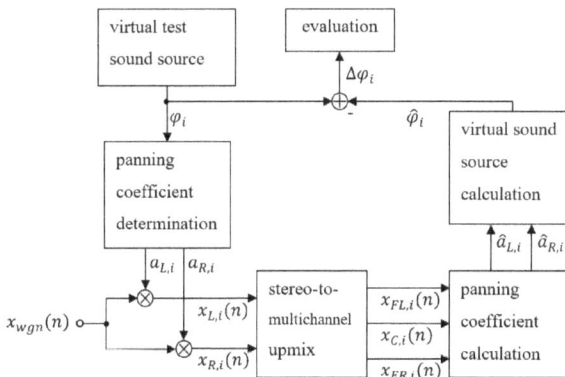

Figure 2. Block diagram: Time- and frequency-independent panning test.

In the case of the time- and frequency-dependent panning test, the angle of the virtual test sound source is randomly generated in the range of $-30°$ to $30°$. This is done at any time n and frequency k in a time–frequency representation with the help of a short-time Fourier transform (STFT). So, the ability of the upmix to respond to fast changes of the virtual sound source should be tested. The score is calculated from the mean of the normalized absolute deviation across all N times and K frequencies with φ_{max} as the maximum absolute value of all angles given by

$$score_{PT2} = 1 - \frac{1}{\varphi_{max}} \cdot \frac{1}{N} \frac{1}{K} \sum_{n=1}^{N} \sum_{k=1}^{K} |\Delta\varphi(n,k)| \tag{5}$$

The evaluation allows the comparison of the angle of the virtual sound source of the stereo input signal with the angle of the virtual sound source of the multichannel output signal (see Figure 3).

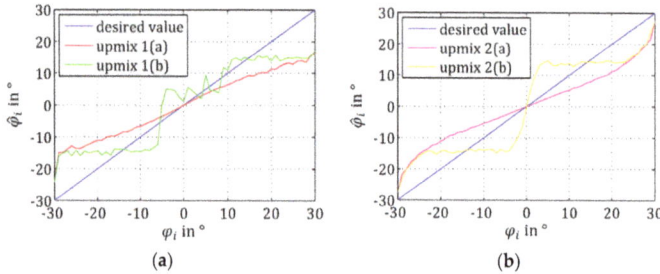

Figure 3. Panning test—angle of the virtual test sound sources compared with the determined virtual sound sources of the upmix algorithms: (**a**) algorithms upmix 1(a) and 1(b); (**b**) algorithms upmix 2(a) and 2(b).

Upmix 1: The more the angle of the virtual sound source in the stereo configuration diverges from $0°$ in mode (a), the larger are the discrepancies in the multichannel configuration. As a consequence, the spatial extent of the initial stereo configuration is partly reduced significantly. At the same time, the majority of sound events is perceived from a small spot around the center. In mode (b) the direction of the virtual sound source in the multichannel configuration does not even tendentially comply with the direction of the virtual sound source in the stereo configuration. That is because of the aim of mode (b) to enhance speech intelligibility. So, only a small range around the center speaker ($\varphi_i = 0°$) is emphasized and parts straight beyond this area are already located considerably further away.

Upmix 2: In mode (a) the angle of the virtual sound source in the multichannel configuration complies tendentially with the angle of the virtual sound source in the stereo configuration. The more φ_i diverges from $0°$, the larger are the discrepancies in the multichannel configuration until the angle $\hat{\varphi}_i$ converges fast towards the angle φ_i. The spatial extent admittedly nearly remains, but the majority of the virtual sound sources is located closer to the center speaker. In mode (b) the spatial extent admittedly nearly remains, but within a certain area beyond the center ($\varphi_i = 0°$) all sound events are solely located in one direction. To enhance speech intelligibility, sound events in the center are emphasized because parts straight beyond this area are located considerably further away.

2.2. Direct Signal Test

Criterion: The remaining amount of direct signal in the surround channels of the stereo-to-multichannel upmix could result in undesired virtual sound sources, which could interfere with the spatial representation of sound events true to original. Although it could lead to a higher subjective perceived spatial quality, the remaining amount of direct signal in the surround channels would be against the assumptions made for upmix algorithms, and should therefore be as low as possible.

Again, a special test signal is defined and used as input signal for the tested upmix algorithm. Different direct signals were taken from the database MedleyDB [12]. These audio recordings were then convolved with room impulse responses and mixed to a test signal (see Appendix A). The procedure of the direct signal test is shown in Figure 4.

The generated surround channels x_{SL} (surround left) and x_{SR} (surround right) of the upmix are analyzed by determining their remaining amount of direct signal $\hat{S}_A$. This is compared with the known test signal S and serves as the basis for evaluation. The score of the direct signal test

$$score_{DT} = 1 - \frac{1}{N} \cdot \frac{1}{K} \sum_{n=1}^{N} \sum_{k=1}^{K} \frac{\hat{S}_{A,env}(n,k)}{S_{env}(n,k)} \tag{6}$$

is calculated from the mean of the quotient of the spectral envelopes [13] $\hat{S}_{A,env}$ and S_{env} of the amounts of direct signals, across all N times and K frequencies, representing their relative deviation. To ensure a comparative evaluation, the summed power of the extracted surround channels must be equivalent to the summed power of the input signals. That is because signals before and after the upmixing process are considered. Only through using normalized surround signals is a comparative evaluation possible. This ensures, among others, that surround signals are considered correctly, which are identically equal to the input signals but reduced in power. That is because they would have the same relative amount of direct signal. The evaluation allows the comparison of the remaining surround channel direct signal with the known direct signal of the used test signal (see Figure 5).

Upmix 1: Mode (a) contains a reduced remaining direct signal in the surround channels. Mode (b) contains a remaining direct signal which is slightly lower or greater.

Upmix 2: The remaining direct signals are almost identical in modes (a) and (b), but slightly increased relative to the direct signal of the test signal. These proportionally increased remaining surround channel direct signals can occur because of input signal level adjustment or positive feedback of the upmix output signals. It should be noted that other upmix algorithms could also exhibit obviously reduced remaining surround channel direct signals.

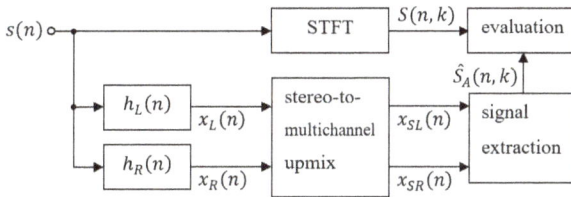

Figure 4. Block diagram: Direct signal test. Note that this is a simplified illustration of room impulse responses as a summary. See Appendix A for details of used room impulse responses and the calculation of x_L and x_R. STFT: short-time Fourier transform.

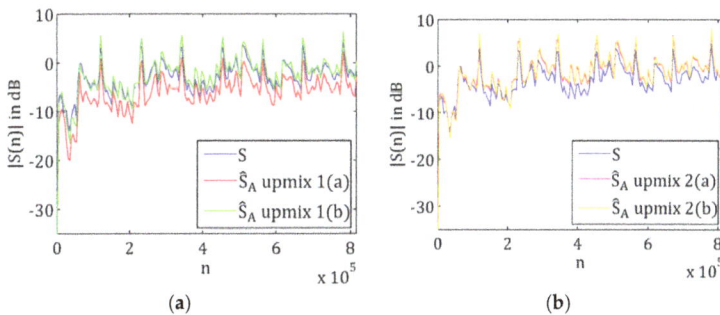

Figure 5. Direct signal test—remaining surround channel direct signal: (**a**) Algorithms upmix 1(a) and 1(b); (**b**) algorithms upmix 2(a) and 2(b). Note that the index n in the plots represents discrete values of time. $|S(n)|$ represents the magnitude of the respective complex direct signal averaged over all frequencies.

2.3. Volume Test

Criterion: Power and loudness of the surround channels of the stereo-to-multichannel upmix should not be greater than the ones of the front channels. No unnatural or unexpected spatial sound should occur because the volume of the surround sound lateral or behind the listener is perceived louder than the volume of the sound events in front of the listener.

The volume test is therefore subdivided into the power test and the loudness test. The evaluation score of the volume test $score_{LT}$ results from the evaluation scores of the power test $score_{LT1}$ and the loudness test $score_{LT2}$, weighted with g_{LT1} and g_{LT2}, and leads to

$$score_{LT} = \frac{g_{LT1} \cdot score_{LT1} + g_{LT2} \cdot score_{LT2}}{g_{LT1} + g_{LT2}} \tag{7}$$

In each case, five-second-long extracts were taken from twelve popular pieces of music from various genres to create a sixty-second-long test signal (see Appendix B, Table A3).

2.3.1. Power Test

The procedure of the power test is shown in Figure 6.

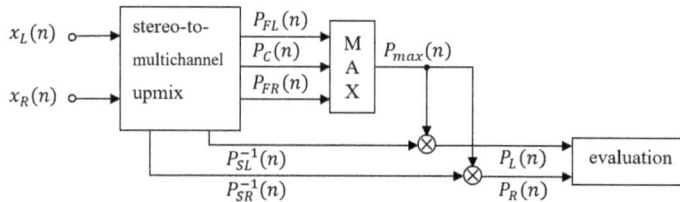

Figure 6. Block diagram: Power test.

The defined stereo test signal is used as input signal for the tested upmix. The power P_{FL}, P_C, P_{FR}, P_{SL} and P_{SR} of the generated upmix output signals x_{FL}, x_C, x_{FR}, x_{SL} and x_{SR} are considered. The maximum power of the three front signals P_{max} is compared with the power of both surround signals P_{SL} and P_{SR} each. The ratios P_L and P_R serve as the basis for evaluation. The evaluation scores of the left and right surround signal, $score_{LT1,L}$ and $score_{LT1,R}$, are calculated from the means of the quotients across all N_L and N_R times in which the power of the particular surround signals is greater than the power of the front channels, with $P_{max}(n) = max[P_{FL}(n), P_C(n), P_{FR}(n)]$:

$$score_{LT1,L} = 1 - \frac{1}{N_L} \sum_{n=1}^{N_L} P_L(n) \ \forall \ P_{SL}(n) \geq P_{max}(n)$$
$$score_{LT1,R} = 1 - \frac{1}{N_R} \sum_{n=1}^{N_R} P_R(n) \ \forall \ P_{SR}(n) \geq P_{max}(n) \tag{8}$$

The evaluation scores $score_{LT1,L}$ and $score_{LT1,R}$ represent the relative deviations of the considered power P_{SL} respectively P_{SR} from the maximum power of the three front signals P_{max}. The evaluation score of the power test $score_{LT1}$ results from the evaluation scores of the power test for the left and right surround signal, $score_{LT1,L}$ and $score_{LT1,R}$, weighted with $g_{LT1,L}$ and $g_{LT1,R}$, and is given by

$$score_{LT1} = \frac{g_{LT1,L} \cdot score_{LT1,L} + g_{LT1,R} \cdot score_{LT1,R}}{g_{LT1,L} + g_{LT1,R}} \tag{9}$$

The evaluation allows the comparison of front with surround channel power (see Figure 7). For reasons of clarity only the left surround channel power is used in the following figures.

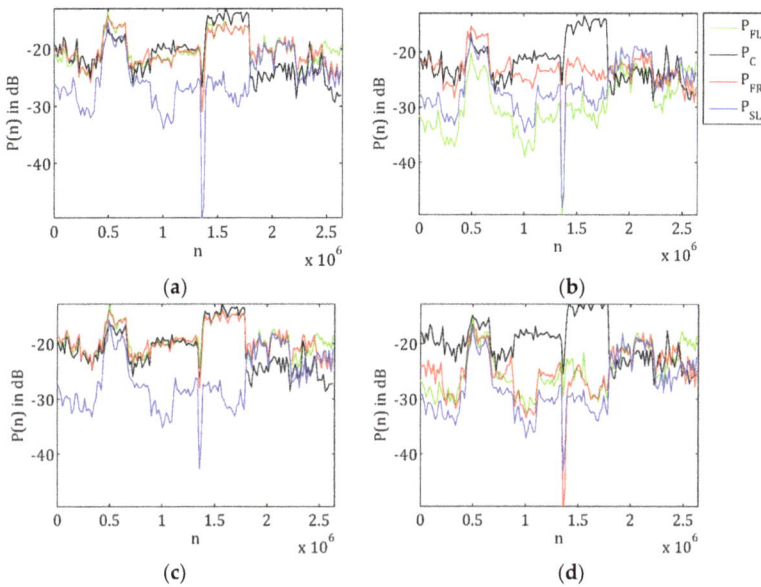

Figure 7. Power test: (**a**) Upmix 1(a); (**b**) upmix 1(b); (**c**) upmix 2(a); (**d**) upmix 2(b).

Upmix 1: In mode (a), the power of the left surround channel is basically lower and in some areas partly as high as the power of the front channel with the greatest power. In mode (b), the power of the left surround channel is mostly lower and in some areas partly higher than the power of the front channel with the greatest power. While in mode (a), the power of each front channel is more or less relatively similar, they are mostly considerably different in mode (b). The strong emphasis on the center channel can especially be recognized.

Upmix 2: In mode (a), the power of the left surround channel is basically lower and in some areas partly as high as the power of the front channel with the greatest power. In mode (b), the power of the left surround channel is lower and in some areas partly as high as or slightly higher than the power of the front channel with the greatest power. While in mode (a), the power of each front channel is more or less relatively similar, they are mostly considerably different in mode (b). The strong emphasis on the center channel can especially be recognized.

2.3.2. Loudness Test

The procedure of the loudness test is shown in Figure 8.

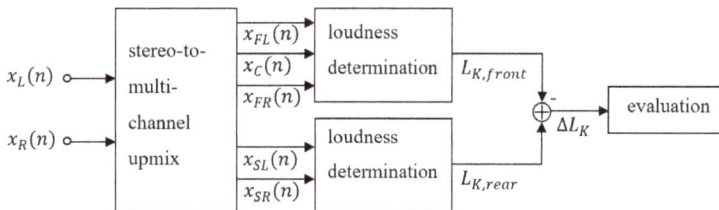

Figure 8. Block diagram: Loudness test.

The defined stereo test signal is used as input signal for the tested upmix. The generated upmix output signals x_{FL}, x_C, x_{FR}, x_{SL} and x_{SR} are considered. The loudness $L_{K,front}$ of the three front channels x_{FL}, x_C and x_{FR} is compared with the loudness $L_{K,rear}$ of both surround channels x_{SL} and x_{SR}. This serves as the basis for the evaluation score of the loudness test $score_{LT2}$. The determination of the loudness is done blockwise according to [14], but separately for the loudness of the front channels $L_{Kj,front}$ and the loudness of the surround channels $L_{Kj,rear}$. The evaluation score is calculated from the mean of the absolute deviations $\Delta L_{Kj} = L_{Kj,rear} - L_{Kj,front}$ across all J blocks in which the loudness of the surround channels is greater than the loudness of the front channels, and is given by

$$score_{LT2} = 1 - \frac{1}{J} \sum_j^J \Delta L_{Kj} \ \forall \ L_{Kj,rear} \geq L_{Kj,front} \tag{10}$$

The evaluation allows the comparison of the loudness of the front channels with the loudness of the surround channels (see Figure 9).

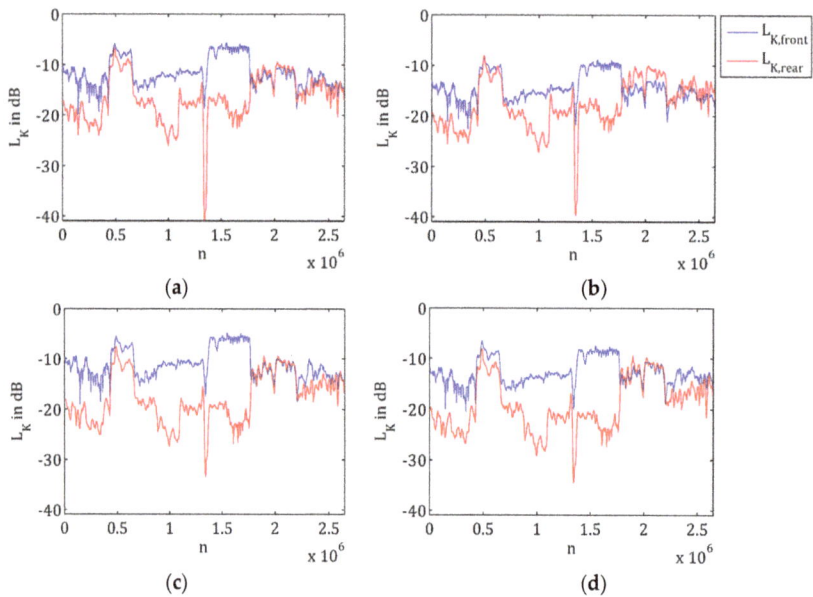

Figure 9. Loudness test: (**a**) Upmix 1(a); (**b**) upmix 1(b); (**c**) upmix 2(a); (**d**) upmix 2(b).

Upmix 1: In mode (a), the loudness of the surround channels is basically lower and in some areas partly similar or slightly greater than the loudness of the front channels. In mode (b), the loudness of the surround channels is basically lower and in some areas partly higher than the loudness of the front channels.

Upmix 2: In mode (a) as well as in mode (b), the loudness of the surround channels is basically lower and in some areas partly similar or slightly greater than the loudness of the front channels.

2.4. Phase Test

Criterion: The surround channels of the stereo-to-multichannel upmix should have a certain correlation to prevent uncomfortable perception. If the surround channels would be completely correlated, a mono sound source would be created, which could be perceived as uncomfortable. If the surround channels would be completely decorrelated, two independent sound sources would be created, which could be perceived as uncomfortable, too [15–18].

The procedure of the phase test is shown in Figure 10. The test signal is identically equal to the test signal of the volume test and is used as input signal for the tested upmix.

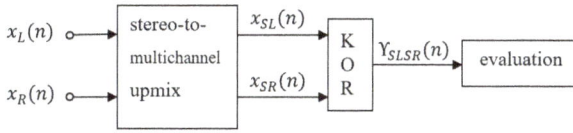

Figure 10. Block diagram: Phase test. Correlation degree is calculated in the block "KOR".

The correlation degree Υ_{SLSR} results from the normalized cross-correlation of the generated upmix output signals x_{SL} and x_{SR} for $\tau = 0$, and leads with

$$\Upsilon_\varepsilon(n) = \begin{cases} \dfrac{|\Upsilon_{SLSR}(n)| - \varepsilon_o}{1 - \varepsilon_o}, & |\Upsilon_{SLSR}(n)| > \varepsilon_o \\ \varepsilon_u - |\Upsilon_{SLSR}(n)|, & |\Upsilon_{SLSR}(n)| < \varepsilon_u \\ 0, & \varepsilon_u \leq |\Upsilon_{SLSR}(n)| \leq \varepsilon_o \end{cases} \tag{11}$$

to the evaluation score of the phase test $score_{PhT}$ according to

$$score_{PhT} = 1 - \frac{1}{N} \sum_{n=1}^{N} \Upsilon_\varepsilon(n) \tag{12}$$

With the requirement that the surround channels x_{SL} and x_{SR} should not be either completely correlated or completely decorrelated, two evaluation limits, $\varepsilon_o = 0.5$ and $\varepsilon_u = 0.2$, were defined within which a certain correlation is supposed to be optimal. The tendency for complete correlation and thus the creation of a mono sound source is higher weighted in the evaluation score than the tendency for complete decorrelation. The evaluation allows the comparison of the correlation degrees of the surround channels (see Figure 11).

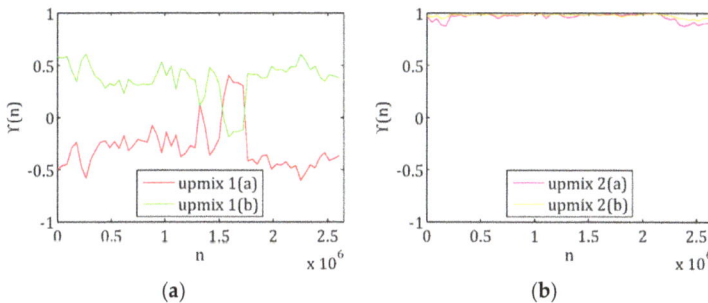

(a) (b)

Figure 11. Correlation degree of surround channels: (a) Upmix 1(a) and 1(b); (b) upmix 2(a) and 2(b).

Upmix 1: In mode (a), the correlation degree of the surround channels is basically negative, in mode (b), basically positive. It is notable that the surround signals of the one mode are a phase-inverted version of the surround signals of the other mode.

Upmix 2: In both modes, the correlation degree of the surround channels is basically approximately one. So, there is the danger that the correlated surround signals are decomposed into a mono signal.

2.5. Perception Test

Criterion: The stereo-to-multichannel upmix should generate a high subjectively perceived spatial quality. This is accompanied with the result that the listening experience is improved compared to the initial stereo configuration, and that the listener feels projected in the middle of the sound events.

The procedure of the perception test is shown in Figure 12. The test signal is identically equal to the test signal of the volume and phase test and is used as input signal for the tested upmix.

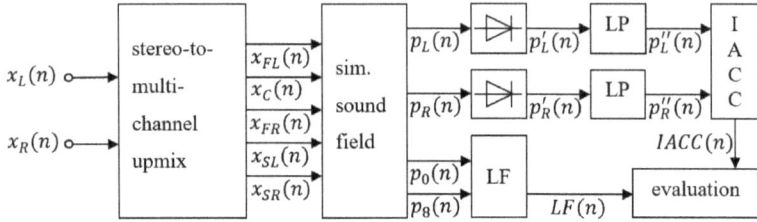

Figure 12. Block diagram: Perception test. IACC: interaural cross-correlation coefficient; LF; lateral energy fraction; LP: low-pass.

The interaural cross-correlation coefficient (IACC) describes the subjectively perceived spatial quality of sound events, and is a measure for apparent source width (ASW). The lateral energy fraction (LF) describes the impression of spatial quality, and is also a measure for listener envelopment (LEV) [7,19,20]. For the determination of IACC, the generated upmix output signals x_{FL}, x_C, x_{FR}, x_{SL} and x_{SR} are used to create a simulated sound field on the basis of head-related impulse responses (HRIR). The binaural signals:

$$p_L(n) = \sum_i x_i(n) * h_{i,L}(n)$$
$$p_R(n) = \sum_i x_i(n) * h_{i,R}(n)$$

(13)

result from the summed generated upmix output signals x_i across all channels i of the multichannel configuration convolved with the particular head-related impulse responses $h_{i,L}$ and $h_{i,R}$ for the left and right ear (with $i \in I$ and $I = \{FL, C, FR, SL, SR\}$).

IACC results from the normalized cross-correlation of the half-wave rectified and with a third-order Butterworth filter ($f_c = 1$ kHz) low-pass filtered binaural signals p_L and p_R. The prefiltering ensures that the results correspond better to the subjectively perceived spatial quality [21–23].

For the determination of LF, the generated upmix output signals x_{FL}, x_C, x_{FR}, x_{SL} and x_{SR} are used to create another simulated sound field. The signal p_0, which is recorded from a virtual omnidirectional microphone, results from the summed generated upmix output signals x_i across all channels i of the multichannel configuration, and is given by

$$p_0(n) = \sum_i x_i(n)$$

(14)

The signal p_8, which is recorded by a virtual bidirectional microphone, results from the summed generated upmix output signals x_i across all channels i of the multichannel configuration weighted with the respective loudspeaker directions φ_i, and can be written as

$$p_8(n) = \sum_i x_i(n) \cdot cos(\varphi_i)$$

(15)

LF results, with the signals p_8 and p_0, from the ratio of acoustic waves, which are arriving at the listening position laterally and from all directions [7,23], given by

$$LF = \frac{\sum_{n=1}^{N} p_8^2(n)}{\sum_{n=1}^{N} p_0^2(n)} \qquad (16)$$

Due to using a simulated sound field, the differentiation between early-and late-arriving signal components is omitted [7].

The evaluation score of the perception test based on IACC results in

$$score_{WT1} = 1 - IACC \qquad (17)$$

and is a direct measure for ASW. The subjectively perceived spatial quality is the higher, the lower IACC is. The evaluation score $score_{WT2}$ of the perception test based on LF results in

$$score_{WT2} = LF \qquad (18)$$

and is a direct measure for LEV. The impression of spatial quality is the higher, the higher LF is. The evaluation score of the perception test $score_{WT}$ results from the evaluation scores of the perception test based on IACC and LF, $score_{WT1}$ and $score_{WT2}$, weighted with g_{WT1} and g_{WT2}, given by

$$score_{WT} = \frac{g_{WT1} \cdot score_{WT1} + g_{WT2} \cdot score_{WT2}}{g_{WT1} + g_{WT2}} \qquad (19)$$

Use of simulated sound fields within the scope of the perception test ensures simplicity because of independence from the properties of room, speakers, microphones, etc., which had to be considered for the determination of IACC and LF based on costly recordings. The evaluation allows the comparison of IACC and LF (see Figure 13).

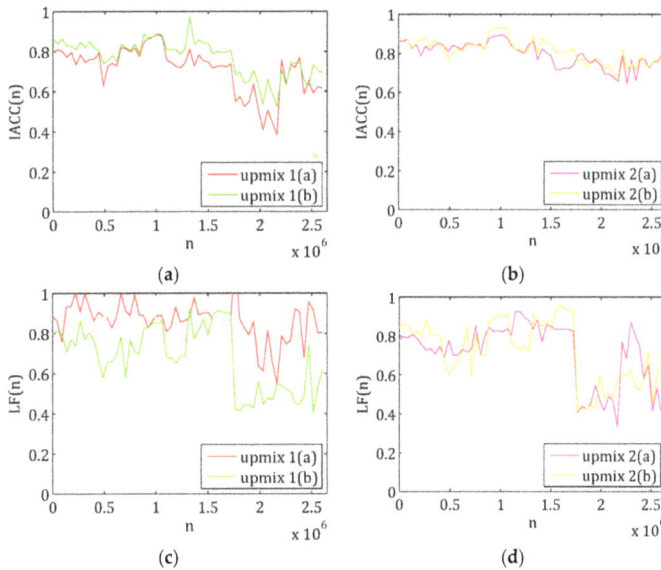

Figure 13. Perception test: (**a**) IACC for upmix 1(a) and 1(b); (**b**) IACC for upmix 2(a) and 2(b); (**c**) LF for upmix 1(a) and 1(b) IACC; (**d**) LF for upmix 2(a) and 2(b).

Upmix 1: In mode (a), IACC assumes middle to high values, LF assumes high values (see Figure 13). According to that, middle to low subjectively perceived spatial quality and a high impression of spatial quality occurs. In mode (b), IACC and LF assume middle to high values. According to that, middle to low subjectively perceived spatial quality and middle to high impression of spatial quality occurs.

Upmix 2: In both modes, IACC assumes high and LF middle to high values (see Figure 14). According to that, low subjectively perceived spatial quality and middle to high impression of spatial quality occurs.

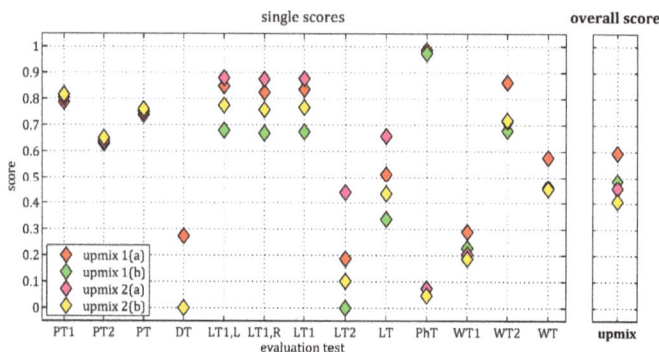

Figure 14. Evaluation scores of the upmix algorithms—graphical overview.

3. Results

Table A4 (Appendix C) summarizes the single scores of all evaluation tests for the exemplarily tested stereo-to-multichannel upmix algorithms, the used weighting factors and the resulting overall evaluation score. The higher a score, the better the test result of the tested stereo-to-multichannel upmix according to the defined criteria. Zero is the worst, one the best evaluation score. With the help of weighting factors, significance of single tests can be adjusted (see Appendix D). Figure 14 illustrates the evaluation scores according to Table A4. All in all, upmix 1(a) has the best overall evaluation score by far, upmix 1(b) the second best. Upmix 2(b) has the worst overall evaluation score, upmix 2(a) the second worst. Note that two commercial upmix algorithms in two different modes were used to demonstrate the functional principle of the proposed evaluation test and to illustrate how possible results can be visualized. The aim of this paper was not to compare existing upmix algorithms but to introduce an objective evaluation test to gain the possibility of objective comparison. So, an overall evaluation of an upmix algorithm with the proposed evaluation test is appropriate in comparison with other upmix algorithms as references. Therefore, Figure 14 provides an appropriate graphical overview for the comparative evaluation of different upmix algorithms. Note that corresponding results of the single evaluation tests were presented in the end of each section.

For the proposed evaluation test several assumptions about stereo-to-multichannel upmix algorithms were made. Since upmix algorithms are also based on assumptions, the evaluation test will measure how well the assumptions made here were met. Furthermore, a self-contained evaluation of a single upmix algorithm should focus on panning test, direct signal test and volume test. That is because the effects of the correlation of the surround channels (phase test) are perceived subjective. In addition, the impacts of lateral energy fraction and interaural cross-correlation (perception test) on perceived spatiality are subjective, too.

4. Conclusions

In this paper, we proposed an objective evaluation for stereo-to-multichannel upmix algorithms based on defined objective criteria, special test signals and several single evaluation tests. Two upmix algorithms available on the market were used to demonstrate the single tests exemplarily. The panning test checks whether the direction of the virtual sound source in the stereo-to-multichannel upmix corresponds to the direction of the virtual sound source in the initial stereo configuration. The direct signal test checks whether the remaining direct signal in the surround channels is as low as possible. The volume test checks whether the power and the loudness of the surround channels is not greater than these of the front channels. The phase test checks whether the surround channels of the stereo-to-multichannel upmix are not either completely correlated or completely decorrelated, but have a certain correlation. And the perception test checks whether the stereo-to-multichannel upmix generates a high subjectively perceived spatial quality.

The introduced objective evaluation test enables an objective comparative evaluation, which can now provide a measurable quantity for the quality of stereo-to-multichannel upmix algorithms. In addition, the objective evaluation test could be used for the optimization of upmix algorithms and also for the clarification and illustration of the impacts and influences of different modes and parameters. The proposed objective evaluation test is assumed as an appropriate alternative or supplement for time-consuming and expensive subjective listening tests.

Nevertheless, a comparison of the proposed objective evaluation test with subjective test results will be a focus of future work as part of appropriate validation.

Acknowledgments: This research was funded by Helmut Schmidt University, Hamburg, Germany.

Author Contributions: Martin Mieth is the first author, the developer of this research and wrote the paper. Udo Zölzer is the corresponding author, managed the overall research, supervised the complete work and edited the manuscript.

Conflicts of Interest: The authors declare no conflict of interest.

Appendix A

The direct signal used for the direct signal test results (according to Table A1) in

$$s(n) = \sum_{m} s_m(n) \tag{A1}$$

with $m \in M$ and $M = \{vocals, bass, drums, rattle, guitar\}$. The two-channel test signal results in

$$
\begin{aligned}
x_L(n) &= \sum_{m} s_m(n) * h_{m,L} \\
x_R(n) &= \sum_{m} s_m(n) * h_{m,R}
\end{aligned}
\tag{A2}
$$

The room impulse responses $h_{m,L}$ und $h_{m,R}$ were taken from the *Aachen Impulse Response (AIR) Database* [24] with the following parameters:

- type of room impulse response: Binaural; recorded with a dummy head.
- type of room: Stairway.
- distance of sound event to listening position: 3 m.
- angle of sound events: see Table A2.

Table A1. Used direct signals from *MedleyDB*.

Sound Event	File from Database	Variable
vocals	*AClassicEducation_NightOwl_RAW_08_01.wav*	s_{vocals}
bass	*AClassicEducation_NightOwl_RAW_01_02.wav*	s_{bass}
drums	*AClassicEducation_NightOwl_RAW_02_01.wav*	s_{drums}
rattle	*AClassicEducation_NightOwl_RAW_11_01.wav*	s_{rattle}
guitar	*AClassicEducation_NightOwl_RAW_05_01.wav*	s_{guitar}

Table A2. Angle of sound events.

Sound Event	Angle of Sound Events [1]
vocals	90°
bass	0°
drums	135°
rattle	45°
guitar	180°

[1] Different directions were provided for the single sound events. Here, it is: 0° left, 90° front, 180° right.

Appendix B

Table A3. Used pieces of music.

Genre	Interpreter	Title	Time of Title	Time of Test Signal
pop	The Human League	Don't you want me	1:18–1:23	0:00–0:05
pop	Michael Jackson	Thriller	1:27–1:32	0:05–0:10
pop	Little Talks	Of Monsters and Men	0:04–0:09	0:10–0:15
rock	Metallica	Enter Sandmann	0:57–1:02	0:15–0:20
rock	Motörhead	Ace of Spades	0:13–0:18	0:20–0:25
rock	Black Sabbath	Paranoid	0:12–0:17	0:25–0:30
dance	The Prodigy	Omen	0:09–0:14	0:30–0:35
dance	The Chainsmokers	Selfie	0:38–0:43	0:35–0:40
classic	Antonio Vivaldi	The Four Seasons	1:10–1:15	0:40–0:45
classic	Carl Orff	O Fortuna	0:08–0:13	0:45–0:50
jazz	Louis Armstrong	What a wonderful world	1:29–1:34	0:50–0:55
jazz	Quincy Jones	Pink Panther Theme	1:48–1:53	0:55–1:00

Appendix C

Table A4. Evaluation scores of the upmix algorithms—tabular overview.

Test Score	Upmix 1(a)	Upmix 1(b)	Upmix 2(a)	Upmix (2b)	Weighting
$score_{PT1}$	0.7876	0.8165	0.8045	0.8155	$g_{PT1} = 2$
$score_{PT2}$	0.6393	0.6296	0.6373	0.6510	$g_{PT2} = 1$
$score_{PT}$	0.7382	0.7542	0.7488	0.7607	$g_{PT} = 3$
$score_{DT}$	0.2727	0	0	0	$g_{DT} = 2$
$score_{LT1,L}$	0.8484	0.6794	0.8796	0.7751	$g_{LT1,L} = 1$
$score_{LT1,R}$	0.8241	0.6686	0.8742	0.7581	$g_{LT1,R} = 1$
$score_{LT1}$	0.8362	0.6740	0.8769	0.7666	$g_{LT1} = 1$
$score_{LT2}$	0.1866	0	0.4387	0.1019	$g_{LT2} = 1$
$score_{LT}$	0.5114	0.3370	0.6578	0.4342	$g_{LT} = 2$
$score_{PhT}$	0.9851	0.9729	0.0737	0.0468	$g_{PhT} = 1$
$score_{WT1}$	0.2894	0.2283	0.2044	0.1867	$g_{WT1} = 1$
$score_{WT2}$	0.8626	0.6788	0.7141	0.7178	$g_{WT2} = 1$
$score_{WT}$	0.5760	0.4536	0.4593	0.4522	$g_{WT} = 1$
$score_{Upmix}$	0.5938	0.4848	0.4550	0.4055	

Appendix D

The proposed objective evaluation test uses weighting factors to determine an overall evaluation score from the weighted single-test evaluation scores. With the help of these weighting factors, significance of single tests can be adjusted.

The panning test is conducted in two versions. The aim of the time- and frequency-independent panning test is to test the ability of an upmix algorithm to reproduce the virtual sound sources true to original. Therefore, time- and frequency-independent virtual test sound sources are defined. The aim of the time- and frequency-dependent panning test is to test the ability of the upmix algorithm to respond to fast changes of the virtual sound source. Therefore, time- and frequency-dependent virtual test sound sources are generated independently. Since the focus of the panning test is whether the direction of the virtual sound source in the stereo-to-multichannel upmix corresponds to the direction of the virtual sound source in the initial stereo configuration, the time- and frequency-independent panning test is higher weighted than the time- and frequency-dependent panning test (the time- and frequency-independent panning test counts twice). The evaluation score of the panning test results from the double-weighted time- and frequency-independent panning test ($g_{PT1} = 2$) and the single-weighted time- and frequency-dependent panning test ($g_{PT2} = 1$). Due to the assumed high overall importance of the panning test in comparison with the other single tests, the panning test is triple-weighted ($g_{PT} = 3$). That is because a stereo-to-multichannel upmix should enhance and extend the listening experience without adding artificial effects or contents, and provide virtual sound sources true to original.

The phase test and the perception test are single-weighted ($g_{PhT} = 1$; $g_{WT} = 1$) because of the assumed low overall importance in comparison to the other single evaluation tests. That is because the effects of the correlation of the surround channels are perceived subjective. In addition, the impacts of lateral energy fraction and interaural cross-correlation on perceived spatiality are subjective, too. The direct test and the volume test are double-weighted ($g_{DT} = 2$; $g_{LT} = 2$) because of the assumed middle overall importance in comparison to the other single evaluation tests. Thus, they are higher weighted than the phase test and the perception test, but lower weighted than the panning test.

Note that the weighting factors used are not obligatory. They can be used in order to adjust the significance of the single evaluation tests as described above. In addition, they can be used to compare several upmix algorithms with a focus on a special issue.

References

1. Avendano, C.; Jot, J.-M. Frequency Domain Techniques for Stereo to Multichannel Upmix. In Proceedings of the 22nd International Conference on Virtual, Synthetic and Entertainment Audio (AES), Espoo, Finland, 15–17 June 2002.
2. Faller, C. Multiple-Loudspeaker Playback of Stereo-Signals. *Jt. Audio Eng. Soc.* **2006**, *54*, 1051–1064.
3. Goodwin, M.M.; Jot, J.-M. Primary-Ambient Signal Decomposition and Vector-Based Localization for Spatial Audio Coding and Enhancement. In Proceedings of the 2007 IEEE International Conference on Acoustics, Speech and Signal Processing, Honolulu, HI, USA, 16–20 April 2007; pp. 9–12.
4. Vickers, E. Frequency-Domain Two-to-Three-Channel Upmix for Center Channel Derivation and Speech Enhancement. In Proceedings of the AES 127th Convention, New York, NY, USA, 9–12 October 2009.
5. Jeon, S.-W.; Park, Y.-C.; Lee, S.-P.; Youn, D.-H.H. Robust Representation of Spatial Sound in Sterei-to-Multichannel Upmix. In Proceedings of the AES 128th Convention, London, UK, 22–25 May 2010.
6. Usher, J.S. Subjective Evaluation and Electroacoustic Theoretical Validation of a New Approach to Audio Upmixing. Ph.D. Thesis, McGill University, Montreal, QC, Canada, September 2006.
7. Choisel, S.; Wickelmaier, F. Relating Auditory Attributes of Multichannel Sound to Preference and to Physical Parameters. In Proceedings of the AES 120th Convention, Paris, France, 20–23 May 2006.
8. Barry, D.; Kearney, G. Localization Quality Assessment in Source Separation-Based Upmixing Algorithms. In Proceedings of the AES 35th International Conference, London, UK, 11–13 February 2009.

9.   ITU-R BS 1116-0. *Methods for the Subjective Assessment of Small Impairments in Audio Systems Including Multichannel Sound Systems*; Recommendation ITU-R BS 1116-0; International Telecommunication Union: Geneva, Switzerland, July 1994.

10. ITU-R BS.775-3. *Multichannel Stereophonic Sound System with and without Accompanying Picture*; Recommendation ITU-R BS.775-3; International Telecommunication Union: Geneva, Switzerland, August 2012.

11. Pulkki, V. Virtual sound source positioning using vector base amplitude panning. *Jt. Audio Eng. Soc.* **1997**, *45*, 457.

12. Bittner, R.; Salamon, J.; Tierney, M.; Mauch, M.; Cannam, C.; Bello, J.P. MedleyDB: A Multitrack Dataset for Annotation-Intensive MIR Research. In Proceeding of the 15th International Society for Music Information Retrieval Conference, Taipei, Taiwan, 27–31 October 2014.

13. Röbel, A.; Rodet, X. Efficient Spectral Envelope Estimation and Its application to Pitch Shifting and Envelope Preservation. In Proceedings of the 8th International Conference on Digital Audio Effects (DAFx05), Madrid, Spain, 20–22 September 2005; pp. 30–35.

14. ITU-R BS.1770-3. *Algorithms to Measure Audio Programme Loudness and True-Peak Audio Level*; Recommendation ITU-R BS.1770-3; International Telecommunication Union: Geneva, Switzerland, August 2012.

15. Riekehof-Böhmer, H.; Wittek, H. Prediction of perceived width of stereo microphone setups. In Proceedings of the AES 130th Convention, London, UK, 13–16 May 2011.

16. Theile, G. Multichannel Natural music Recording Based on Psychoacoustic Principles. In Proceedings of the AES 19th International Conference, Schloss Elmau, Germany, 21–24 June 2001.

17. Damaske, P. Subjektive Untersuchung von Schallfeldern. *Acta Acust.* **1967**, *19*, 68.

18. Usher, J.S. Design Criteria for High Quality Upmixers. In Proceedings of the AES 28th International Conference, Piteå, Sweden, 30 June–2 July 2006.

19. Hirst, J.M. Spatial Impression in Multichannel Surround Sound Systems. Ph.D. Thesis, University of Salford, Salford, UK, 2006.

20. Bradley, J.S.; Soulodre, G.A. The influence of late arriving energy on spatial impression. *J. Acoust. Soc. Am.* **1995**, *97*, 2590–2597. [CrossRef]

21. Mason, R.; Brooks, T.; Rumsey, F. Frequency dependency of the relationship between perceived auditory source width and the interaural cross-correlation coefficient for time-invariant stimuli. *J. Acoust. Soc. Am.* **2005**, *117*, 1337–1350. [CrossRef] [PubMed]

22. Neher, T.; Brookes, T.; Mason, R. Musically representative test signals for interaural cross-correlation coefficient measurement. *Acta Acoust.* **2006**, *92*, 787–796.

23. Bradley, J.S. Comparison of concert hall measurements of spatial impression. *J. Acoust. Soc. Am.* **1994**, *96*, 3525–3535. [CrossRef]

24. Jeub, M.; Schäfer, M.; Vary, P. A Binaural Room Impulse Response Database for the Evaluation of Dereverberation Algorithms. In Proceedings of the 2009 16th International Conference on Digital Signal Processing (DSP), Santorini, Greece, 5–7 July 2009.

applied sciences

MDPI

Article

Auditory Distance Control Using a Variable-Directivity Loudspeaker †

Florian Wendt *, Franz Zotter, Matthias Frank and Robert Höldrich

Institute of Electronic Music and Acoustics, University of Music and Performing Arts Graz, Inffeldgasse 10/III, 8010 Graz, Austria; zotter@iem.at (F.Z.); frank@iem.at (M.F.); hoeldrich@iem.at (R.H.)
* Correspondence: wendt@iem.at; Tel.: +43-316-389-3520
† This paper is an extended version of the paper published at the International Conference on Digital Audio Effects (DAFx-16), Brno, Czech Republic, 5–9 September 2016; pp. 295–300.

Academic Editors: Woon-Seng Gan and Jung-Woo Choi
Received: 19 April 2017; Accepted: 22 June 2017; Published: 29 June 2017

Abstract: The directivity of a sound source in a room influences the D/R ratio and thus the auditory distance. This study proposes various third-order beampattern pattern designs for a precise control of the D/R ratio. A comprehensive experimental study is conducted to investigate the hereby achieved effect on the auditory distance. Our first experiment auralizes the directivity variations using a virtual directional sound source in a virtual room using playback by a 24-channel loudspeaker ring. The experiment moreover shows the influence of room, source-listener distance, signal, and additional single-channel reverberation on the auditory distance. We verify the practical applicability of all the proposed beampattern pattern designs in a second experiment using a variable-directivity sound source in a real room. Predictions of experimental results are made with high accuracy, using room acoustical measures that typically predict the apparent source width.

Keywords: icosahedral loudspeaker; variable-directivity source; auditory distance; D/R ratio; apparent source width

---

1. Introduction

Studies on sound localization mainly focus on the directional aspect and auditory distance perception receives substantially less scientific attention. However, a recent review of localization studies could show that when listeners are asked to describe the location of perceived auditory objects, the most commonly attribute used is *distance* [?]. The several distance cues assessed by the auditory system vary in their effective ranges and can be divided into two group, cf. [? ?]. While the first group of cues yields an *absolute* distance perception based on internal references, distance perception obtained from cues of the second group is a *relative* judgment. The most-studied indicator from the relative cues is amplitude. In the free field, where only the direct sound is present, the amplitude decreases with distance. As the auditory system is exquisitely sensitive to small changes in amplitude, it permits fine relative distance discrimination. For simple implementation electro-acoustic applications can use gain modifications to shape auditory scenes with regard to distance. The so-called D/R ratio (direct-to-reverberant energy ratio) is a cue providing coarse but absolute distance information, as shown in several studies [? ? ? ?]. In a room, the D/R ratio is inversely related to the distance of the sound source and characterizes the energy ratio of direct and reflected sound.

Laitinen [?] proposed an elegant solution to control the D/R ratio from a single point in the room. He employed a sound source controlled to approximate an omni-directional directivity and a second-order cardioid pattern steering away from or towards the listener. This variation of the directivity achieved control of the D/R ratio in a relatively dry and small room.

Our contribution extends Laitinen's approach by an auralization-based listening experiment of (i) a directional source in various ideal higher-order beampattern designs/beam constellations, (ii) two different rooms, (iii) two different source-listener distances, (iv) including single-channel reverberation. In addition to Laitinen's work, our paper establishes models of the hereby achieved auditory distance using simple acoustical measures. In extension to our previous study presented at DAFx [?], this article also includes the results of a second experiment using (v) the icosahedral loudspeaker (IKO), with a more elaborate third-order beampattern control [?] to synthesize the various beampatterns in a room.

The paper is arranged as follows: It outlines the first experiment based on auralized rooms and directivities and presents detailed results with discussions of the influence of room, signal, and reverberation. The second section presents models of the experimental results. The last section presents the second experiment that verifies the practical applicability of the beampattern designs to the directivity synthesis by the IKO in a real room.

2. Experiment I: Directivity-Controlled Auditory Distance in Auralized Rooms

Considered beampatterns up to the third order are based on frequency-independent beampattern designs by a combination of Legendre polynomials $g_i(\vartheta) = \sum_{n=0}^{i}(2n + 1)\, a_n\, P_n(\cos\vartheta)$, using max-$r_E$ weights that are common in Ambisonics (cf. [? ?])

$$a_n = \frac{P_n[\cos(\frac{137.9°}{i+1.151})]}{\sqrt{\sum_{n=0}^{i}(2n+1)\,[P_n(\cos\frac{137.9°}{i+1.151})]^2}}. \tag{1}$$

This exhibits a relatively narrow main lobe and sufficiently suppressed side lobes for any beam order i.

The proposed beampattern designs vary:

A   the beam order i from three to zero for $g_i(\vartheta)$ and $g_i(\pi - \vartheta)$;
B   the ratio a/b of two opposing beams: $a\, g_3(\vartheta) + b\, g_3(\pi - \vartheta)$;
C   the angle α of a beam pair: $g_3(\vartheta - \alpha/2) + g_3(\vartheta + \alpha/2)$.

Table ?? lists all tested beampattern designs in particular, which differently modify the amount of diffuse, lateral, and direct energy, thus the D/R ratio. Each beampattern indicated by the index 1 and 7 corresponds to a 3rd-order beam facing towards and away from the listening position ($A_1 = B_1 = C_1$, $A_7 = B_7 = C_7$). Furthermore, beam pairs indicated by indices 1/7, 2/6, and 3/5 of each design are identical in their shape but horizontally rotated by $180°$. Figure ?? shows the beampatterns $A_{1...4}$, $B_{1...4}$, and $C_{1...4}$ normalized to constant energy.

Table 1. Properties of tested beampattern designs A, B, and C.

	$A_{1/7}$	3rd-order max-r_E beam to/off listener
A	$A_{2/6}$	2nd-order max-r_E beam to/off listener
	$A_{3/5}$	1st-order max-r_E beam to/off listener
	A_4	omnidirectional beampattern
B	$B_{1...7}$	3rd-order max-r_E beams to and off listener linearly blended at $[\infty, 6, 3, 0, -3, -6, -\infty]$ dB
C	$C_{1...7}$	two 3rd-order max-r_E beams horizontally arranged at $\pm 30°\ [0, 1, \ldots 6]$ with respect to the listener

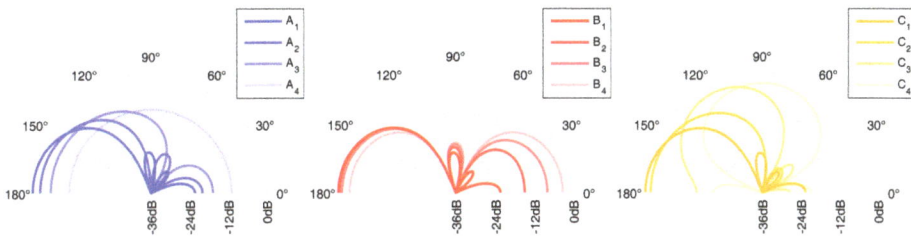

Figure 1. Beampattern designs *A*, *B*, *C* controlling the D/R ratio.

2.1. Experimental Setup

The effect is evaluated in a first listening experiment, in which the variable-directivity source in a room is auralized using the image source method. The room is shoebox-shaped with a frequency-independent absorption coefficient $\bar{a}$. Specular reflections up to 3rd order are considered [?] and diffuse reflections are simulated as spherical harmonics using the software tool MCRoomSim [?]. For simplicity, diffuse reverberation of an omni-directional excitation is considered.

Playback employed a ring of 24 equally-distributed Genelec 8020 loudspeakers with a radius of $r = 1.5\,\text{m}$ placed in an anechoic laboratory. Each listener was sitting in the center of the arrangement with ear height adjusted to the loudspeaker ring, cf. Figure ??.

Figure 2. Experimental setup in the anechoic laboratory.

On the circular setup each specular reflection is auralized by the loudspeaker with the closest azimuth angle. This avoids timbral effects of amplitude panning [?]. Elevated specular reflections are attenuated in the auralization by the cosine of their elevation. Diffuse reflections are played back in Ambisonics format. The impulse response $h_l(t)$ of the *l*-th loudspeaker is obtained after superimposing specular and diffuse reflections using MATLAB.

Obviously, a two-dimensional representation of a three-dimensional sound field is not optimal, but findings in [?] indicate that reflections from floor and ceiling do not have a significant influence on the auditory distance.

Each impulse response was convolved with the signals $S_{1...3}$, yielding a 24-channel audio file for each condition. Audio playback was controlled by the open source software Pure Data on a standard PC with RME MADI audio interface and DirectOut D/A converters.

To monitor the influence of room acoustics, three different layouts were tested, including two rooms and two source-listener distances, cf. $R_{1...3}$ in Table ??.

Table 2. Properties of tested rooms R and signals S.

	R_1	IEM CUBE,	$10.3\,\text{m} \times 12\,\text{m} \times 4.8\,\text{m}$,	$T_{60} = 700\,\text{ms}$,	$d_1 = 1.7\,\text{m}$
room	R_2	IEM CUBE,	$10.3\,\text{m} \times 12\,\text{m} \times 4.8\,\text{m}$,	$T_{60} = 700\,\text{ms}$,	$d_2 = 2.9\,\text{m}$
	R_3	IEM Lecture Room,	$7.6\,\text{m} \times 6.8\,\text{m} \times 3\,\text{m}$,	$T_{60} = 570\,\text{ms}$,	$d_3 = 1.7\,\text{m}$
	S_1	female speech taken from *Music for Archimedes*, CD Bang and Olufsen 101 (1992)			
signal	S_2	sequence of irregular artificial bursts			
	S_3	speech-spectrum noise with increased kurtosis			

Geometry and reverberation time of the auralized rooms are based on two rooms at our institute, namely the IEM CUBE, a $10.3\,\text{m} \times 12\,\text{m} \times 4.8\,\text{m}$ large room with $T_{60} = 700\,\text{ms}$, and the IEM Lecture Room, $7.6\,\text{m} \times 6.8\,\text{m} \times 3\,\text{m}$ with $T_{60} = 570\,\text{ms}$. Both rooms were chosen as they are typical venues for concerts or experiments with the IKO as a variable-directivity source [? ?].

The simulated sound source was placed near the corners of the room at a distance of 2 m and 3 m (IEM CUBE) and 1 m and 2 m (IEM Lecture Room). The listening position was chosen at a virtual distance of $d = 1.7\,\text{m}$ to the sound source. Additionally, for the IEM CUBE an increased source-listener distance of $d = 2.9\,\text{m}$ was tested.

The listener was facing the sound source simulated at height of 1.8 m above the floor with an angular offset of $\Delta\phi = 15°$ with regard to the sidewalls. Figure ?? shows the setup of the auralized room using the 24-channel loudspeaker ring and Table ?? lists rooms and source-listener distances tested in the experiment.

The signals fed into auralization were chosen to investigate the influence of speech versus noise, noise spectrum, and noise envelope to the effect: female speech (S_1), a sequence of irregular artificial bursts (S_2), and Gaussian white noise shaped to speech spectrum (S_3) as listed in Table ??. For S_3, envelope fluctuations were slightly accentuated by multiplying the noise with its Hilbert envelope and by restriction to its original bandwidth, cf. [?]. By this procedure, S_1 and S_3 have similar spectra and kurtosis, which measures the envelope fluctuation, whereas S_2 is more transient with more energy at frequency above 1 kHz. All signals were normalized to their RMS value for level equalization.

The above signals are anechoic. To monitor potential influence of additional reverberation for some conditions, signa   vels of reverberation were tested, of which level 1 corre   verberation time of $T_{60} = 0.5\,\text{s}$,

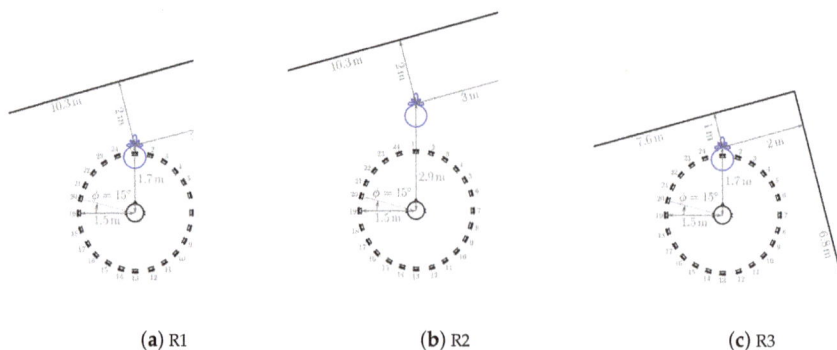

(a) R1   (b) R2   (c) R3

Figure 3. Room and source configuration for R_1, R_2, and R_3 together with loudspeaker ring used for auralization. R_1 and R_2 are based on the IEM CUBE differing in the source-listener distance and room R_3 is based on the IEM Lecture Room.

The listening experiment was carried out as a multi-stimulus test where listeners had to comparatively rate multiple conditions, denoted as sets. Their task was to indicate the distance of auditory objects on a graphical user interface displaying a continuous slider for each condition of a set along the ordinal scale *very close* (vc), *close* (c), *moderate* (m), *distant* (d), and *very distant* (vd). The listeners were allowed to repeat each condition at will, and audio files were played back in loop. Fifteen listeners participated in the experiment (three female, twelve male; age 23–54). All of them were experienced listeners in 3D audio and experienced participants in psychophysical studies of hearing; all reported normal hearing acuity.

Tested sets (set 1 to 12, see Table **??**) comprise 7 conditions, each representing a beampattern, room, signal, and reverberation level. Under a varied beampattern design, e.g., $A_{1...7}$, the influence of room (set 1, 10, 11), signal (set 1, 2, 3), and reverberation level (set 1, 12) was only examined separately, yielding responses $x^{\mathrm{I}}_{1...7}$ for each subject. These separate multi-stimulus sets do not yet permit cross comparison due to the absence of a common reference. As a solution, maintaining a limited testing time, the additional 9-stimulus comparison sets (13...15) were tested with fewer beampatterns $A_{1,4,7}$ and instead involving cross-comparisons with regard to signal (13), room (14), and reverberation level (15). They yield cross-comparison responses $x^{\mathrm{II}}_{1,4,7}$ that enable a comparison involving a fine-grained directivity variation in Figures **??**–**??**.

In these figures, responses $x^{\mathrm{I}}_{2,3}$ and $x^{\mathrm{I}}_{4,5}$ were re-mapped for each listener by linear scaling and shifting to match $x^{\mathrm{I}}_{1,4}$ with $x^{\mathrm{II}}_{1,4}$, and $x^{\mathrm{I}}_{4,7}$ with $x^{\mathrm{II}}_{4,7}$, respectively:

$$
x_i = \begin{cases}
x^{\mathrm{II}}_i & \text{for } i \in \{1,4,7\}, \\[2mm]
\dfrac{x^{\mathrm{II}}_4 - x^{\mathrm{II}}_1}{x^{\mathrm{I}}_4 - x^{\mathrm{I}}_1}(x^{\mathrm{I}}_i - x^{\mathrm{I}}_1) + x^{\mathrm{II}}_1 & \text{for } i \in \{2,3\}, \\[2mm]
\dfrac{x^{\mathrm{II}}_7 - x^{\mathrm{II}}_4}{x^{\mathrm{I}}_7 - x^{\mathrm{I}}_4}(x^{\mathrm{I}}_i - x^{\mathrm{I}}_4) + x^{\mathrm{II}}_4 & \text{for } i \in \{5,6\},
\end{cases} \tag{2}
$$

i.e., a complete response set $x_{1...7}$ per listener, signal, room, and reverberation level.

During the listening session, the listeners were requested to face loudspeaker 1 ($\phi = 0°$, cf. Figure **??**), which corresponds to the direction of the auralized sound source.

At the beginning of the experiment, each listener was given a short training to familiarize with the evaluation scale. The training set included expected extreme values with regard to the auditory distance. Listeners were asked to rate along the whole scale and use extremes as an internal reference for further evaluations.

After the training phase, multi-stimulus tasks were presented. Each time a multi-stimulus set was displayed, the arrangement of its stimuli was an individual random permutation. The listener could have the stimuli sorted by own ratings to facilitate comparative rating. The first part of the experiment consisted of the sets with 7 stimuli (set 1 to 12) in an individual random permutation, and the second part of the sets consisting of 9 conditions (set 13 to 15) in an individual random permutation.

None of the listeners reported that they perceived the auralization as unnatural or confusing; some emphasized the naturalness of the auralization.

Table 3. Composition of tested sets, consisting of 7 and 9 samples.

Set No.	Design	Index	Signal	Room	Reverb. Level
1	A	1...7	S_1	R_1	0
2	A	1...7	S_2	R_1	0
3	A	1...7	S_3	R_1	0
4	B	1...7	S_1	R_1	0
5	B	1...7	S_2	R_1	0
6	B	1...7	S_3	R_1	0
7	C	1...7	S_1	R_1	0
8	C	1...7	S_2	R_1	0
9	C	1...7	S_3	R_1	0
10	A	1...7	S_1	R_2	0
11	A	1...7	S_1	R_3	0
12	A	1...7	S_1	R_1	1
13	A	1,4,7	$S_{1...3}$	R_1	0
14	A	1,4,7	S_1	$R_{1...3}$	0
15	A	1,4,7	S_1	R_1	0,1,2

2.2. Influence of Beampattern Design

Figure **??** shows a detailed analysis of the auditory distance for the beampatterns $A_{1...7}$, $B_{1...7}$, and $C_{1...7}$ according to Table **??** and Figure **??**, based on the responses to the sets $1...3$, $4...6$, and $7...9$ of Table **??**, using all signals $S_{1...3}$ and the room R_1. The direct comparability of all curves in Figure **??** is feasible as all designs were determined to include reference patterns corresponding to a 3rd-order beam facing to ($A_1 = B_1 = C_1$) and off ($A_7 = B_7 = C_7$) the listening position, respectively. This allowed to linearly re-map the responses gathered in the sets $1...9$ to fill out the entire interval $[0; 1]$ for each listener. Figure **??** shows the medians and the corresponding 95% confidence intervals.

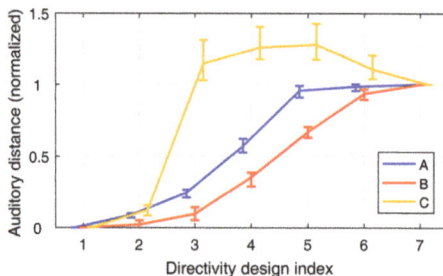

Figure 4. Medians and corresponding 95% confidence intervals for all beampattern designs A, B, and C, pooled over all signals and normalized individually on directivities indicated by 1 and 7.

Both designs A and B yield monotonic curves. A pairwise analysis of variance (ANOVA) of the data pooled over all signals reveals the beampattern to be significant factor ($p \ll 0.01$) for $A_{1...5}$. For the design B, all directivities are (weakly) significant ($B_{1...7}$, $p < 0.08$).

A comparison of the curve obtained for $A_{2,4,6}$ to the results of Laitinen [?], reveals a similar linear mapping to auditory distance.

By contrast, the curve obtained for $C_{1...7}$ is not monotonic in the proposed sequence. If we compare strength and angle of direct sound and specular reflections arriving at the listener for directivities C_4 and C_7, cf. Figure **??**, we see more energy coming from lateral directions for C_4. The more diffuse sound field explains the significantly greater auditory distance ($p \leq 0.04$) for $C_{2...6}$ compared to C_7.

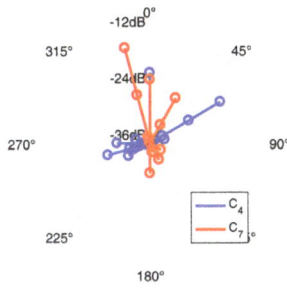

Figure 5. Direct sound and specular reflections arriving at the listening position for C_4 and C_7, normalized with respect to C_1.

For conditions with more energy coming from lateral directions, e.g., $C_{3...6}$, major intersubjective differences are found affecting the size of respective 95% confidence intervals. Therefore we conclude that these conditions lead to an ambiguous distance percept.

2.3. Influence of the Signal

The influence of the signal $S_{1...3}$ on the auditory distance of the design A in R_1 is evaluated by supplementing responses of set 13 with re-mapped responses of set 1 to 3 using Equation (??).

Figure ?? shows the median values and corresponding 95% confidence intervals of the auditory distance for the room R_1 and beampattern design A. Along the indices, the distance impression exhibits a monotonic increase for all signals until A_5. The ANOVA of neighboring values reveals beampatterns $A_{2...5}$ as a significant factor ($p < 0.03$). By contrast, beampatterns $A_{5...7}$ do not yield a significant change ($p \geq 0.45$), despite continuously reducing the D/R ratio. This seems to comply with a general tendency to auditorily underestimate the physical distance [?].

A signal-wise comparison of the obtained data reveals the significantly smaller auditory distance for S_2 than for S_1 or S_3 ($p_{S_2/S_1} \ll 0.01$, $p_{S_2/S_3} = 0.02$). This seems to comply with the finding in [? ?] that the auditory distance of broadband signals decreases with the relative amount of high-frequency energy.

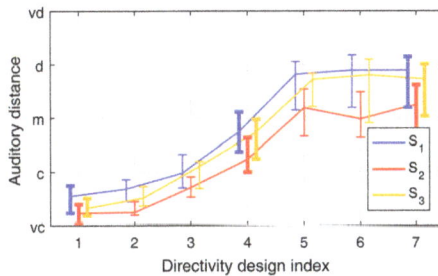

Figure 6. Medians and 95% confidence intervals for tested signals $S_{1...3}$ in R_1 with beampattern design A.

2.4. Influence of the Room

The influence of the room and the source-listener distance ($R_{1...3}$) is evaluated by the data of the set 14. Figure ?? shows the median values and corresponding 95% confidence intervals, regarding signal S_1 and beampattern design A, supplemented by the linearly and individually re-mapped responses of the sets 1, 10, and 11 using Equation (??).

A smaller room with shorter T_{60} and sound source closer to adjacent walls but with the same source-listener distance (R_3) leads to a flatter curve. Similar flattening accompanied by an additional offset to bigger auditory distances is achieved by extending the source-listener distance (R_2). Interestingly, for all tested rooms R the beampattern is a significant factor ($p_{R_1} < 0.09, p_{R_2} < 0.03, p_{R_3} < 0.04$) in the range of $A_{1...5}$. This significance is similar to the values obtained with pooled signals $S_{1...3}$ ($p \ll 0.01$, see Figure **??**).

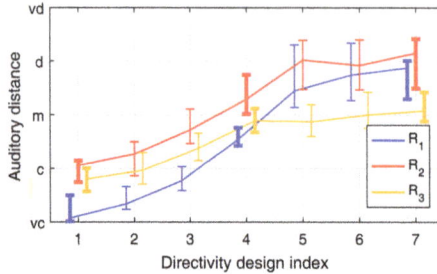

Figure 7. Medians and corresponding 95% confidence intervals for tested rooms $R_{1...3}$ with beampattern design A and signal S_1.

2.5. Influence of Single-Channel Reverberation

In audio playback reverberation effects are often used to control the auditory distance. To get an idea how this effect contributes to the proposed effect, artificial reverberation is added to signal S_1 and tested with beampattern patterns $A_{1,4,7}$ in room R_1. Figure **??** shows respective median values together with corresponding 95% confidence intervals. According to the ANOVA, the influence of reverberation on the auditory distance is significant ($p < 0.05$).

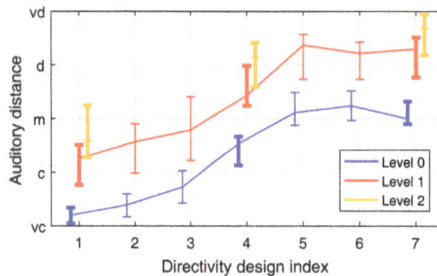

Figure 8. Medians and corresponding 95% confidence intervals for reverberation levels 0, 1, 2 in R_1 with S_1 and beampattern design A.

Individually and linearly re-mapped responses from the sets 1 and 12 were used supplementing the responses from set 15 to provide a more detailed analysis for the reverberation levels 0, 1 in terms progression over the 7 design indices. Both reverberation levels yield a similar progression with the known saturation for $A_{>5}$. The beampattern is a (weakly) significant factor ($p < 0.09$) for the dry signal (rev. level 0) in the range of $A_{1...5}$, and by adding reverberation (rev. level 1), differences between the neighboring conditions $A_{1,2}$ and $A_{2,3}$ are no longer significant ($p \geq 0.16$).

3. Modeling the Auditory Distance

This section discusses linear auditory distance models for the presented effect, based on characteristic metrics of the spatial sound field and their regression to the experimental data.

3.1. Direct-To-Reverberant Energy Ratio

The most obvious predictor in this context is the D/R ratio. It is widely accepted for prediction of auditory distance [?] and is defined as

$$D/R = 10 \log_{10} \frac{\int_{0ms}^{T} s^2(t)dt}{\int_{T}^{\infty} s^2(t)dt}. \tag{3}$$

By using $s(t) = \sum_l h_l(t)$, the D/R ratio can be calculated based on the loudspeaker impulse responses, with a time constant T regarding only direct sound.

Regression analysis fits a linear regression function $f(D/R) = kD/R + d$ depending on the D/R ratio to the normalized experimental data and yields $k = -0.049$ and $d = 0.11$. Figure ??a shows the pooled data compared with $f(D/R)$. Although the D/R ratio and the median values of the pooled data are highly correlated ($R^2 = 0.93$) their progression along the beampattern indices tends to underestimate the distance.

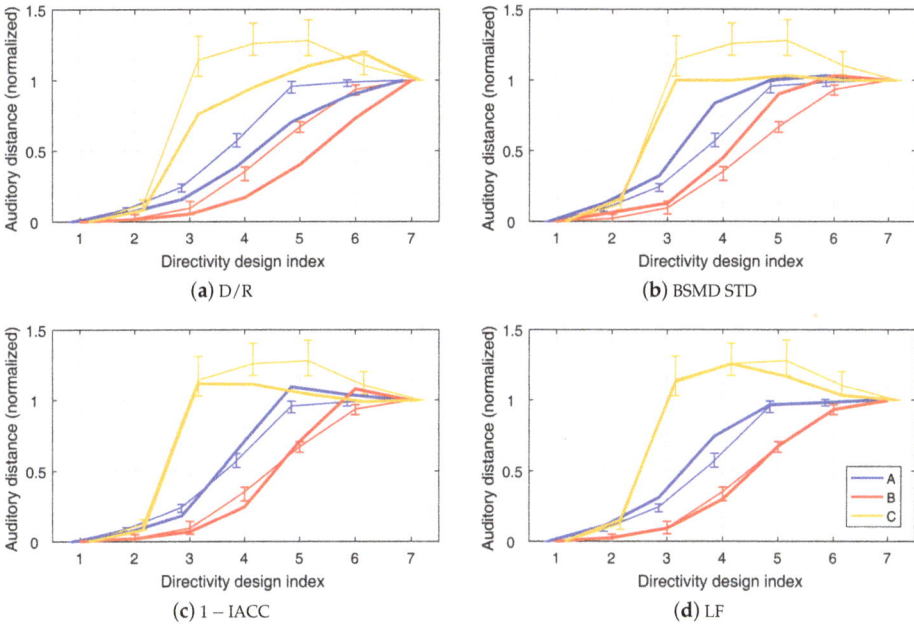

Figure 9. Comparison of medians and 95% confidence intervals for all conditions (thin lines) with predictors (thick lines): D/R, BSMD STD, LF, and IACC.

3.2. Binaural Spectral Magnitude Difference Standard Deviation

In [?] a feature is introduced related to the standard deviation of the magnitude spectrum of the room transfer function. Similar to the D/R ratio, this feature, noted as BSMD STD, represents a distance-dependent behavior and is implemented to model the source-listener distance within the freely available Auditory Modeling Toolbox (http://amtoolbox.sourceforge.net/). For calculating the BSMD STD, any binaural signal is sufficient.

Binaural input signals are generated by firstly convolving the signal of each propagation path arriving at the listener with respective HRTF measurements of a KEMAR dummy-head microphone (freely accessible measurements of the MIT available at http://sound.media.mit.edu/resources/KEMAR.html) and then summing up obtained signals for each ear respectively. The linear regression yields the same correlation as the D/R ratio ($R^2 = 0.93$ with $k = 0.32$ and $d = -1.52$), although their progression along the beampattern index is qualitatively different, cf. Figure ??b.

3.3. Inter-Aural Cross Correlation Coefficient

As reverberation caused by the room simulation introduces binaural cues by altering the sound attributes at the two ears differentially, the inter-aural cross correlation coefficient (IACC) is used as an additional measure for auditory distance. The IACC is based on the inter-aural cross correlation function (IACF):

$$\text{IACF}(\tau) = \frac{\int_{t_1}^{t_2} s_{\text{left}}(t)s_{\text{right}}(t+\tau)dt}{\sqrt{\int_{t_1}^{t_2} s_{\text{left}}^2(t)dt \int_{t_1}^{t_2} s_{\text{right}}^2(t)dt}}, \tag{4}$$

with $s_{\text{left}}(t) = h_{\text{left}}(t) * s(t)$ and $s_{\text{right}}(t) = h_{\text{right}}(t) * s(t)$.

The binaural impulse response $h(t)$ corresponds to responses for left and right ear at $\phi = 0°$. The IACC is defined as the maximum absolute value within $\tau = \pm 1$ ms:

$$\text{IACC} = \max_{\forall \tau \in [-1\text{ms};1\text{ms}]} |\text{IAFC}(\tau)|. \tag{5}$$

The early IACC, considering a time window of $t_1 = 0$ ms to $t_2 = 80$ ms, is commonly used in room acoustics as an objective measure for apparent source width (ASW). It is widely accepted that a lower IACC value leads to a bigger ASW, and therefore $1-$ IACC is positively correlated with the magnitude of perceived width. With the IACC binaurally measured in the experimental setup, linear regression yields $f(1 - \text{IACC}) = 1.52(1 - \text{IACC}) - 0.20$ to model the experimental data ($R^2 = 0.97$, cf. Figure ??c).

3.4. Lateral Energy Fraction

The lateral energy fraction (LF) is another acoustic measure quantifying the spatial impression. Similaraly then the IACC, considering a time window up to 80 ms, it has been accepted as a measure of the effect of source broadening [? ?]. Simply stated, the LF is the ratio of the sum of the early lateral energy to the sum of the early total energy:

$$\text{LF} = \frac{\int_{5\text{ms}}^{80\text{ms}} s_{\text{lat}}^2(t)dt}{\int_{0\text{ms}}^{80\text{ms}} s^2(t)dt}, \tag{6}$$

with $s_{\text{lat}}(t) = \sum_l h_l(t) \sin(\phi_l)$ and ϕ_l as azimuthal angle of the l-th loudspeaker.

Linear regression yields $f(\text{LF}) = 7.3\text{LF} - 0.54$, cf. Figure ??d. This LF-based linear model delivers the best matching results underlined by a sublime correlation of $R^2 = 0.99$.

4. Experiment II: Directivity-Controlled Auditory Distance in a Real Room

The findings of the first experiment are evaluated with a real variable-directivity source in a room. Considering the good performance of models that were actually developed to predict the apparent source width (ASW), the second experiment evaluates the ASW in addition to the auditory distance. This enables us to examine the inter-relation of the two attributes.

4.1. Experimental Setup

In this experiment the effect is proven with a sound source able to vary its directivity namely the icosahedral loudspeaker (IKO, http://iko.sonible.com/). This 20-sided, 20-channel playback device employs spherical beamforming as developed in [? ?] and allows to steer beams up to third order into freely adjustable directions.

The directivity of the IKO was controlled using the freely available ambiX plug-in suite [?] with Reaper as DAW. Firstly, auralized signals were encoded using the ambiX encoder, then converted using ambiX converter, and lastly filtered according to [?] using mcfx convolver. This yields twenty-channel audio files used as conditions for the experiment. Audio playback was controlled by Pure Data on a standard PC with RME MADI audio interface to drive a Sonible d:24 power amplifier.

The size controller of the ambiX encoder allows to vary the beam width from third to zeroth order (SIZE $= 0 \ldots 1$). To create a beampattern design representing the design A of the first experiment, different settings of the size controller were used. By informal listening of the author, size values and orientations were determined, including a zeroth order beam and to opposed third order beams. Directivities $A^*_{1\ldots4}$ are facing to the listener with values of SIZE $= (0, 0.27, 0.47, 1)$ and conditions $A^*_{5\ldots7}$ are rotated by $180°$ with SIZE $= (0.47, 0.27, 0)$. Thus, design A^* can be seen as modified version of A, adjusted by the ear. Figure **??** shows calculated beampattern patterns $A^*_{1\ldots4}$ normalized to constant energy.

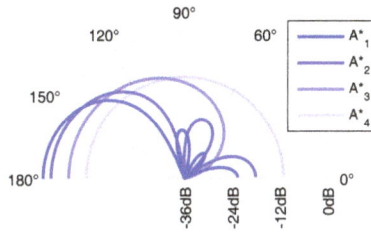

Figure 10. Beampattern design A^*.

Other designs evaluated in the experiment are known from the first experiment. While design B and C are identical (see Figure **??**), design D is composed of directivities $B_{1,3,4\ldots7}$ and C_4 in order to achieve the most distinct effect. A horizontal cross-section through measured frequency-dependent beampatterns of the IKO is shown in Figure **??**.

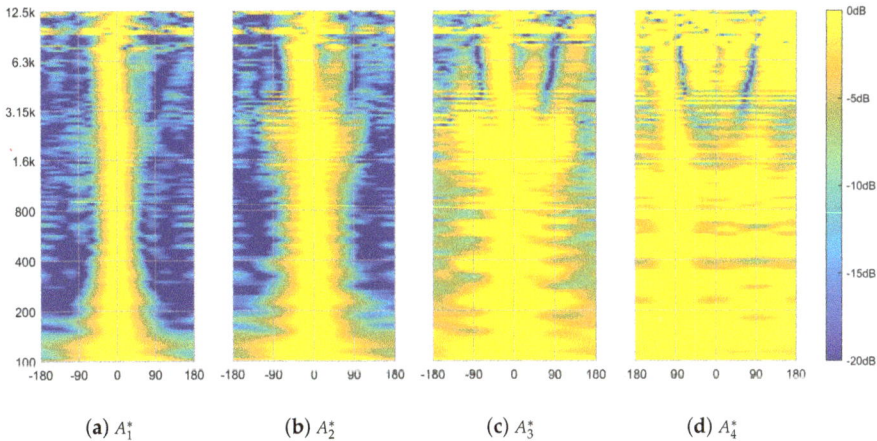

(a) A^*_1 (b) A^*_2 (c) A^*_3 (d) A^*_4

Figure 11. *Cont.*

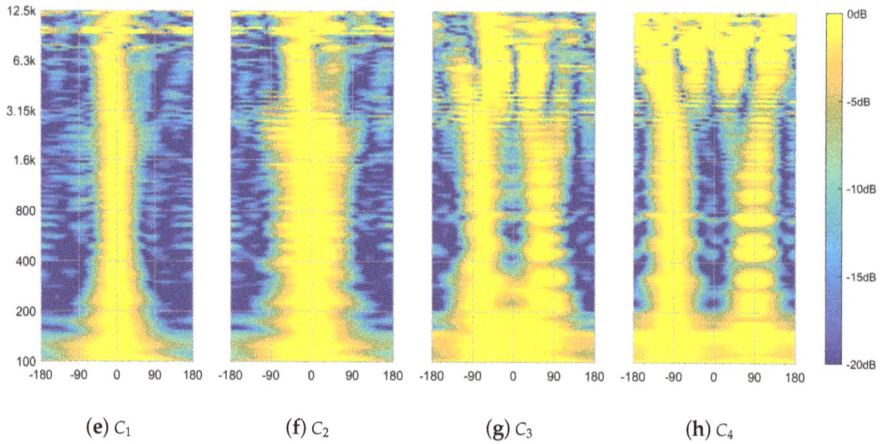

(**e**) C_1 (**f**) C_2 (**g**) C_3 (**h**) C_4

Figure 11. Horizontal cross-section through measured frequency-dependent beampatterns of the IKO normalized the maximum of each resulting directivity in each cross section. Decibel values are color coded over frequency in Hertz and azimuth angle in degree.

Room and positioning of the sound source corresponded to the condition R_2, cf. Table **??**. Similarly to the first experiment, listeners were asked to rate the distance of the auditory object on a graphical user interface. On a screen, the sketch of the setup was displayed and listeners had to adjust 7 randomly sorted markers to the auditory distance, where each marker represented a beampattern of the designs under test. Markers could either be moved directly (drag and drop) or, for fine adjustments, steered with a slider. Each condition could be repeated at will until listeners were satisfied with the match between marker placement and what they heard. To facilitate the task a fine grid indicating distances of 0.5 m was displayed on the screen, cf. Figure **??**a. In the room microphone stands marked distances of $(1, 2, 4)$ m, cf. Figure **??**b.

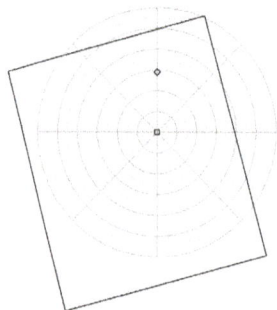

(**a**) Sketch of the setup as represented in the GUI. The square represtents the listenener and a diamond represents the IKO.

(**b**) Experimental setup in the IEM CUBE.

Figure 12. Conducting the listening experiment II.

Listeners were asked to provide an honest report of what they actually perceived. This instruction had to do with the fact that there was no time limit to provide answers. It aimed specifically at asking listeners to avoid developing theories about which condition they were presented, as some listeners were aware of results from the first experiment.

Additional to distance, the second experiment also examined the apparent source width (ASW) of auditory objects created by beampattern designs A^*, B, and C in a separate task. The procedure was the same as in the first experiment, so that rating was done on a graphical user interface displaying a continuous slider for each condition of a set to permit comparative rating. Listeners were asked to rate using the whole scale *very narrow* (vn), *narrow* (n), *moderate* (m), *broad* (b), and *very broad* (vb).

The signal fed into auralization was female speech (S_1, see Table **??**). All conditions were normalized in loudness and were played back in loop at comfortable level of 70 dB(A).

During the listening session, listeners were sitting on a chair with ear height adjusted to the IKO (1.3 m) and while listening to conditions, they were requested to face the IKO. Both tasks were performed consecutively with a short break in between. Half of the listeners started with the distance rating task and the other half with the rating of the ASW. Ten listeners participated in this experiment (all male; age 28–54), nine of them performed already the first experiment.

4.2. Auditory Distance

Figure **??** shows the results for the distance rating task of the second experiment.

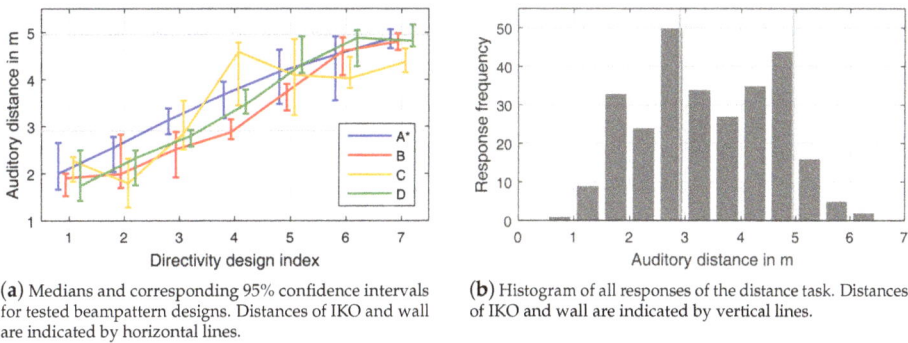

(**a**) Medians and corresponding 95% confidence intervals for tested beampattern designs. Distances of IKO and wall are indicated by horizontal lines.

(**b**) Histogram of all responses of the distance task. Distances of IKO and wall are indicated by vertical lines.

Figure 13. Experimental results of the distance task for signal S_1 and room R_2 with use of the IKO.

A pairwise ANOVA of the data reveals the beampattern to be a significant factor for $A^*_{2...4}$ ($p \leq 0.02$). Although medians of the design A^* form a linear curve, the significant range of the first experiment using the design A in room R_2 is not achieved, cf. Figure **??**. For design B, significance is attained by directivities $B_{3...6}$ ($p \leq 0.02$) and for design C by directivities $C_{2...4}$ ($p \leq 0.01$).

Although comparisons of these results with significances of the first experiment shown in Figure **??** should be interpreted with caution, indications are found that the IKO yields less pronounced distance impressions. Even with design D, corresponding to a combination of directivities that should yield the most pronounced effect, not more then 4 significantly different distances are obtained ($D_{2...5}$, $p \leq 0.01$).

Medians of design B remain sigmoid-shaped as they are in the first experiment, whereas medians obtained by design C show major differences. While simulated conditions $C_{3...6}$ of the first experiment created significantly different impressions that are localized more distant compared to C_7, they are either localized closer to C_7, e.g., C_3, or are no longer significantly different ($C_{4...6}$, $p \geq 0.17$) when auralized with the IKO. Informal notes of listeners indicate that the spectral coloration of some conditions led to an impression as if the auditory object is right behind the IKO and the incoming sound is filtered due to acoustic shadowing. Similarly to the first experiment these conditions yield major intersubjective differences as indicated by the size of the 95% confidence intervals.

If we take a look at the measured beampattern C_3 in Figure **??**h we find evidence that shifted results can be explained rather by the spectrum of the IKO into the listener's direction as the image source behind the IKO ($180°$) receives highly attenuated signals and the direct sound has a low-pass character (400 Hz).

Visual cues could explain the less pronounced ratings for large distances, because responses of most listeners (7/10) are within the feasible space limited by the wall at approximately 5 m, leading to a high response frequency in the interval right in front of the wall, cf. Figure ??b.

The high frequency of responses within the interval of the IKO are due to another effect caused by visual cues. Studies could show that seeing only one possible sound source biases the perceived distance towards it, e.g., [?].

Interestingly, in the first experiment visual cues were available similarly, but no influence thereof was obtained. Therefore we conclude that in the laboratory environment, in which visual cues do not comply with auditory cues, the former play a minor role. This agrees with findings in [? ?] showing that sensory interactions, e.g., vision vs. audition, include a weighting process where the most reliable cue contributes the most to the multi-sensory percept.

4.3. Apparent Source Width

Figure ?? shows the results for the ASW rating task of the second experiment. Assessed ASWs of all tested designs form monotone curves and resemble respective distance curves shown in Figure ??. The correlation of medians is high ($R^2_{A*} = 0.99$, $R^2_B = 0.98$, $R^2_C = 0.94$) and the significant range for width is the same as it is for distance, except for neighboring conditions $A^*_{4/5}$ whose differences were found to be significant for ASW.

The correlation of ASW to auditory distance is not surprising, if we consider the model predictions of the first experiment. Both best predicting models for distance, $1-$IACC and LF are measurements that are typically used in room acoustics to quantify the ASW. This is in contrast to the inverse relation between the physical source-listener distance and the ASW found in [?]. In contrast to our study, the ASW decreased almost linearly as the distance is doubled and $1-$IACC and LF predicted results opposing ASW. Thus, it seems that the ASW of auditory objects created by reflections is larger then for real loudspeakers facing the listener.

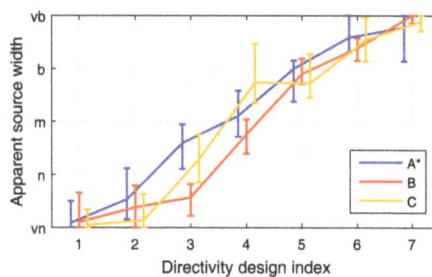

Figure 14. Medians and corresponding 95% confidence intervals for beampattern designs of assessed width for signal S_1 with use of the IKO.

5. Conclusions

In this contribution, an investigation was carried out into the influence of various beampatterns on the auditory distance. Two-dimensional simulation of a variable-directivity sound source at a single point in the room was shown to provide control of the auditory distance. Different beampattern designs were proposed that cause pronounced and graduated distance impressions. Additionally, the influence of the auralized room, source-listener distance, signal, and single-channel reverberation was studied.

The mapping of beampatterns $A_{1...7}$ and $B_{1...7}$ to auditory distance curves is sigmoid-shaped. It resembles the compressive power functions described in [?], characterizing the relation between physical and auditory distance. Moreover, agreeing with [? ?], signals with an increased relative amount of high-frequency energy appeared to be closer in the study.

Both decreasing the auralized room and increasing the source-listener distance yield a more compressed curve, which is slightly offset in case of the increased source-listener distance. Despite this, the range of discriminability is persistent.

The use of single-channel reverberation is effective at increasing the auditory distance, however, it narrows the directivity-controllable range of distinguishable distance impressions.

Successful modeling of the experimental results was presented and all models yield curves that are highly correlated with the experimental data. Interestingly, spatial measures used to quantify the ASW provide very accurate predictions.

In addition to the findings obtained by loudspeaker-based auralization in the anechoic chamber, we could also present an evaluation of the designs synthesized by a variable-directivity sound source in a room. A listening experiment could show that in real environments the distance perception is biased due to visual cues leading to less pronounced distance impressions. In addition to the auditory distance the apparent source width was evaluated and we could show that in contrast to the natural environments of the study [?], the width highly correlates with distance impressions caused by the directivity of the sound source. This finding explains the performance of spatial measures in the first experiment and enhances the robustness of this new effect in real environments.

Possible directions of further research are to investigate whether and to which extend larger performance venues (e.g., Ligeti Hall) affect the mapping to perceived distance to answer the question if this effect is preserved for a larger audience.

Acknowledgments: The authors thank all listeners for their participation in the listening experiment. This work was funded by the Austrian Science Fund (FWF) project nr. AR 328-G21, Orchestrating Space by Icosahedral Loudspeaker. The paper is an extended version of [?].

Author Contributions: Florian Wendt, Franz Zotter, Matthias Frank, and Robert Höldrich conceived and designed the experiments; Florian Wendt performed the experiments; Florian Wendt, Franz Zotter, Matthias Frank, and Robert Höldrich analyzed the data; Florian Wendt wrote the paper with periodic contributions by the other authors.

Conflicts of Interest: The authors declare no conflict of interest.

References

. Mason, R. How Important Is Accurate Localization in Reproduced Sound? In Proceedings of the 142th Convention of the Audio Engineering Society, Berlin, Germany, 20–23 May 2017.

. Zahorik, P.; Brungart, D.S.; Bronkhorst, A.W. Auditory distance perception in humans: A summary of past and present research. *Acta Acust. United Acust.* **2005**, *91*, 409–420.

. Kolarik, A.J.; Moore, B.C.J.; Zahorik, P.; Cirstea, S.; Pardhan, S. Auditory distance perception in humans: A review of cues, development, neuronal bases, and effects of sensory loss. *Atten. Percept. Psychophys.* **2016**, *78*, 373–395.

. Mershon, D.H.; King, L.E. Intensity and reverberation as factors in the auditory perception of egocentric distance. *Percept. Psychophys.* **1975**, *18*, 409–415.

. Zahorik, P. Assessing auditory distance perception using virtual acoustics. *J. Acoust. Soc. Am.* **2002**, *111*, 1832–1846.

. Larsen, E.; Iyer, N.; Lansing, C.R.; Feng, A.S. On the minimum audible difference in direct-to-reverberant energy ratio. *J. Acoust. Soc. Am.* **2008**, *124*, 450–461.

. Kolarik, A.; Cirstea, S.; Pardhan, S. Discrimination of virtual auditory distance using level and direct-to-reverberant ratio cues. *J. Acoust. Soc. Am.* **2013**, *134*, 3395–3398.

. Laitinen, M.V.; Politis, A.; Huhtakallio, I.; Pulkki, V. Controlling the perceived distance of an auditory object by manipulation of loudspeaker directivity. *J. Acoust. Soc. Am.* **2015**, *137*, EL462–EL468.

. Wendt, F.; Frank, M.; Zotter, F.; Höldrich, R. Directivity patterns controlling the auditory source distance. In Proceedings of the 19th International Conference on Digital Audio Effects (DAFx-16), Brno, Czech Republic, 5–9 September 2016; pp. 295–300.

. Zotter, F.; Zaunschirm, M.; Frank, M.; Kronlachner, M. A Beamformer to Play with Wall Reflections: The Icosahedral Loudspeaker. *Comput. Music J. (Accept. Publ.)* **2017**, *41*.

. Daniel, J. Représentation de Champs Acoustiques, Application à la Transmission et à la Reproduction de Scènes Sonores Complexes Dans un Contexte Multimédia. Ph.D. Thesis, Université Paris 6, Paris, France, 2001.

. Zotter, F.; Frank, M. All-round ambisonic panning and decoding. *AES J. Audio Eng. Soc.* **2012**, *60*, 807–820.

. Allen, J.B.; Berkley, D.A. Image Method for Efficiently Simulating Small-room Acoustics. *J. Acoust. Soc. Am.* **1979**, *65*, 943–950.

. Wabnitz, A.; Epain, N.; Jin, C.T.; Van Schaik, A. Room acoustics simulation for multichannel microphone arrays. In Proceedings of the International Symposium on Room Acoustics, Melbourne, Australia, 29–31 August 2010.

. Tervo, S.; Pätynen, J.; Kuusinen, A.; Lokki, T. Spatial decomposition method for room impulse responses. *J. Audio Eng. Soc.* **2013**, *61*, 17–28.

. Guski, R. Auditory localization: Effects of reflecting surfaces. *Perception* **1990**, *19*, 819–830.

. Wendt, F.; Sharma, G.K.; Frank, M.; Zotter, F.; Höldrich, R. Perception of Spatial Sound Phenomena Created by the Icosahedral Loudspeaker. *Comput. Music J.* **2017**, *41*, 76–88.

. Zaunschirm, M.; Frank, M.; Zotter, F. An Interactive Virtual Icosahedral Loudspeaker Array. *Fortschritte der Akusitk* **2016**, 1331–1334.

. Kohlrausch, A.; Kortekaas, R.; van der Heijden, M.; van de Par, S.; Oxenham, A.J.; Püschel, D. Detection of Tones in Low-noise Noise: Further Evidence for the Role of Envelope Fluctuations. *Acta Acust. United Acust.* **1997**, *83*, 659–669.

. Coleman, P. Dual Role of Frequency Spectrum in Determination of Auditory Distance. *J. Acoust. Soc. Am.* **1968**, *44*, 631–632.

. Little, A.D.; Mershon, D.H.; Cox, P.H. Spectral content as a cue to perceived auditory distance. *Perception* **1992**, *21*, 405–416.

. Georganti, E.; May, T.; Van De Par, S.; Mourjopoulos, J. Sound source distance estimation in rooms based on statistical properties of binaural signals. *IEEE Trans. Audio Speech Lang. Process.* **2013**, *21*, 1727–1741.

. Marshall, A.H. A note on the importance of room cross-section in concert halls. *J. Sound Vib.* **1967**, *5*, 100–112.

. Barron, M.; Marshall, A.H. Spatial impression due to early lateral reflections in concert halls: The derivation of a physical measure. *J. Sound Vib.* **1981**, *77*, 211–232.

. Lösler, S. MIMO-Rekursivfilter für Kugelarrays. Master's Thesis, University of Music and Performing Arts Graz, Graz, Austria, 2014.

. Zotter, F. Analysis and Synthesis of Sound-Radiation with Spherical Arrays. Ph.D. Thesis, University of Music and Performing Arts Graz, Graz, Austria, 2009.

. Kronlachner, M. Plug-in Suite for Mastering the Production and Playback in Surround Sound and Ambisonics. In Proceedings of the 136th Convention of the Audio Engineering Society, Berlin, Germany, 26–29 April 2014.

. Anderson, P.W.; Zahorik, P. Auditory/visual distance estimation: Accuracy and variability. *Front. Psychol.* **2014**, *5*, 1–11.

. Ernst, M.O.; Banks, M.S. Humans integrate visual and haptic information in a statistically optimal fashion. *Nature* **2002**, *415*, 429–433.

. Mendonça, C.; Mandelli, P.; Pulkki, V. Modeling the Perception of Audiovisual Distance: Bayesian Causal Inference and Other Models. *PLoS ONE* **2016**, *11*, e0165391.

. Lee, H. Apparent Source Width and Listener Envelopment in Relation to Source-Listener Distance. In Proceedings of the 52nd Audio Engineering Society Conference, Guildford, UK, 2–4 September 2013.

applied
sciences

MDPI

Article

Late Reverberation Synthesis Using Filtered Velvet Noise [†]

Vesa Välimäki *, Bo Holm-Rasmussen [‡], Benoit Alary and Heidi-Maria Lehtonen [‡]

Acoustics Lab, Department of Signal Processing and Acoustics, Aalto University, 02150 Espoo, Finland;
bo.holm.rasmussen@gmail.com (B.H.-R.); benoit.alary@aalto.fi (B.A.); hmleht@gmail.com (H.-M.L.)
*   Correspondence: vesa.valimaki@aalto.fi; Tel.: +358-50-5691-176
[†]   This paper is a revised and extended version of a paper published in the International Conference on Digital Audio Effects (DAFX), Maynooth, Ireland, 2–5 September 2013.
[‡]   Current addresses: Bo Holm-Rasmussen—The Royal Danish Academy of Fine Arts, Laboratory for Sound, 1050 Copenhagen, Denmark. Heidi-Maria Lehtonen—Dolby Sweden, 11330 Stockholm, Sweden.

Academic Editors: Woon-Seng Gan and Jung-Woo Choi
Received: 15 March 2017; Accepted: 3 May 2017; Published: 6 May 2017

Abstract: This paper discusses the modeling of the late part of a room impulse response by dividing it into short segments and approximating each one as a filtered random sequence. The filters and their associated gain account for the spectral shape and decay of the overall response. The noise segments are realized with velvet noise, which is sparse pseudo-random noise. The proposed approach leads to a parametric representation and computationally efficient artificial reverberation, since convolution with velvet noise reduces to a multiplication-free sparse sum. Cascading of the differential coloration filters is proposed to further reduce the computational cost. A subjective test shows that the resulting approximation of the late reverberation often leads to a noticeable difference in comparison to the original impulse response, especially with transient sounds, but the difference is minor. The proposed method is very efficient in terms of real-time computational cost and memory storage. The proposed method will be useful for spatial audio applications.

Keywords: audio systems; digital filters; digital signal processing; room acoustics

---

1. Introduction

Artificial reverberation research started in the 1960s, when Schroeder developed the first methods to simulate the room effect with a computer [1,2]. His methods plus numerous other approaches, which were introduced thereafter, have been reviewed by Gardner [3] and recently in a series of two papers by Välimäki et al. [2,4].

Concert halls and listening rooms are often considered to be linear and time-invariant systems. Therefore, it should be possible to fully reproduce their sonic characteristics by replicating the impulse response, which is measured between a source and a listening point. A room impulse response (RIR) is often divided into three phases: the direct (or dry) sound, early reflections, and the late reverberation. This paper focuses on the modeling of the late reverberation, which is noise-like and contains the contribution of a large number of reflections.

Convolution with a measured RIR is a popular technique resulting in very realistic reverberation [2,4,5]. However, convolution is computationally intensive, and modification or parameterization of the measured RIR can be cumbersome. Partitioned fast convolution methods [6–9] reduce the computational complexity considerably compared to direct convolution and avoid most of the delay introduced by the basic fast convolution, which corresponds to a full-scale FFT(Fast Fourier transform)-based implementation. Moorer suggested that the late part of the RIR can be well characterized as exponentially decaying white noise [10]. This observation led to useful applications when

Rubak and Johansen used a finite-impulse response (FIR) filter with random coefficients in a recursive reverberation algorithm [11,12]. Karjalainen and Järveläinen developed an improved algorithm in which a random coefficient FIR filter is cascaded with a lowpass comb filter [13]. They also introduced velvet noise, which is smooth-sounding ternary random noise [13]. Later, Lee et al. [14] and Oksanen et al. [15] investigated alternative recursive reverberator structures using velvet noise.

This paper focuses on room reverberation modeling using velvet noise, extending our previous work [16,17]. The RIR is divided into short segments and each of them is approximated as a filtered velvet noise (FVN) sequence. The coloration filters and their associated gain account for the spectral shape and level of each RIR segment, so together they enable the approximation of a given frequency-dependent decay behavior in the time domain. Finally, cascaded Schroeder allpass filters are used to obtain a smooth, wideband, noise-like response. This approach is thus orthogonal to the modal filter bank idea, which divides the RIR into slices in the frequency dimension [18,19], and to Jot's idea of estimating the reverberation time across frequency bands [20] and calibrating a feedback delay network reverberator [21,22]. Such methods are best suited for exponentially decaying responses.

This FVN approach leads to a parametric representation and computationally efficient RIR synthesis, since convolution with velvet noise is economical to implement. A novel idea is proposed to cascade the coloration filters, so that the effect of all filtering operations of the previous stages are accounted for by using differential filters in the subsequent stages.

The rest of this paper is organized as follows: Sections 2 and 3 discuss velvet noise and the basic version of the FVN method, respectively, and Section 4 describes a new differential filtering strategy and an impulse response segmentation strategy for it. Section 5 shows how well the algorithm can synthesize the impulse response of a real concert hall, and how the synthetic response can be modified. Section 6 compares the computational complexity and memory usage with other algorithms, and Section 7 presents a subjective evaluation of the proposed method. Section 8 concludes this paper.

2. Velvet Noise

Velvet noise is a special kind of random noise, which was discovered by Karjalainen and Järveläinen [13]. It consists of sample values -1, 0, and 1 only. The most surprising attribute of velvet noise is that even when 95% of its samples are zero, it sounds smoother than Gaussian random noise, which is generally thought to be the prototype of white noise [13,23]. Velvet noise is of interest in this work, because it provides a computationally efficient way to convolve an arbitrary signal with white noise [16].

2.1. Generation of Velvet Noise

Velvet noise can be interpreted as a randomly jittered impulse train in which the sign of each impulse is chosen randomly to be positive or negative [23]. To generate velvet noise, one should first select the pulse density N_d, i.e., the number of impulses per second. It yields the main design parameter, the average distance between impulses T_d, as:

$$T_d = \frac{f_s}{N_d},$$ (1)

where f_s is the sample rate. Other randomization techniques have also been proposed, for example the totally random ternary sequence by Rubak and Johansen [11], which does not include any rule to limit how close to or far away from each other two neighboring impulses can occur. However, it is not perceived to be as smooth as velvet noise at low pulse densities [23]. The restriction of having only one impulse within every T_d samples appears to be an economical choice, which minimizes roughness [13,23].

In velvet noise, the impulse locations $k(m)$ are determined as:

$$k(m) = \text{round}[mT_d + r_1(m)(T_d - 1)], \tag{2}$$

where $m = 0, 1, 2, \ldots$ is the pulse counter and $r_1(m)$ is a value produced with a random-number generator with uniform distribution $(0,1)$. The term -1 at the end of Equation (2) helps to avoid coinciding pulses [23].

The complete velvet-noise sequence can then be written as:

$$s(n) = \begin{cases} 2\,\text{round}[r_2(m)] - 1, & \text{when } n = k(m), \\ 0, & \text{otherwise,} \end{cases} \tag{3}$$

where n is the sample index, $k(m)$ are the impulse locations determined using Equation (2), and $r_2(m)$ is the value of a second random-number generator with uniform distribution $(0,1)$ used to select the sign of each impulse [23].

When the sample rate of 44,100 Hz is used, the choice of N_d = 2205 pulses/s, according to Equation (2), leads to a convenient integer value of T_d = 20 samples for the average pulse distance. Figure 1a shows the first 500 samples of an example velvet-noise sequence with these parameters. There is only one non-zero sample seen between any two grid boundaries. The autocorrelation function of the velvet-noise sequence shown in Figure 1b is close to a unit impulse, as its maximum occurring at $n = 0$ is 1.0 and at other lags the correlation is smaller than about 0.01. The power spectrum of the velvet-noise sequence shown in Figure 1c is fairly flat.

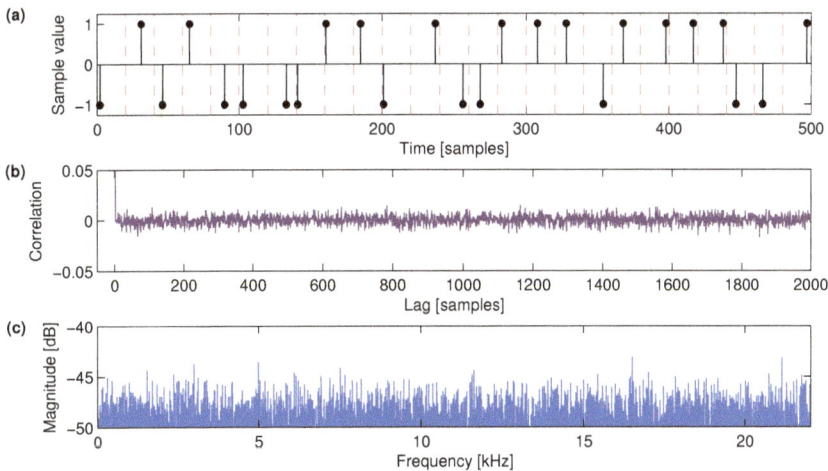

Figure 1. (a) Non-zero sample values, (b) the autocorrelation function, and (c) the estimated spectrum of a velvet-noise sequence. In (a), the vertical dashed lines indicate the grid boundaries. In (b), the value of autocorrelation at zero lag is 1.0, but this first value is truncated in the figure.

2.2. Velvet-Noise Convolution

Time-domain convolution of a signal with velvet noise can be highly economical computationally. The samples of the velvet-noise sequence $s(n)$ are used as FIR filter coefficients. Velvet-noise convolution (VNC) is very fast to compute, because all multiplications by zero can be dispensed as their locations in the sequence are known. Additionally, as the non-zero samples contained in the velvet noise are either -1 or 1, multiplications are not needed. Thus, convolution with velvet noise reduces to a sparse multiplication-free convolution.

In practice, then, the input signal is propagated in the delay line of the filter, and only those input signal samples which coincide with the non-zero coefficients of the velvet-noise sequence are added together to produce the output. One idea is to separately run through the indices of coefficient values $+1$ and -1, add the corresponding sample values taken from the delay line, and subtract the two sums. This VNC process can be formulated as:

$$x(n) * s(n) = \sum_{m_+} x[n - k(m_+)] - \sum_{m_-} x[n - k(m_-)], \tag{4}$$

where $x(n)$ is the input signal, $*$ denotes the convolution, and $k(m_+)$ and $k(m_-)$ contain the indices of the positive and negative impulses, respectively, in the velvet-noise sequence $s(n)$. This multi-tap delay-line implementation of VNC is illustrated in Figure 2.

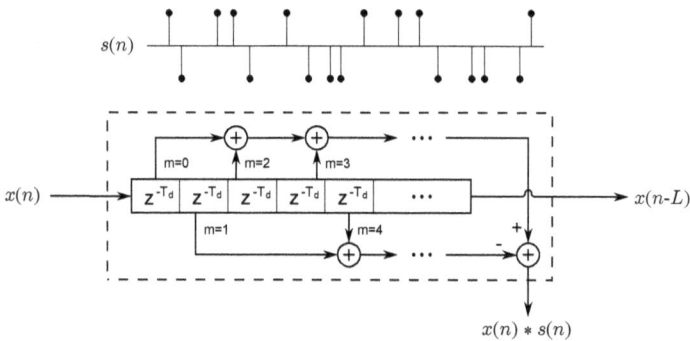

Figure 2. Velvet-noise convolution: Convolving the signal $x(n)$ with a velvet-noise sequence $s(n)$ reduces to the multiplication-free process of computing two sparse sums of delayed input signal samples and their difference. Blocks containing z^{-T_d}, where z is the complex variable of the Z transformation, refer to delay lines of T_d samples. The output tap of each delay-line element is located at the sample point determined by sequence $s(n)$.

For example, when 5% of the velvet noise coefficients are non-zero ($+1$ or -1) and the filter length is L samples, computing an output sample requires $0.05L$ additions and no multiplications. For a $1 - s$ noise sequence at the 44.1-kHz sample rate, the filter length is $L = 44{,}100$, and this yields 2205 additions per output sample. For comparison, a regular FIR filter of the same length requires $L - 1 = 44{,}099$ additions and $L = 44{,}100$ multiplications, or 88,199 operations, to compute each output sample, which is 40 times more than using VNC.

3. Filtered Velvet Noise Reverberation Algorithm

The key idea of the FVN reverberation algorithm is to divide the RIR into short non-overlapping segments and to approximate each segment as filtered white noise. Velvet noise is used instead of standard white noise, such as Gaussian noise, since then the convolution with the input signal is fast to compute.

Figure 3 shows the block diagram of the basic FVN reverberation algorithm. The delay lines of each VNC block serve two purposes: they delay the input signal appropriately for the next stage, as indicated by the right-hand-side output signal $x(n - L)$ in Figure 2, and they provide the state variables of the sparse multi-tap delay line used to implement the VNC, i.e., a very efficient multiplication-free convolution of input signal with the velvet-noise sequence. The sparse sum of each segment is next filtered by its own spectral coloration filter $H_m(z)$ and attenuated appropriately by the gain term G_m, as shown in Figure 3.

Uniform segmentation of an RIR should not be used, as the constant frame rate causes a periodic disturbance in the synthetic response. This is reminiscent of the flutter echo effect, which is a common problem in room acoustics. Much effort has been made to reduce this effect in recursive reverberation algorithms that use a pseudo-random noise sequence [13,14]. Thus, it makes sense to use a non-uniform segmentation scheme in the FVN algorithm, as suggested in [16]. Another motivation to use a non-uniform framing is that the filter for each segment would be sufficiently different. In a typical RIR in which the exponential decay is faster at high frequencies than at low, a constant decrease in bandwidth, such as a 1-kHz narrowing, takes place non-uniformly in time—quickly in the beginning and slower towards the end of the RIR. This also motivates the use of longer segments at the end than at the beginning of the RIR. Figure 4a shows an example of a RIR and its segmentation. The impulse response has been measured in the concert hall in Pori, Finland (this impulse response of the concert hall is available online at http://legacy.spa.aalto.fi/projects/poririrs/).

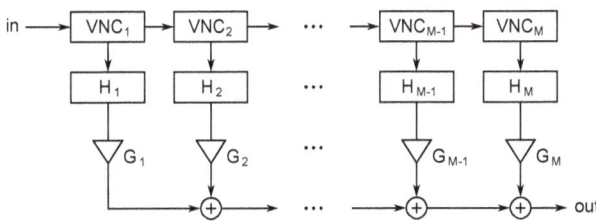

Figure 3. Basic principle of the filtered velvet noise (FVN) algorithm [16]. The delay lines between the filtering branches of length L_m are in practice combined with velvet-noise convolution (VNC) blocks, cf. Figure 2. Blocks H_m and G_m, for $m = 1, 2, ..., M$, represent the spectral coloration filters and gain factors for each segment, respectively.

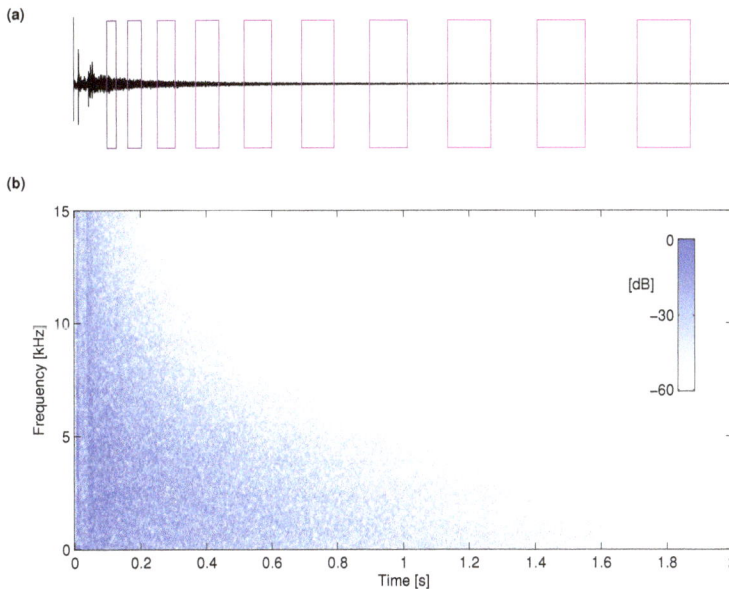

Figure 4. (**a**) Measured room impulse response (RIR) of the concert hall in Pori, Finland, with the boxes indicating every second segment used for modeling, and (**b**) its spectrogram showing frequency-dependent decay.

3.1. Coloration Filters

To design the spectral coloration filter $H_m(z)$ linear prediction (LP) can be used for each segment [24]. The coloration filters should match the overall lowpass characteristic of each short segment. For this reason, low-order LP is sufficient in this application. Prediction order 10 is used in this work, which leads to 10th-order all-pole coloration filters. Figure 5 shows examples of coloration filters estimated for the RIR of the Pori concert hall. The overall shape of the responses follows the frequency-dependent decay, as expected.

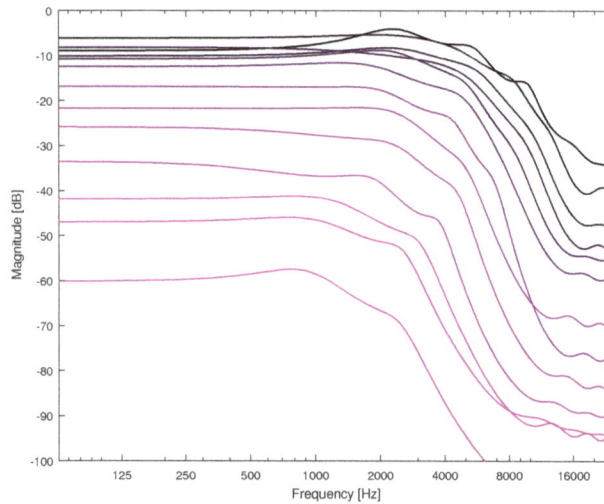

Figure 5. Magnitude responses of coloration filters of order 10 for every second segment of the impulse response of the Pori concert hall. The same color codes as in Figure 4a are used such that the darker lines correspond to the beginning of the RIR and the color gets lighter towards the end of the RIR.

Since only one lowpass filter and one gain coefficient are required per segment, the computation of the VNC becomes the most demanding part of the structure. For this reason, ways to reduce the pulse density without sacrificing the sound quality were investigated. Karjalainen and Järveläinen [13] showed that the sufficient pulse density is lower for lowpass-filtered velvet noise than in the full audio band: in particular, for a cutoff frequency of $f_c = 1.5\,\text{kHz}$, the lowpass-filtered velvet noise sounds smoother than Gaussian white noise even with the lowest pulse density they tested, 600 pulses/s. Since the bandwidth of the RIR becomes narrower towards its end, the pulse density of velvet noise may also be decreased from one segment to another. Figure 4b clearly shows the narrowing of the bandwidth (blue area) of a measured RIR over time.

3.2. Schroeder Allpass Filters

In order to further smooth the synthetic RIR, a cascade of Schroeder allpass (SAP) filters is used. This allows further reduction of the pulse density in VNC. Each SAP filter has the following transfer function [1]:

$$A(z) = \frac{a + z^{-N}}{1 + az^{-N}},\tag{5}$$

where $-1 < a < 1$ is the allpass filter coefficient and N is the delay-line length. Figure 6 shows the structure of the FVN algorithm when the total sum of all branches is further processed with a cascade of filters, SAP_1 to SAP_K.

Figure 7 shows the spectrogram of a velvet-noise sequence having only 44 non-zero samples per second and that of a SAP filter consisting of four cascaded filters. The delay line lengths of the SAP filters are 225, 341, 441, and 556 samples, and their filter coefficient is $a = 0.7$. The rightmost spectrogram is the result of convolving the velvet-noise sequence with the SAP filter's response, showing a wideband noise-like behavior. This example shows that the gaps in velvet noise can be filled by cascading SAP filters. The spectrograms in Figure 7 were generated using a 600-sample Hann window with 500 samples of overlap.

By experimenting with different pulse densities and listening to the outcome, it was decided that $N_d = 100$ pulses/s is sufficient in the very beginning of the late reverberation, where segments are very short, whereas $N_d = 40$ pulses/s can be enough at the end where the bandwidth gets narrow. Between these extremes, the density is decreased linearly as a function of the segment index m. The selected pulse density for each segment is shown in Figure 8.

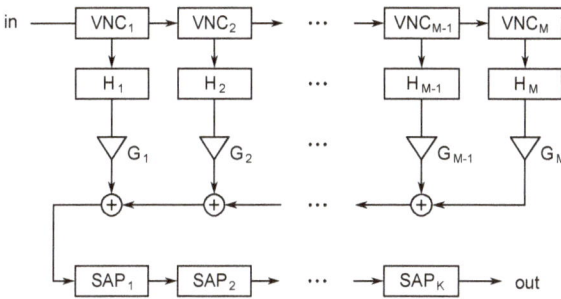

Figure 6. FVN algorithm with Schroeder allpass filters (SAP) [16].

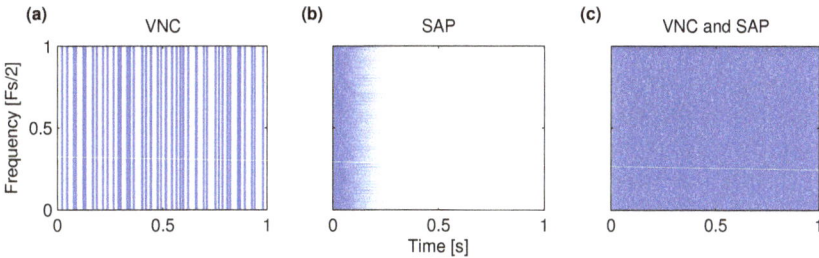

Figure 7. Spectrograms of (**a**) a velvet-noise sequence, (**b**) the impulse response of four cascaded SAP filters, and (**c**) their convolution. White corresponds to 60 dB lower level than blue.

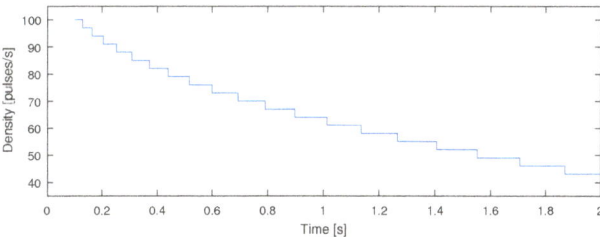

Figure 8. Pulse density and length of each segment for the Pori hall. The pulse density can be decreased towards the end of the RIR.

3.3. Segment Gains

Finally, the gain G_m for each segment, as shown in Figure 6, must be determined so that the overall decay rate of the RIR model is preserved. To ensure that this is the case, an analysis–synthesis approach is used. Each RIR segment is first whitened with the LP inverse filter obtained using the 10th-order LP, and the average signal power of this filtered signal segment is calculated to establish a reference. Then a long sequence (e.g., one second) of velvet noise with the pulse density assigned to that segment is processed with the all-pole coloration filter and with the cascade of SAP filters. The average signal power of this filtered velvet noise is then calculated, and the gain of this segment, G_m, is set based on the ratio of this signal power to the reference signal power. This routine ensures that the gain of each segment is adjusted accurately.

4. Advanced FVN Algorithm

In this section we elaborate on the basic FVN method: coloration filters are redesigned so that they can be cascaded, which helps reduce the filter order for each segment.

4.1. Differential Coloration Filters

Since the cutoff frequency of the filters in each segment usually decreases towards the end of the RIR, it is possible to exploit the previous filters in the subsequent filtering stages. The basic idea is to design the first lowpass coloration filter $H_1(z)$ but to construct the other filters by cascading differential filters $\Delta H_m(z)$, for $m \geq 2$. This structure is illustrated at the top of Figure 9.

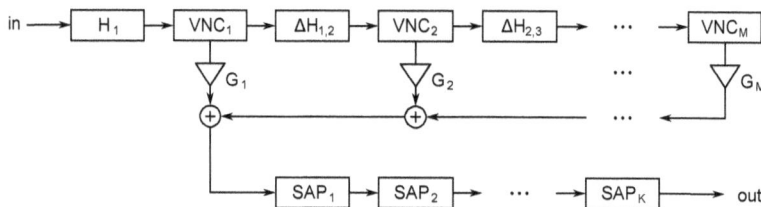

Figure 9. Advanced FVN algorithm with cascaded differential coloration filters $\Delta H_m(z)$.

The first filter can be designed manually to imitate the spectral shape of the initial RIR segment, which has a fairly flat spectrum. Here we use a 10th-order all-pole filter obtained with linear prediction. The magnitude response of this filter is shown in Figure 10a.

The differential filters are second-order notch filters with the transfer function $H(z) = 1 + (V_0 - 1)$ $[1 - A_2(z)]/2$ with

$$A_2(z) = \frac{-c + d(1-c)z^{-1} + z^{-2}}{1 + d(1-c)z^{-1} - cz^{-2}}, \tag{6}$$

where $c = [\tan(\pi f_b/f_s) - V_0]/[\tan(\pi f_b/f_s) + V_0]$, $d = -\cos(2\pi f_c/f_s)$ for $0 < V_0 < 1$ is the attenuation at the center frequency f_c and f_b is the bandwidth of the notch (Hz) [25]. The differential filters can be designed to match the difference between the neighboring coloration filters. Figure 10b shows responses of the notch filters designed from the family of 10th-order coloration filters. Figure 10c shows the total effect of cascading 1 to $M - 1$ of these filters with the first filter $H_1(z)$. The overall shapes and cutoff points are very similar to the responses shown in Figure 5.

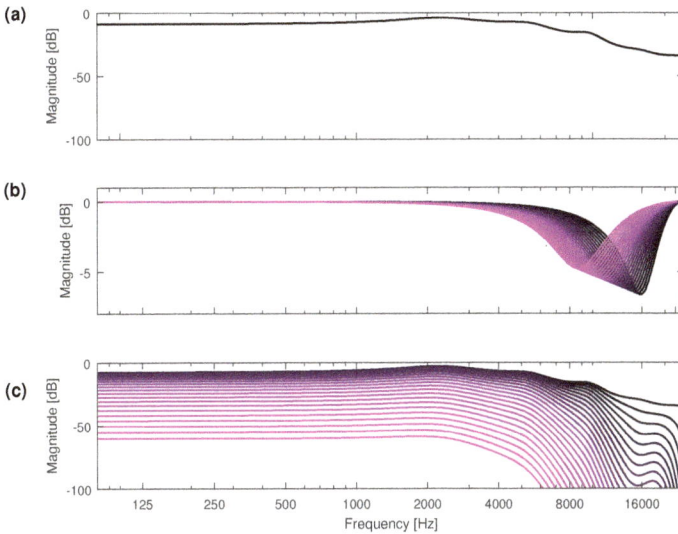

Figure 10. Magnitude responses of (**a**) the first-segment coloration filter $H_1(z)$; (**b**) differential filters; and (**c**) cascaded differential filters with the first-segment filter.

4.2. Revised Segmentation Method

The differential filtering technique was found to benefit more from a different segmentation method than what was used in the basic FVN method. The main idea here is to start a new segment when the difference in the spectrum from the start of the previous segment becomes sufficiently large. The RIR was analyzed in short windows (2048 samples) using low-order linear prediction (order 6 was used). Based on the magnitude responses of the corresponding all-pole filters, which provide an approximation of the spectral envelope of the windowed signals, a bandwidth for each segment was estimated. The bandwidth estimate was determined as the frequency at which the spectral envelope estimate decreased 20 dB from its maximum.

Using a linearly decaying threshold function, the segment boundaries were chosen based on reaching a sufficiently large change in bandwidth in the estimated spectral envelope. Therefore, a larger difference is required at the beginning than at the end of the RIR before starting a new segment. This led to the segmentation of the Pori RIR shown in Figure 11. The main difference compared to the previous method, shown in Figure 8, is that the revised segmentation reflects the significant changes in the magnitude response of the RIR.

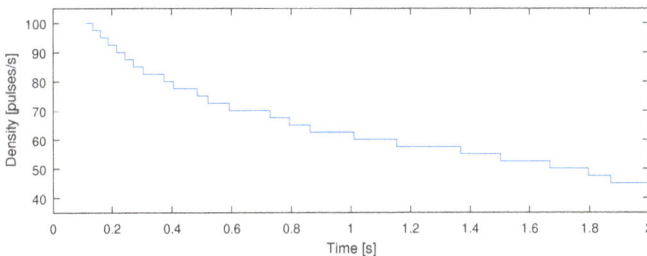

Figure 11. Segment lengths and density based on the revised segmentation strategy, which is used with differential coloration filters.

5. Design Examples

This section shows an example of modeling an RIR and modifying it. We show and analyze here the approximation of the Pori RIR implemented using the advanced method. An example of modeling this RIR using the basic FVN method has been presented earlier [16].

5.1. RIR Modeling Using Advanced FVN

The synthetic RIR produced using the advanced FVN model and its spectrogram are shown in Figure 12. As an objective comparison, Figure 13 shows the reverberation time T_{30} against octave bands for three RIRs (original, basic FVN, and advanced FVN). We decided to use T_{30} instead of T_{60}, because the signal-to-noise ratio near the end of the RIR does not sufficiently measure 60-dB decay; T_{30} is the measured time of a 30-dB decay multiplied by two.

All three RIRs in Figure 13 show the same tendency of lower reverberation time for higher frequencies than low frequencies. The octave-band reverberation times for the basic FVN algorithm stay within ±7% of the reference in all octave bands. For the second algorithm this spread is within ±12%. The increased deviation is in accordance with the assumption that the second algorithm is a rougher approximation due to the lower coloration filter order.

An informal listening test comparing the two new reverb algorithms with a reference convolution reverb has been carried out using headphones. The reference RIR and its approximation with the basic FVN algorithm sound very similar even when comparing the impulse responses themselves. The approximation produced by the advanced FVN algorithm has a slightly more unnatural sound when listening to its impulse response. Results of a subjective test comparing the audio signal processed with the original RIRs and their FVN approximations are presented in Section 7 of this paper.

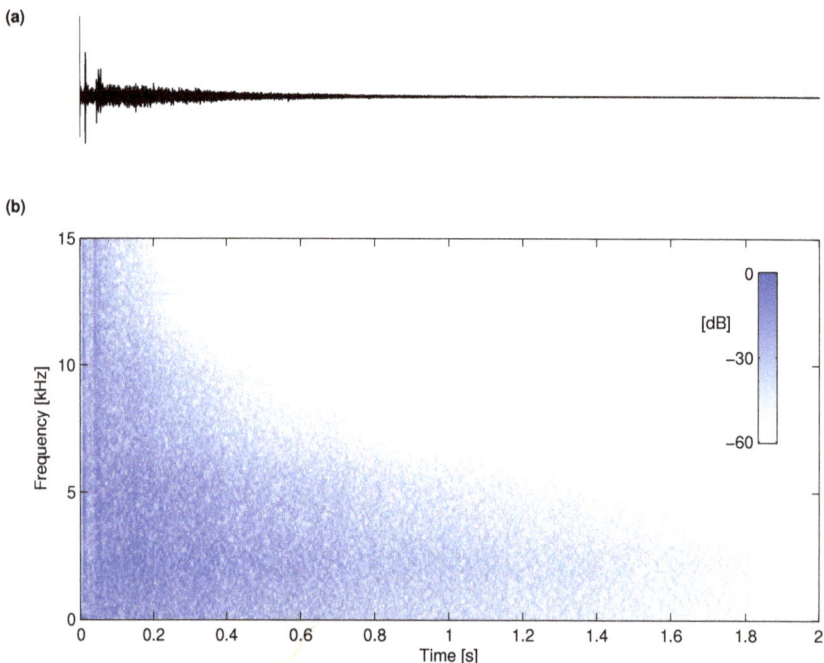

Figure 12. (a) Synthetic RIR produced using the advanced FVN method and (b) its spectrogram. Cf. Figure 4.

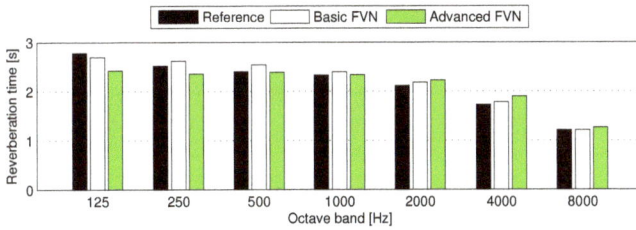

Figure 13. Reverberation time, T_{30}, for the original RIR (reference), its basic FVN synthesis, and advanced FVN synthesis.

5.2. Modification of the Approximated RIR

The parametric representation used in the FVN method allows modifying the modeled RIR in various ways. We have previously shown that the RIR can be dramatically shaped simply by modifying the gain term G_m [16]. In this way it is possible, for example, to increase or decrease the decay rate of the RIR. Here we show another option, time-stretching of the RIR.

Figure 14 shows the result of shortening the RIR by 50%. The number of segments, velvet noise density, coloration filters, or gains have not been changed, but the lengths of the VNC filters have been shortened to half. The early part of the RIR has not been modified, however. The overall shape of the RIR and the spectrogram are both seen to be preserved with respect to Figure 12, but the time scale has been modified. Another option to change the decay rate would be to modify the coloration filters and gains in the FVN model. Figure 15 shows an example in which the VNC filters have been lengthened by 100%, which leads to a twice-longer and, thus, more slowly decaying RIR. These examples demonstrate the possibilities for meaningful parametric modifications allowed by the FVN method.

The modeled impulse responses and test signals are available online at http://research.spa.aalto.fi/publications/papers/applsci-fvn/.

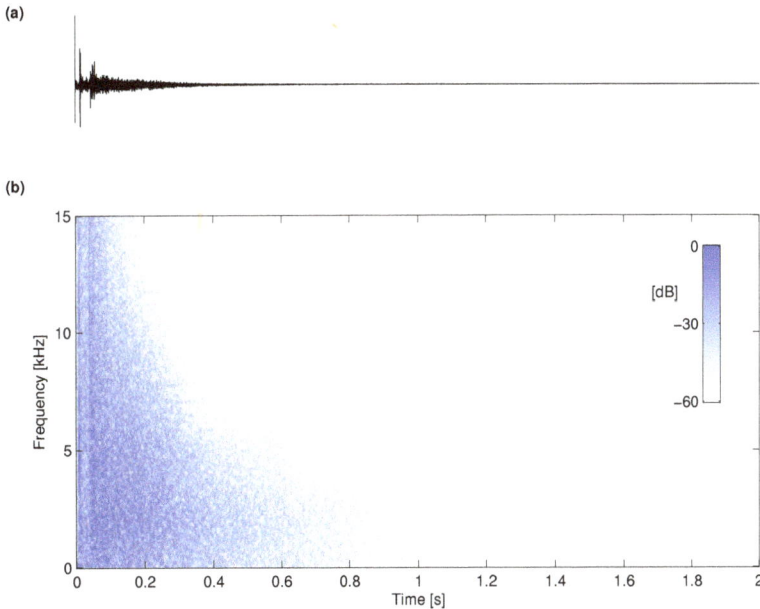

Figure 14. (a) 50% shortened synthetic RIR and (b) its spectrogram.

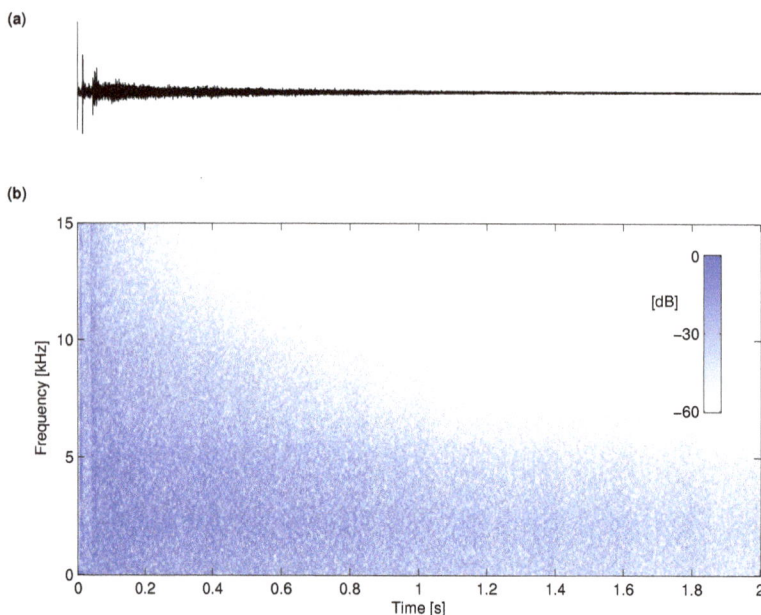

Figure 15. (**a**) 100% stretched RIR and (**b**) its spectrogram.

6. Computation and Memory Costs

The computational efficiency of reverberation algorithms is of great importance when they are used for real-time audio processing. Reverberation algorithms are also known to require a considerable amount of fast memory for storing past signal-sample values, which can be critical in implementations on limited hardware. Additionally, multichannel RIRs must be stored in spatial audio, which may require a considerable amount of memory storage. In this section, these implementation costs of the two versions of the FVN algorithm are compared with direct convolution and with partitioned fast convolution. The implementation cost of the early reflections is not included in the calculations, but it is assumed that the late part of the RIR lasts for 2 s.

6.1. Costs of the Basic FVN Algorithm

The number of floating-point operations (FLOPs) per processed sample required by the basic FVN algorithm are listed in Table 1. The numbers given are for the RIR modeling example of the Pori concert hall (see Figure 4). The FLOPs are specified as the number of additions and multiplications for each module of the algorithm. Note that the VNC filters only require additions and no multiplications. In Table 1, 'H' and 'G' are the coloration filters and gain adjustments, respectively, for each signal segment, and 'Sum' refers to the addition of output signals of the 20 branches before they are fed to the SAP filters (see Figure 6). In Table 1, note that the SAP filters only take 4% of total operations, but the coloration filters take 64%. This proves that efforts to reduce the cost of the coloration filtering stage are well motivated.

Table 1. Operations required to process one sample in each module of the basic FVN algorithm. The largest number in each column is in bold.

Module	Additions	Multiplications	Percentage
VNC	160	0	26%
H	200	200	64%
G	0	20	3%
Sum	19	0	3%
SAP	14	14	4%
Total	393	234	100%

6.2. Costs of the Advanced FVN Algorithm

Table 2 dissects the operations of each module in the advanced FVN, which uses the differential coloration filtering approach. Each differential coloration filter is implemented as a direct-form second-order IIR (infinite impulse response) filter, which requires five multiplications and four additions per sample. The VNC and SAP filters used for the two versions of the FVN algorithm are the same, and hence the same numbers of operations appear for these modules in Table 2 as in Table 1. The differential coloration filters 'ΔH' possess about half of the total arithmetic instructions, showing the advantage of collaborative cascaded filtering.

Table 2. Operations of the advanced FVN algorithm. Note that ΔH also includes the first coloration filter $H_1(z)$. The largest number in each column is in bold.

Module	Additions	Multiplications	Percentage
VNC	**139**	0	31%
ΔH	131	**106**	52%
ΔG	0	25	6%
Sum	24	0	5%
SAP	14	14	6%
Total	308	145	100%

6.3. Comparison Against Other Algorithms

Next, we compare the computational and memory costs of the proposed methods to other convolution reverberation approaches. We enumerate the number of FLOPs and the number of signal memory samples required for a 88,200 samples-long impulse response, as in the previous section.

The values listed in Table 3 for the direct convolution are based on the direct-form FIR implementation, which leads to the same number of multiplications as the number of RIR samples (88,200) and one less addition (88,199). In direct convolution, the required amount of fast memory is the same as the RIR length, since it defines the delay-line length (88,200 samples). The values for the partitioned fast convolution are taken from the recent improvement of the algorithm by Wefers and Vorländer (see Table 1 in [8]).

Table 3 shows that the proposed algorithms are over 100 times more efficient computationally than the direct convolution and approximately as efficient as the best partitioned fast convolution algorithm, which is only 12% more efficient than the advanced FVN. The memory consumption of the new method is the same as that of the direct convolution and 50% smaller than that of the partitioned convolution algorithm.

Table 3 also shows that the FVN method is useful for compression of RIR data: whereas the direct and partitioned convolution algorithms must store all RIR samples, the FVN methods only store two arrays of pointers, which give the locations of the positive and negative impulses, an array of segment lengths (20 in this case), plus 12 filter parameters per segment (a gain factor and 11 feedback coefficients of the 10th-order all-pole filter). The advanced FVN method is even more efficient in this respect, as there are less impulses in the VCN block and the differential filters only require five parameters each. This yields a total of 294 parameters to be stored. The amount of data is only 0.33% compared to the original RIR samples. This implies that the FVN approach enables very efficient storage of multichannel RIR data.

Table 3. Operation count, fast memory and storage memory consumption of various reverberation algorithms for modeling a 2-s RIR at a 44.1-kHz sample rate. The smallest numbers are in bold. FLOPS: floating-point operations.

Algorithm	FLOPs	Delay-Line Memory	Storage Memory
Direct convolution	176,401	**88,200**	88,200
Partitioned fast convolution	**399**	176,400	88,200
Basic FVN	627	90,442	420
Advanced FVN	453	**88,200**	**294**

7. Subjective Evaluation

The proposed advanced FVN method was evaluated using a subjective test. Three different concert halls impulse responses were approximated from pre-recorded RIR [26]. One of the RIRs was the Pori concert hall response used in the examples above, which has a reverberation time of 2.3 s at middle frequencies. The second hall was the Cologne Philharmonie, which has a shorter mid-frequency reverb time (1.9 s). Its RIR is quite dry, containing mainly the direct sound, a few reflections, and a relatively short reverberation tail. The third hall was the Vienna Musikverein, which has the longest reverberation time (3.2 s) of the selected halls. Its RIR sounds very reverberant, having a lot of early reflections soon after the direct sound.

The beginning of each RIR approximation was taken from the measured RIR. Thus, the early-reflection part of the impulse responses remained the same as the original, and only the tail of the RIR was modified by the basic and advanced FVN approximations. The duration of each early-reflection segment was adjusted manually based on preliminary testing as follows: 110 ms for the Pori Concert Hall, 119 ms for the Cologne Philharmonie, and 52 ms for the Vienna Musikverein.

Three different sound files were processed with the three reference (original) RIRs and their approximations produced using the advanced FVN method, which yielded altogether 18 (3×6) sound files. The three test sounds contained drumming, slowly changing chords played with a synthesizer, and a cappella singing (the first 10 s of "Tom's Diner" by Suzanne Vega).

The test type was ABX [27], which refers to a pair-wise test in which the subject always compares three sound files, A, B, and X, and is asked to identify whether sound X is the same as A or B. Additionally, in our test, the subjects had to evaluate the perceived difference between A and B on a five-point scale, a variant of the mean-opinion score. Figure 16 shows the user interface used in the listening test. The 18 test sounds were played in pseudo-random order, and they all appeared twice during the test, leading to 36 cases to be evaluated. Additionally, four extra cases were played in the beginning of the test, the answers of which were deleted from the data, since learning was assumed to occur during the first few cases, and only after this are the persons able to carefully and objectively evaluate the sounds. Thus, the total number of cases presented to the subjects in the listening test was 40.

Twelve subjects with no reported hearing problems participated in the listening test. Their age varied between 23 and 41 years. All subjects had previously participated in listening tests. It took typically 30 to 40 min for the subjects to finalize the test. The test can be assumed not to have been too difficult or tiresome.

Table 4 summarizes how the test subjects identified the synthetic reverberation from the original for different sound types. Since the subjects were allowed to listen to all sounds several times, detecting even the smallest differences turned out to be easy. Thus, in 86% of all cases, the persons identified the approximated RIR from the original. Detecting the difference in drumming, which contains transients, was the easiest, and the identification score was 99%. Chords were the most difficult case, as the sounds are mostly stationary and the synthetic sounds had a slow attack. The difference was still detected in about three cases out of four. The difficulty in detecting the differences in singing was between the two extreme cases, and the recognition was successful in 84% of cases. After the test, the test subjects commented that it was fairly easy to find the different items in drum samples, but for the other two sounds it felt more difficult. However, the average rating for the differences was 3.1, which corresponds to a "small difference". This implies that although it was often possible to discriminate between the original RIR and its approximation, the perceived difference was not considered to be very large.

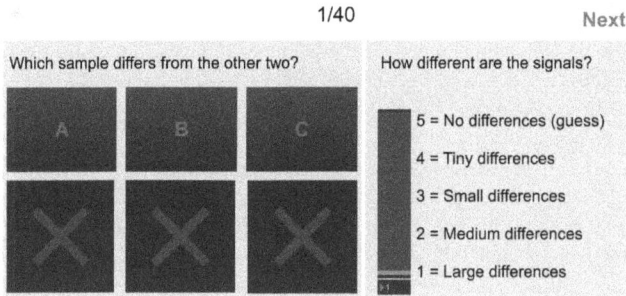

Figure 16. User interface of the ABX test with the 5-point difference rating used in the listening test. The verbal descriptions associated with each quality level appear on the right.

Table 4. Identification of FVN approximation of reverberated sounds in the listening test.

Sound Type	Drums	Chords	Singing	Average
Identification	99%	76%	84%	86%
Quality rating	2.1	3.9	3.3	3.1

Table 5 shows the listening test results for the three different halls. Interestingly, there was no significant difference between the different RIR types, but the identification of all approximations was close to the average, or 86%. The quality rating was, similarly, close to the average for all concert halls. Thus, the FVN method appears to be equally well suited to both short and long RIRs.

Table 5. Identification of the FVN approximation of different RIRs in the listening test.

Concert Hall	Pori	Cologne Philharmonie	Vienna Musikverein	Average
Identification	84%	85%	88%	86%
Quality rating	3.3	3.0	3.0	3.1

8. Conclusion and Future Prospects

This paper discussed the modeling of the late part of a measured room impulse response using filtered velvet-noise sequences. The idea here is to divide the impulse response into many non-overlapping segments of variable length and to approximate each segment using a spectral coloration filter and a sparse FIR filter having its coefficients taken from a velvet-noise sequence. The summed output of these filtering stages is further processed with a cascade of a few Schroeder allpass filters to increase the density and to smooth out the transitions between the segments. In this configuration, velvet-noise convolution can provide a smooth response even with very low pulse densities. Moreover, the sparsity of the velvet noise may vary along the reverberation tail so that towards the end, where the bandwidth gets narrow, the sequences are sparser. To obtain a realistic model of a target RIR, the coloration filters can be designed by applying linear prediction to the variable-length RIR segments.

Additionally, this paper contributed a method to improve the computational efficiency of the FVN reverberation algorithm: the idea is to link the coloration filters so that each of them receives as the input the output of the previous stage. This way each segment only requires a differential coloration filter, which reduces the bandwidth sufficiently with respect to the previous stages. Instead of being a high-order IIR filter, each differential coloration filter is a second-order notch filter.

The performance of the proposed algorithm was demonstrated with a modeling example, and the results showed that the algorithm is able to accurately model the overall characteristics of the target concert hall impulse response. The design procedure yields a flexible parametric approximation of the late part of the target impulse response, allowing for variations such as time-scale modification. Furthermore, the proposed reverberation algorithm is computationally efficient, providing a major advantage over the direct convolution: in the example case of 2-s RIR modeling, the proposed method reduces the computational cost by over 99.6% compared to direct convolution, and it is in this respect comparable to the best FFT-based partitioned convolution methods.

Results of a subjective test were also reported, showing that the FVN approximations are often perceptually different from the original, but that the difference between the original RIR and its FVN approximation is considered small. The difference is easiest to observe when the audio signal contains transients, such as in drum sounds. However, the FVN method was observed to be equally well-suited for approximating long and short RIRs, as there was not much difference in the identification of different RIRs.

The proposed method can be used to implement convolution reverberation in which instead of directly using the measured impulse response, its FVN model is implemented. This allows the possibility for parametric control of the impulse response characteristics. The proposed FVN method has a computational complexity that is comparable to the partitioned fast convolution method, but with a far reduced memory storage, which is important in spatial audio, where multichannel sound reproduction requires a large set of multidirectional impulse responses.

Acknowledgments: This work has been funded in part by the Academy of Finland (ICHO project, Aalto University project No. 13296390). The authors would like to thank Prof. Tapio Lokki for his help in providing and selecting the room acoustic data used in this work.

Author Contributions: Vesa Välimäki wrote the paper and contributed to the original idea; Bo Holm-Rasmussen contributed to the original idea, programmed the methods, produced audio examples, and produced most figures; Benoit Alary contributed to the advanced method, produced Figures 5, 10, 11, and 16, validated the method, and produced audio examples; Heidi-Maria Lehtonen contributed to the original idea, was the advisor of the work of Bo Holm-Rasmussen, and contributed to writing, when she was a postdoctoral researcher at Aalto University.

Conflicts of Interest: The authors declare no conflict of interest.

References

1. Schroeder, M.R.; Logan, B.F. Colorless artificial reverberation. *J. Audio Eng. Soc.* **1961**, *9*, 192–197.
2. Välimäki V.; Parker, J.D.; Savioja, L.; Smith, J.O.; Abel, J.S. Fifty years of artificial reverberation. *IEEE Trans. Audio Speech Lang. Process.* **2012**, *20*, 1421–1448.
3. Gardner, W.G. Reverberation algorithms. In *Applications of Digital Signal Processing to Audio and Acoustics*; Kahrs, M., Brandenburg, K., Eds.; Kluwer: New York, NY, USA, 2002; pp. 85–131.
4. Välimäki, V.; Parker, J.D.; Savioja, L.; Smith, J.O.; Abel, J.S. More than 50 years of artificial reverberation. In Proceedings of the Audio Engineering Society 60th International Conference on Dereverberation and Reverberation of Audio, Music, and Speech, Leuven, Belgium, 3–5 February 2016.
5. Shelley, S.B.; Murphy, D.T.; Chadwick, A.J. B-format acoustic impulse response measurement and analysis in the forest at Koli national park, Finland. In Proceedings of the International Conference on Digital Audio Effects (DAFX), Maynooth, Ireland, 2–5 September 2013; pp. 351–355.
6. Kulp, B.D. Digital equalization using Fourier transform techniques. In Proceedings of the Audio Engineering Society 85th Convention, Los Angeles, CA, USA, 3–6 November 1988.

Appl. Sci. **2017**, *7*, 483

7.   Gardner, W.G. Efficient convolution without input-output delay. *J. Audio Eng. Soc.* **1995**, *43*, 127–136.

8.   Wefers, F.; Vorländer, M. Optimal filter partitions for non-uniformly partitioned convolution. In Proceedings of the Audio Engineering Society 45th International Conference on Applications of Time-Frequency Processing in Audio, Helsinki, Finland, 1–4 March 2012.

9.   Wefers, F. Partitioned Convolution Algorithms for Real-Time Auralization. Ph.D. Thesis, RWTH Aachen University, Institute of Technical Acoustics, Aachen, Germany, 2014.

10. Moorer, J.A. About this reverberation business. *Comput. Music J.* **1979**, *3*, 13–28.

11. Rubak, P.; Johansen, L.G. Artificial reverberation based on a pseudo-random impulse response. In Proceedings of the Audio Engineering Society 104th Convention, Amsterdam, The Netherlands, 16–19 May 1998.

12. Rubak, P.; Johansen, L.G. Artificial reverberation based on a pseudo-random impulse response II. In Proceedings of the Audio Engineering Society 106th Convention, Munich, Germany, 8–11 May 1999.

13. Karjalainen, M.; Järveläinen, H. Reverberation modeling using velvet noise. In Proceedings of the Audio Engineering Society 30th International Conference on Intelligent Audio Environments, Saariselkä, Finland, 15–17 March 2007.

14. Lee, K.S.; Abel, J.S.; Välimäki, V.; Stilson, T.; Berners, D.P. The switched convolution reverberator. *J. Audio Eng. Soc.* **2012**, *60*, 227–236.

15. Oksanen, S.; Parker, J.; Politis, A.; Välimäki, V. A directional diffuse reverberation model for excavated tunnels in rock. In Proceedings of the IEEE International Conference on Acoustics, Speech, and Signal Processing (ICASSP), Vancouver, BC, Canada, 26–31 May 2013; pp. 644–648.

16. Holm-Rasmussen, B.; Lehtonen, H.M.; Välimäki, V. A new reverberator based on variable sparsity convolution. In Proceedings of the International Conference on Digital Audio Effects (DAFX), Maynooth, Ireland, 2–5 September 2013; pp. 344–350.

17. Holm-Rasmussen, B. Velvet Noise in Reverberation Algorithms. MSc Thesis, Technical University of Denmark, Kgs. Lyngby, Denmark, October 2013.

18. Karjalainen, M.; Järveläinen, H. More about this reverberation science: Perceptually good late reverberation. In Proceedings of the Audio Engineering Society 111th Convention, New York, NY, USA, 30 November–3 December 2001.

19. Abel, J.S.; Coffin, S.A.; Spratt, K.S. A modal architecture for artificial reverberation with application to room acoustics modeling. In Proceedings of the Audio Engineering Society 137th Convention, Los Angeles, CA, USA, 9–12 October 2014.

20. Jot, J.M. An analysis/synthesis approach to real-time artificial reverberation. In Proceedings of the IEEE International Conference on Acoustics, Speech, and Signal Processing (ICASSP), San Francisco, CA, USA, 23–26 March 1992; pp. 221–224.

21. Jot, J.M.; Chaigne, A. Digital delay networks for designing artificial reverberators. In Proceedings of the Audio Engineering Society 90th Convention, Paris, France, 19–22 February 1991.

22. Schlecht, S.J.; Habets, E.A.P. Feedback delay networks: Echo density and mixing time. *IEEE/ACM Trans. Audio Speech Lang. Process.* **2017**, *25*, 374–383.

23. Välimäki, V.; Lehtonen, H.M.; Takanen, M. A perceptual study on velvet noise and its variants at different pulse densities. *IEEE Trans. Audio Speech Lang. Process.* **2013**, *21*, 1481–1488.

24. Makhoul, J. Linear prediction: A tutorial review. *Proc. IEEE* **1975**, *63*, 561–580.

25. Dutilleux, P.; Holters, M.; Disch, S.; Zölzer, U. Filters and delays. In *DAFX: Digital Audio Effects*, 2nd ed.; Zölzer, U., Ed.; Wiley: Hoboken, NJ, USA, 2011; pp. 47–81.

26. Pätynen, J. A Virtual Symphony Orchestra for Studies on Concert Hall Acoustics. Ph.D. Thesis, Aalto University, Espoo, Finland, November 2011.

27. Clark, D. High-resolution subjective testing using a double-blind comparator. *J. Audio Eng. Soc.* **1982**, *30*, 330–338.

![applied sciences logo] *applied sciences*

MDPI

Article

Solution Strategies for Linear Inverse Problems in Spatial Audio Signal Processing

Mingsian R. Bai [1,*], Chun Chung [1], Po-Chen Wu [1], Yi-Hao Chiang [1] and Chun-May Yang [2]

[1] Department of Power Mechanical Engineering, National Tsing Hua University, No. 101, Section 2, Kuang-Fu Road, Hsinchu 30013, Taiwan; pups0409@yahoo.com.tw (C.C.); ricky82826@gmail.com (P.-C.W.); precious199382@gmail.com (Y.-H.C.)

[2] Department of Electrical Engineering, National Chiao Tung University, No. 1001, Ta-Hsueh Road, Hsinchu 30013, Taiwan; cmyang@cn.nctu.edu.tw

* Correspondence: msbai63@gmail.com; Tel.: +886-3-5742915

Academic Editors: Woon Seng Gan and Jung-Woo Choi
Received: 30 March 2017; Accepted: 26 May 2017; Published: 5 June 2017

Abstract: The aim of this study was to compare algorithms for solving inverse problems generally encountered in spatial audio signal processing. Tikhonov regularization is typically utilized to solve overdetermined linear systems in which the regularization parameter is selected by the golden section search (GSS) algorithm. For underdetermined problems with sparse solutions, several iterative compressive sampling (CS) methods are suggested as alternatives to traditional convex optimization (CVX) methods that are computationally expensive. The focal underdetermined system solver (FOCUSS), the steepest descent (SD) method, Newton's (NT) method, and the conjugate gradient (CG) method were developed to solve CS problems more efficiently in this study. These algorithms were compared in terms of problems, including source localization and separation, noise source identification, and analysis and synthesis of sound fields, by using a uniform linear array (ULA), a uniform circular array (UCA), and a random array. The derived results are discussed herein and guidelines for the application of these algorithms are summarized.

Keywords: inverse problems; Tikhonov regularization; compressive sensing (CS); convex optimization (CVX); focal underdetermined system solver (FOCUSS); steepest descent (SD); Newton's method (NT); conjugate gradient (CG); golden section search (GSS)

1. Introduction

Numerous inverse problems exist in the field of acoustics. For example, nearfield acoustic holography (NAH) is a noise source identification method that reconstructs a surface field of the source on the basis of sound pressure measured in the nearfield of the source [1–5]. The deconvolution approach for the mapping of acoustic sources (DAMAS) is also a method for noise source identification [6]. Another example is the source signal separation problem, where an individual source signal is to be extracted from a mixed array of signals [7]. Inverse problems can also be found in source sound field synthesis (SFS) problems, where the sound field produced by secondary sources is to be matched with a target field [8,9]. Other examples include sound field control [10,11], crosstalk cancellation in binaural audio rendering [12], noise reduction in speech enhancement [13], room response equalization, and dereverberation [14,15]. In linear range acoustics, each of these problems can be formulated as a linear system ($\mathbf{Ax} = \mathbf{b}$). The current study focused on the solutions of farfield noise source identification problems, sound source localization and separation problems, and sound field analysis (SFA) and synthesis (SFS) problems. Although inverse solutions of acoustic problems have long been investigated by researchers, according to our review of the literature, no conclusive results that give solution strategies and parameter choice guidelines can be found in the literature. Furthermore, although audio quality is the chief concern in practical audio reproduction, most previous research has examined general numerical accuracy and stability. This study explored these problems from the perspective of reproduced signal quality. Solution strategies were compared in a unified framework, and guidelines of parameter choice are summarized herein.

In general, inverse problems can be divided into two categories: overdetermined and square systems, and underdetermined systems. To solve overdetermined systems, least-squares methods, Tikhonov regularization (TIKR) [16], and truncated singular value decomposition (TSVD) are commonly used. Traditionally, Morozov's discrepancy principle, generalized cross-validation (GCV), and the L-curve method can be used to choose the regularization parameter in the TIKR method [17–20]. However, solution methods that are better suited to audio applications than conventional approaches are proposed in this work. In particular, golden section search (GSS) [21] is applied to find optimal regularization parameters. To solve underdetermined problems, compressive sampling (CS) [22,23] solved by using convex optimization (CVX) [24–26] is a widely used approach that is known to be computationally expensive. In the present study, computationally efficient iterative approaches that incorporate sparsity constraints, including the focal underdetermined system solver (FOCUSS) [27], steepest descent (SD), Newton's (NT) method, and conjugate gradient (CG), were developed. These algorithms were compared for several audio application scenarios. The first scenario was sound source localization and separation using a uniform linear array (ULA) and a uniform circular array (UCA). To assess the separation quality, perceptual evaluation of audio quality (PEAQ), perceptual evaluation of speech quality (PESQ), and segmental signal-to-noise ratio (segSNR) were adopted [13,28]. The second scenario was concerned with analyses and syntheses of sound fields. Recently, an integrated array system was developed on the basis of a freefield model for spatial audio recording and reproduction [8,9]. This study extended the previous work to a reverberant environment; a live room was fitted with reflective walls. For the SFA, a 24-element circular microphone array was utilized to encode the sound field based on plane-wave decomposition, whereas in the SFS, a 32-element rectangular loudspeaker array was employed to decode the encoded sound field using three approaches. The third scenario was sound source localization and separation using a random array.

2. Inverse Solution Algorithms

In this section, an array model is given, along with its assumptions. Assume that the sound sources are at the farfield of the microphone array such that sound waves impinging on the array can be regarded as plane waves. In the following array model, time-harmonic dependence $e^{j\omega t}$, $j = \sqrt{-1}$ and ω as angular frequency, is assumed so that the model is essentially formulated in the frequency domain. M microphones and N sources are considered. The array pressure vector can be expressed as [29].

$$\mathbf{p} = \mathbf{As} + \mathbf{v}, \tag{1}$$

where $\mathbf{p} \in \mathbb{C}^M$ is the sound pressure vector received by the microphone, $\mathbf{s} \subset \mathbb{C}^N$ is the source amplitude vector, $\mathbf{v}$ is the noise vector, and $\mathbf{A} \in \mathbb{C}^{M \times N}$ is the steering matrix associated with the sources. Therefore, given the pressure measurement $\mathbf{p}$ and the steering matrix $\mathbf{A}$, solving the problem of Equation (1) for the unknown source amplitude vector $\mathbf{s}$ is a linear inverse problem. Linear inverse problems can be divided into three categories: square systems ($M = N$), overdetermined systems ($M > N$), and underdetermined systems ($M < N$). In the following, solution strategies are presented for these categories.

2.1. Overdetermined or Square Systems

2.1.1. TSVD and Least-Squares Problems

The most basic approach [16–18] to solve linear inverse problems is the least-squares method in which the following cost function is minimized:

$$J = \|\mathbf{e}\|_2^2 = \mathbf{e}^H \mathbf{e}, \tag{2}$$

where $\mathbf{e} = \mathbf{p} - \mathbf{As}$ denotes the error vector and "$\| \cdot \|_2$" denotes the vector 2-norm. If matrix $\mathbf{A}$ is of full-column rank, the least-squares solution can be written as

$$\hat{\mathbf{s}} = [\mathbf{A}^H \mathbf{A}]^{-1} \mathbf{A}^H \mathbf{p}. \tag{3}$$

More generally, by the TSVD of

$$\mathbf{A} = \mathbf{U}\boldsymbol{\Sigma}\mathbf{V}^H$$

$$\hat{\mathbf{s}} = \mathbf{A}^+ \mathbf{p}, \tag{4}$$

where $\boldsymbol{\Sigma}$ represents a diagonal matrix with singular values at its diagonal entries, $\mathbf{U}$ and $\mathbf{V}$ represent unitary matrices, and $\mathbf{A}^+$ represents the pseudoinverse of the matrix $\mathbf{A}$ defined as

$$\mathbf{A}^+ = \mathbf{V}\boldsymbol{\Sigma}^+ \mathbf{U}^H, \tag{5}$$

where $\boldsymbol{\Sigma}^+ = diag\left[\sigma_1^{-1}, \ldots, \sigma_r^{-1}, 0, \ldots, 0\right] \in \mathbb{C}^{N \times M}$, $r = \text{rank}(\mathbf{A})$ [30,31]. Note that the expression of Equation (4) is sufficiently general that it always provides the minimum-norm least-squares solution, even when the matrix $\mathbf{A}$ is rank-deficient. The square of the residual error becomes

$$e_{LS}^2 = \|\mathbf{p} - \mathbf{A}\hat{\mathbf{s}}\|_2^2 = \|(\mathbf{I} - \mathbf{A}\mathbf{A}^+)\mathbf{p}\|_2^2 = \sum_{i=r+1}^{M} \left|\mathbf{u}_i^H \mathbf{p}\right|^2 \tag{6}$$

with $\mathbf{u}_i$ being the ith column of $\mathbf{U}$ and $\mathbf{I}$ being the identity matrix.

In practice, the matrix $\mathbf{A}$ can contain small singular values and can be very ill-conditioned, which leads to numerical instability during the inversion of $\mathbf{A}$. Two common methods to cope with the ill-posedness of the problem are TSVD and TIKR. Briefly, the TSVD method is simply to discard small singular values of the matrix $\mathbf{A}$, whereas TIKR involves attempts to minimize the following cost function [16]:

$$J = \|\mathbf{A}\hat{\mathbf{s}} - \mathbf{p}\|_2^2 + \beta^2 \|\mathbf{s}\|_2^2, \tag{7}$$

where the regularization parameter β weights the residual norm and the solution norm. After some manipulation, we derive with the following optimal solution:

$$\hat{\mathbf{s}} = \left(\mathbf{A}^H \mathbf{A} + \beta^2 \mathbf{I}\right)^{-1} \mathbf{A}^H \mathbf{p} \tag{8}$$

This result can be rewritten in terms of the TSVD of $\mathbf{A}$ as follows:

$$\hat{\mathbf{s}} = \left(\mathbf{A}^H \mathbf{A} + \beta^2 \mathbf{I}\right)^{-1} \mathbf{A}^H \mathbf{p} = \sum_{i=1}^{N} f_i \sigma_i^{-1} \alpha_i \mathbf{v}_i, \tag{9}$$

where $\mathbf{A} = \mathbf{U}\mathbf{S}\mathbf{V}^H = \sum_{i=1}^{N} \sigma_i \mathbf{u}_i \mathbf{v}_i^H$, with $\mathbf{u}_i$ and $\mathbf{v}_i$ being the ith column partition of the matrices $\mathbf{U}$ and $\mathbf{V}$, $\alpha_i = \mathbf{u}_i^H \mathbf{p}$, where

$$f_i(\beta) = \frac{\sigma_i^2}{\sigma_i^2 + \beta^2} = \frac{1}{1 + (\beta/\sigma_i)^2} \tag{10}$$

denotes the filter function.

It can also be shown that the minimum residual vector can be written as

$$\mathbf{p} - \mathbf{A}\hat{\mathbf{s}} = \sum_{i=1}^{N} (1 - f_i)\alpha_i \mathbf{u}_i + \mathbf{r}_\perp \tag{11}$$

where $\mathbf{r}_\perp = (\mathbf{I} - \mathbf{A}\mathbf{A}^+)\mathbf{p} = \sum\limits_{i=M+1}^{M} \alpha_i \mathbf{u}_i$ is the residual vector of the components of $\mathbf{p}$ orthogonal to $\{\mathbf{u}_1 \cdots \mathbf{u}_n\}$. The residual norm can be written as

$$\|\mathbf{p} - \mathbf{A}\hat{\mathbf{s}}\|_2^2 = \sum_{i=1}^{n} (1 - f_i)^2 |\alpha_i|^2 + \sum_{i=n+1}^{m} |\alpha_i|^2 = \sum_{i=1}^{n} (1 - f_i)^2 |\alpha_i|^2 + \|\mathbf{r}_\perp\|_2^2 \tag{12}$$

From Equation (9), the solution 2-norm can be written as

$$\|\hat{\mathbf{s}}\|_2^2 = \sum_{i=1}^{N} f_i^2 \sigma_i^{-2} |\alpha_i|^2 \tag{13}$$

2.1.2. Choice of Regularization Parameter β

In traditional solution strategies for inverse problems, methods are available for choosing regularization parameters, such as Morozov's discrepancy principle, the Generalized Cross-Validation (GCV) method, and the L-curve method [19,20]. The first two methods have been discussed extensively in the literature. Therefore, we subsequently focus on only the L-curve method for brevity.

The L-curve method is widely used for choosing regularization parameters in inverse solutions. In the curve, the solution norm is plotted versus the residual norm by varying the regularization parameter. From Equations (9) and (11), it is straightforward to find the solution norm

$$\|\hat{\mathbf{s}}\|_2^2 = \sum_{i=1}^{N} f_i^2 \sigma_i^{-2} |\alpha_i|^2 \tag{14}$$

and the residual norm

$$\|\mathbf{p} - \mathbf{A}\hat{\mathbf{s}}\|_2^2 = \sum_{i=1}^{N} (1 - f_i)^2 |\alpha_i|^2 + \sum_{i=n+1}^{M} |\mathbf{r}_\perp|^2 \tag{15}$$

Regularization helps to improve robustness against system perturbation and measurement noise. Insights can be gained by writing the solution error as

$$\mathbf{s} - \hat{\mathbf{s}} = \mathbf{s} - \mathbf{A}^\# \mathbf{b} = (\mathbf{s} - \mathbf{A}^\# \overline{\mathbf{p}}) - \mathbf{A}^\# \mathbf{e} = \left[\sum_{i=1}^{N} (1 - f_i) \frac{\mathbf{u}_i^H \overline{\mathbf{p}}}{\sigma_i} \mathbf{v}_i \right] - \sum_{i=1}^{N} f_i \frac{\mathbf{u}_i^H \mathbf{e}}{\sigma_i} \mathbf{v}_i, \tag{16}$$

where $\mathbf{A}^\# = (\mathbf{A}^H \mathbf{A} + \beta^2 \mathbf{I})^{-1} \mathbf{A}^H$, $(\mathbf{s} - \mathbf{A}^\# \overline{\mathbf{p}})$ is the regularization error and $\mathbf{A}^\# \mathbf{e}$ is the perturbation error.

Hence, when $\beta \to 0, f_i \to 1$ for very well-conditioned problems with high signal-to-noise ratio (SNR) measurements, the solution error is dominated by the perturbation error and a few high-order modes are filtered out. In this case, the solution norm is sensitive to the choice of β. The solution tends to be undersmoothed and susceptible to measurement noise. Conversely, when $\beta \to \infty, f_i \to 0$ for very ill-conditioned problems with low SNR measurements, the solution error is dominated by regularization error and numerous high-order modes are filtered out. The solution tends to be oversmoothed and fine details such as resolution are thus lost. In this case, the residual norm is sensitive to the choice of β.

The parameter β acts as a weighting factor between the residual norms and the solution norm. Choosing an appropriate β to strike the balance between these two terms is vital. However, conventional approaches such as GCV and the L-curve method do not always provide satisfactory results in this situation. In this paper, a new method is proposed to facilitate the choice of the regular parameter β for the TIKR method.

Setting the gradient of the cost function of the TIKR method, $J = \|\mathbf{p} - \mathbf{A}\mathbf{s}\|_2^2 + \beta^2 \|\mathbf{s}\|_2^2$, to zero leads to the normal equation

$$(\mathbf{A}^H \mathbf{A} + \beta^2 \mathbf{I}) \mathbf{s} = \mathbf{A}^H \mathbf{p} \tag{17}$$

Without loss of generality, assume $\mathbf{A}$ is of full-column rank. Note that

$$\mathbf{A}^H\mathbf{A} + \beta^2\mathbf{I} = \mathbf{V}\left[\Sigma^H\Sigma + \beta^2\mathbf{I}\right]\mathbf{V}^H = \mathbf{V}diag\left[(\sigma_1^2 + \beta^2),\ldots,(\sigma_N^2 + \beta^2)\right]\mathbf{V}^H$$

which has an effective condition number $\sqrt{(\sigma_1^2 + \beta^2)/(\sigma_N^2 + \beta^2)}$. Therefore, if we want the condition number to be τ after regularization, we must require

$$\tau^2 = \frac{\sigma_1^2 + \beta^2}{\sigma_N^2 + \beta^2} \tag{18}$$

Let κ be the condition number of $\mathbf{A}$; that is, $\kappa = \sigma_1/\sigma_M$. In general, for very ill-conditioned systems, $\kappa \gg \tau \gg 1$, and

$$\beta = \sqrt{\frac{\sigma_1^2 - \tau^2\sigma_M^2}{\tau^2 - 1}} \approx \sqrt{\frac{\sigma_1^2 - \tau^2(\sigma_1^2/\kappa^2)}{\tau^2}} = \sqrt{\frac{\sigma_1^2}{\kappa^2\tau^2}(\kappa^2 - \tau^2)} \approx \frac{\sigma_1}{\tau} \tag{19}$$

Therefore, the regularization parameter β can be chosen to be the maximal singular value σ_1 of $\mathbf{A}$ divided by the regularized condition number τ. For example, one may choose that $\tau = 100$, which causes 40-dB potential loss of SNR in the inverse solution. Normally, $\mathbf{A}$ tends to be ill-conditioned at low frequencies. Choosing a frequency-independent β may suffice for the worst-case scenario. Thus, the regularization parameter β is chosen according to the maximal condition number at a selected low frequency (100 Hz is selected in the following simulation). Next, a coarse search is performed by varying β in orders of 10. A potential interval in which an optimal β may exist is located by observing how an objective function, such as Perceptual Evaluation of Speech Quality (PESQ) [29], varies with β. Finally, a fine-grained search is performed in the potential interval by using the Golden Section Search (GSS) algorithm [21].

GSS is an optimization technique suited for finding the extremum of a unimodal function. It is a simple bracketing method that does not require evaluation of the gradient of the cost function. In each search, a probe point must be selected within the left and right brackets according to the golden ratio. The golden ratio can be defined as

$$\varphi = \frac{1 + \sqrt{5}}{2} = 1.618 \text{ , } \varphi \text{ is the golden ratio}$$

Let $f(\beta)$ be the objective function (PESQ in our case) for which we wish to find the optimal β that maximizes $f(\beta)$. First, $f(\beta)$ has been evaluated at two points, β_1 and β_3. The maximizing value is between β_1 and β_3. The golden ratio can be used to find β_2 and β_4. β_2 and β_4 can be written as

$$\beta_2 = \beta_3 - (\beta_3 - \beta_1)/\varphi, \beta_4 = \beta_1 + (\beta_3 - \beta_1)/\varphi$$

If $f(\beta_2)$ is larger than $f(\beta_4)$, a maximum clearly lies in the interval between β_1 and β_4. Therefore, the new β_3 is equal to β_4. If $f(\beta_2)$ is smaller than $f(\beta_4)$, a maximum lies in the interval between β_2 and β_3. Therefore, the new β_1 is equal to β_2. Figure 1 shows the Schematic of golden section search. The process is repeated until the gap between β_2 and β_4 is small. The iteration process stops when the beta converges within a prespecified tolerance window (0.0001 in our case). The optimal beta β_{opt} can be written as

$$\beta_{opt} = (\beta_2 + \beta_4)/2$$

The algorithm is summarized as the following pseudocode:

$$\beta_2 = \beta_3 - (\beta_3 - \beta_1)/\varphi$$
$$\beta_4 = \beta_1 + (\beta_3 - \beta_1)/\varphi$$
$$while \ |\beta_2 - \beta_4| > \varepsilon$$
$$\quad if \ f(\beta_4) < f(\beta_2)$$
$$\quad\quad \beta_3 = \beta_4$$
$$\quad else$$
$$\quad\quad \beta_1 = \beta_2$$
$$end$$
$$\beta_{opt} = (\beta_2 + \beta_4)/2$$

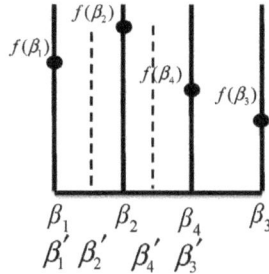

Figure 1. Schematic of golden section search. If $f(\beta_2)$ is higher than $f(\beta_4)$, β_3' is equal to β_4 in the next iteration.

Therefore, this study developed a procedure for choosing optimal regularization parameter beta. The procedure involves four steps as follows:

- Step 1. Select τ according to a condition number threshold.
- Step 2. Select a constant $\beta = \sigma_1/\tau$ as an initial guess. For a frequency-domain design, it may be necessary to choose a frequency-independent β for the worst-case scenario.
- Step 3. Perform a coarse search by running the simulation forward and backward with 10 s powers of β. Locate a potential interval in which an optimal β may exist by observing the trend of an objective function, such as PESQ, with respect to β.
- Step 4. Perform a fine search by using optimization methods such as the GSS algorithm to find the optimal regularization parameter β.

2.2. Underdetermined Systems

In this section, algorithms are presented for underdetermined problems, where the number of microphones (M) is lower than the number of potential sources (N). In this case, the solution is generally not unique unless we impose constraints. Although a pseudoinverse gives a unique minimum-norm least-squares solution, the resolution of the solution is generally not favorable because the solution error tends to be evenly distributed among all entries. Instead, we impose sparsity as the constraint to limit the cardinality (nonzero entries) of the solutions in the study, which suggests that pruning procedures of some sort must be incorporated into the iteration process.

2.2.1. CVX Algorithms

An underdetermined problem with sparse solutions can be written as the following CS problem:

$$\min_{\mathbf{s}} \|\mathbf{s}\|_1 \quad st. \quad \|\mathbf{As} - \mathbf{p}\|_2 \le \varepsilon, \tag{20}$$

where $\| \cdot \|_1$ denotes the vector 1-norm. Numerous methods are available for solving this constrained optimization problem [22,23]. Suppose that the noise energy is constrained within a threshold ε that can be selected with reference to the aforementioned least-squares solution. This problem can be solved numerically by CVX. Freeware was adopted to conduct CVX in this study [24–26].

2.2.2. Iterative Approaches

In the following, we apply four iterative algorithms to solve underdetermined problems. The first method is Focal Underdetermined System Solver (FOCUSS) [27], which is an iterative technique well suited for finding sparse solutions to underdetermined linear systems. The algorithm has two integral parts: a low-resolution initial estimate of the real signal and the iteration process that refines the initial solution to the final localized solution. Because the system is underdetermined, the sensors are more numerous than the sources. In this case, we assume our dictionary contains 36 sources. These sources are located at 5° intervals from the x label. Actually, this case has only three sources. Therefore, if the result is perfect, our final solution has only three nonzero solutions.

The FOCUSS algorithm can be summarized in three steps,

$$\mathbf{W}_k = [\text{diag}(\mathbf{s}_{k-1})] \tag{21}$$

$$\mathbf{q}_k = (\mathbf{AW}_k)^+ \mathbf{p} \tag{22}$$

$$\mathbf{s}_k = \mathbf{W}_k \mathbf{q}_k \tag{23}$$

In Equation (21), $[\text{diag}(\mathbf{s}_{k-1})]$ converts the vector $\mathbf{s}_{k-1}$ into a diagonal weight matrix. The TIKR solution is used as the initial condition. Similar to other fixed-point iteration methods, the algorithms converge within finite numbers of iterations to the sparse solution with appropriate initial conditions.

The large term in the weighting reduces the 2-norm of $\mathbf{q}$

$$\|\mathbf{W}^+ \mathbf{s}_p\|^2 = \|\mathbf{q}\|^2 = \sum_{i=1,\omega_i \neq 0}^{n} \left(\frac{\mathbf{s}_{pi}}{w_i}\right)^2 \tag{24}$$

The relatively large entries in $\mathbf{W}$ reduce the contributions of the corresponding elements of $\mathbf{s}_p$ to the cost, and the solution is nonzero in the source direction. The pseudoinverse in Equation (22) can also be implemented by using the TIKR method. Therefore, it can also be written as

$$\mathbf{q}_k = (\mathbf{W}_{k-1}^H \mathbf{A}^H \mathbf{AW}_{k-1} + \beta^2 \mathbf{I})^{-1} \mathbf{W}_{k-1}^H \mathbf{A}^H \mathbf{p} \tag{25}$$

The FOCUSS-TIKR method is robust to noise because of the regularization parameter β. The iteration process stops when the solution converges within a prespecified tolerance window (0.0001 in our case).

2.2.3. Iterative Approaches: Promote Sparsity by Pruning

CVX algorithms can solve CS problems, but these algorithms are known to be very computationally expensive, which prevents their use in real-time processing. To address this challenge, several iterative approaches are proposed as follows.

In these iterative techniques, the quadratic residual function

$$J(\mathbf{s}) = \frac{1}{2}\|\mathbf{As} - \mathbf{p}\|_2^2 = \frac{1}{2}\left[\mathbf{s}^H \mathbf{A}^H \mathbf{As} - \mathbf{p}^H \mathbf{As} - \mathbf{s}^H \mathbf{A}^H \mathbf{p} + \mathbf{p}^H \mathbf{p}\right] \tag{26}$$

is to be minimized. The key step that executes the "compressive sampling" is a pruning process that must be incorporated into each iteration to promote sparsity, as illustrated in Figure 2. First, several main peaks as well as sidelobes may appear in the source diagram. We reset all elements in the source vector $\mathbf{s}$ below a prespecified threshold ($s_{\max} - D$) to zero. D is initially set to a very small value D_0; it is then increased incrementally by ΔD in each iteration, typically by the same ΔD in every step.

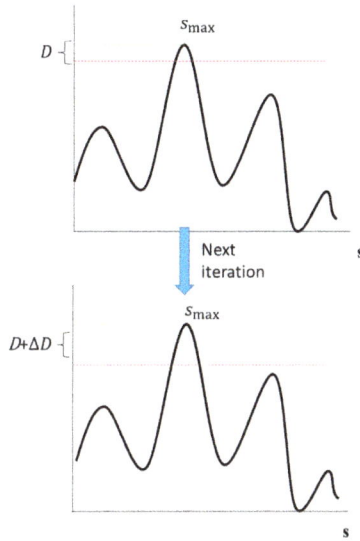

Figure 2. Pruning process to promote sparsity of inverse solution.

The iterative pruning process is summarized with a flowchart in Figure 3. The stopping condition is

$$D > D_{\max} \text{ or } \|\nabla J(\mathbf{s})\|_2^2 \leq 0.001 \tag{27}$$

in which $\nabla J(\mathbf{s})$ is the gradient of the quadratic residual function $J(\mathbf{s})$. Three approaches were employed in this study to update the solutions in the iterative CS algorithms such that each quadratic residual function $J(\mathbf{s})$ is minimized.

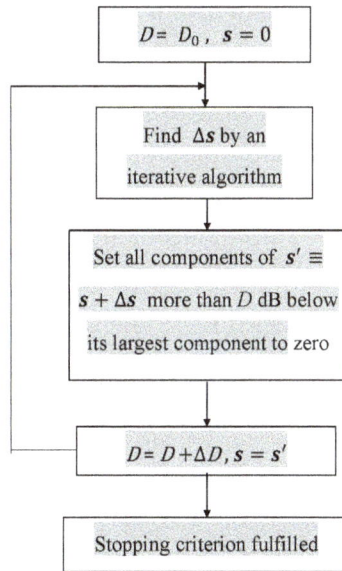

Figure 3. Flowchart of iterative compressive sampling algorithms (adapted from reference [5]).

Steepest Decent (SD) Method

The SD algorithm is based on the notion that the search direction at each iteration is the negative gradient of the cost function in Equation (26) for minimization problems. The gradient vector of the quadratic residual function is given by

$$\mathbf{w} = -\nabla J(\mathbf{s}) = -(\mathbf{A}^H \mathbf{A} \mathbf{s} - \mathbf{A}^H \mathbf{p}) = \mathbf{A}^H \mathbf{r}, \tag{28}$$

where $\mathbf{r} = \mathbf{p} - \mathbf{As}$ is the residual vector.

The new solution $\mathbf{s}'$ is updated as

$$\mathbf{s}' = \mathbf{s} + \Delta\mathbf{s} = \mathbf{s} + \mu\mathbf{w} \tag{29}$$

where μ is the step size. To determine the optimal step size μ, let the vector $\mathbf{g}$ be

$$\mathbf{g} = \mathbf{Aw} \tag{30}$$

It can be shown after some algebraic manipulations that

$$F(\mathbf{s}') = \frac{1}{2}(\mu^2 \mathbf{g}^H \mathbf{g} - 2\mu \mathbf{g}^H \mathbf{r} + \mathbf{r}^H \mathbf{r}) \tag{31}$$

From Equation (31), the step size μ to minimize $F(\mathbf{s}')$ along the direction $\mathbf{w}$ can be found by setting the derivative of $F(\mathbf{s}')$ with respect to μ to zero. Consequently, we obtain

$$\mu = \frac{\|\mathbf{w}\|_2^2}{\|\mathbf{g}\|_2^2} \tag{32}$$

Newton's Method

The NT method is also an iterative method gradient search. Recall the gradient of the cost function is

$$\nabla F(\mathbf{s}) = \mathbf{v} = (\mathbf{A}^H \mathbf{A} \mathbf{s} - \mathbf{A}^H \mathbf{p}) = -\mathbf{A}^H(\mathbf{p} - \mathbf{As}) = -\mathbf{A}^H \mathbf{r}, \tag{33}$$

where $\mathbf{r}$ is as defined before. Setting this gradient to zero leads to the optimal solution

$$\mathbf{s}' = (\mathbf{A}^H \mathbf{A})^{-1} \mathbf{A}^H \mathbf{p} = \mathbf{A}^+ \mathbf{p}, \tag{34}$$

where $\mathbf{A}^+ = (\mathbf{A}^H \mathbf{A})^{-1} \mathbf{A}^H$. Next, multiplying $(\mathbf{A}^H \mathbf{A})^{-1}$ on both sides of $\mathbf{v}$ yields

$$(\mathbf{A}^H \mathbf{A})^{-1} \mathbf{v} = (\mathbf{A}^H \mathbf{A})^{-1}(\mathbf{A}^H \mathbf{A} \mathbf{s} - \mathbf{A}^H \mathbf{p}) = \mathbf{s} - (\mathbf{A}^H \mathbf{A})^{-1} \mathbf{A}^H \mathbf{p} = \mathbf{s} - \mathbf{A}^+ \mathbf{p}$$
$$= -(\mathbf{A}^H \mathbf{A})^{-1} \mathbf{A}^H \mathbf{r} = -\mathbf{A}^+ \mathbf{r}$$

or

$$-\mathbf{A}^+ \mathbf{r} = \mathbf{s} - \mathbf{A}^+ \mathbf{p} = \mathbf{s} - \mathbf{s}'$$

Hence, the solution can be updated as

$$\mathbf{s}' = \mathbf{s} + \mathbf{A}^+ \mathbf{r} \tag{35}$$

In practical implementation, the pseudoinverse $\mathbf{A}^+$ is usually of the form of TIKR, namely $\mathbf{A}^+ = (\mathbf{A}^H \mathbf{A} + \beta^2 \mathbf{I})^{-1} \mathbf{A}^H$, because $\mathbf{A}^H \mathbf{A}$ is singular for underdetermined problems. As a refinement of the algorithm, a step size μ can be used to rewrite the update equation as

$$\mathbf{s}' = \mathbf{s} + \mu \mathbf{A}^+ \mathbf{r} \tag{36}$$

Conjugate Gradient (CG) Method

The CG method is an iterative algorithm well suited for the numerical solution of systems of linear equations, associated with symmetric and positive-definite matrices. Instead of the negative gradient used in the SD method, which occasionally causes zigzag convergence, the search direction of the CG method is a linear combination of the current negative gradient and the previous search direction. Development of the CG algorithm is based on nested Krylov subspaces. For details, we refer the interested readers to Ref. [32–34]. For brevity, we only summarize the CG algorithm with the following pseudocode:

$$\mathbf{r}_0 = \mathbf{p} - \mathbf{A}\mathbf{s}_0$$
$$\mathbf{p}_0 = \mathbf{r}_0$$
$$k = 0$$
$$for$$
$$\alpha_k = \frac{\mathbf{r}_k^T \mathbf{r}_k}{\mathbf{p}_k^T \mathbf{A}\mathbf{p}_k}$$
$$\mathbf{s}_{k+1} = \mathbf{s}_k + \alpha_k \mathbf{p}_k$$
$$\mathbf{r}_{k+1} = \mathbf{r}_k - \alpha_k \mathbf{A}\mathbf{p}_k$$
$$\beta_k = \frac{\mathbf{r}_{k+1}^T \mathbf{r}_{k+1}}{\mathbf{r}_k^T \mathbf{r}_k}$$
$$\mathbf{p}_{k+1} = \mathbf{r}_{k+1} + \beta_k \mathbf{p}_k$$
$$k = k + 1$$
$$end$$

To put the system of equations in Equation (1) into a more tractable form, we multiply by $\mathbf{A}^H$ on both sides, which leads to the normal equations

$$\mathbf{A}^H \mathbf{p} = \mathbf{A}^H \mathbf{A}\mathbf{s}_p \tag{37}$$

This equation is equivalent to finding the vector $\mathbf{s}$ for which the gradient of $F(\mathbf{s})$ equals zero.

$$\nabla F(\mathbf{s}) = -\mathbf{A}^H (\mathbf{p} - \mathbf{A}\mathbf{s}) = 0 \tag{38}$$

3. Comparison of Algorithms

This section presents the application of the preceding algorithms to acoustic source localization and separation problems through numerical simulation. These algorithms were compared for three example problems, with the aid of a uniform linear array (ULA) and a random array. In addition, an inverse solution involved in sound field analysis (SFA) and sound field synthesis (SFS) in spatial audio was investigated. Microphone data are synthetic and generated by the model of Equation (1).

3.1. Uniform Linear Array

In the numerical simulation shown in Figure 4, 10-microphone ULA was utilized to separate the signals emitted by three sources. The sources were located at the far field such that the plane wave assumption was valid. The spacing between adjacent microphones was 10 cm. This simulated underdetermined system contained 36 sources as our dictionary. We separated the signals and localized the signals in one stage. TIKR, FOCUSS, and CS-CVX algorithms were used to solve these inverse problems. Figure 5 shows the condition numbers of different frequencies. The problem is ill-conditioned at low frequencies. Source localization results are shown in Figure 6.

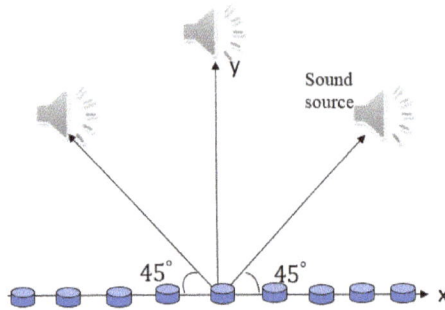

Figure 4. Numerical simulation of 10-microphone uniform linear array utilized to separate the signals emitted by three sources. The first source, located at $\theta = 45^{\circ}$, was broadcasting a male speech signal. The second source, located at $\theta = 90^{\circ}$, was broadcasting a female speech signal. The third source, located at $\theta = 135^{\circ}$, was broadcasting a music signal.

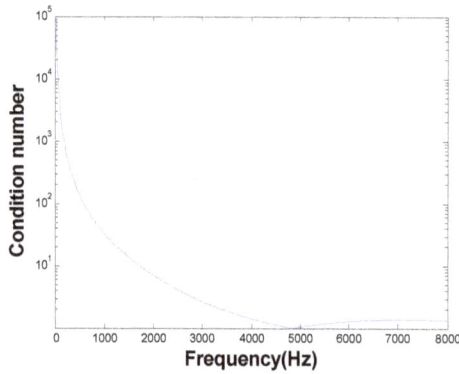

Figure 5. Condition number versus frequency.

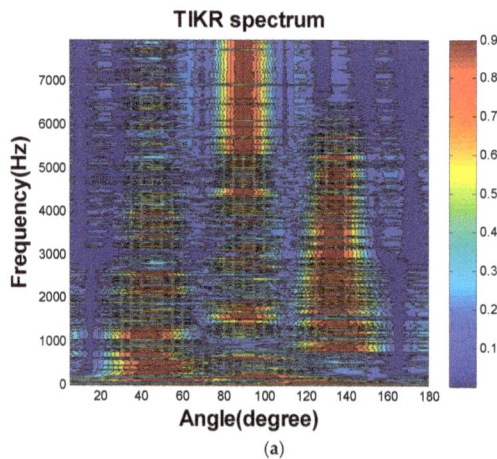

(a)

Figure 6. *Cont.*

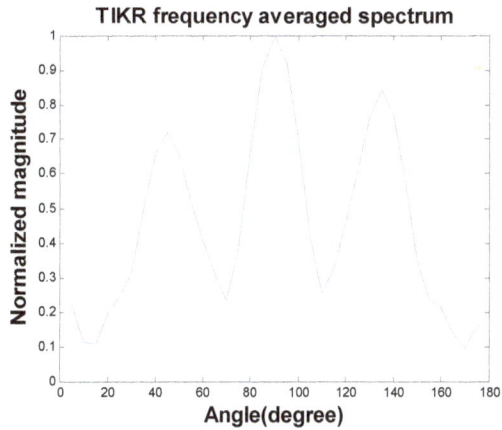

(b)

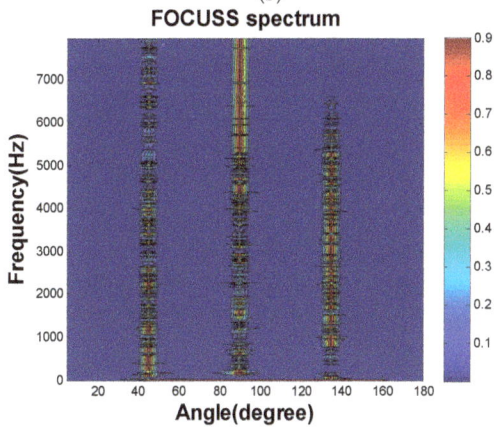

(c)

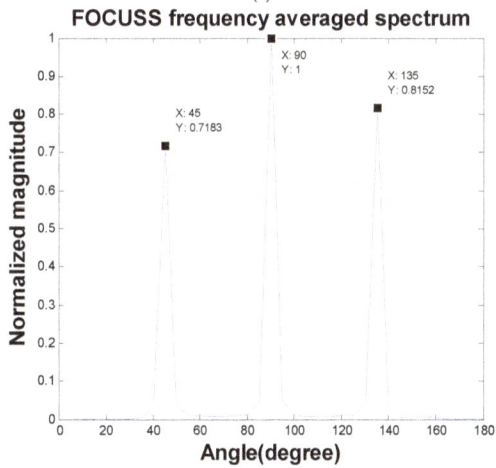

(d)

Figure 6. *Cont.*

(e)

(f)

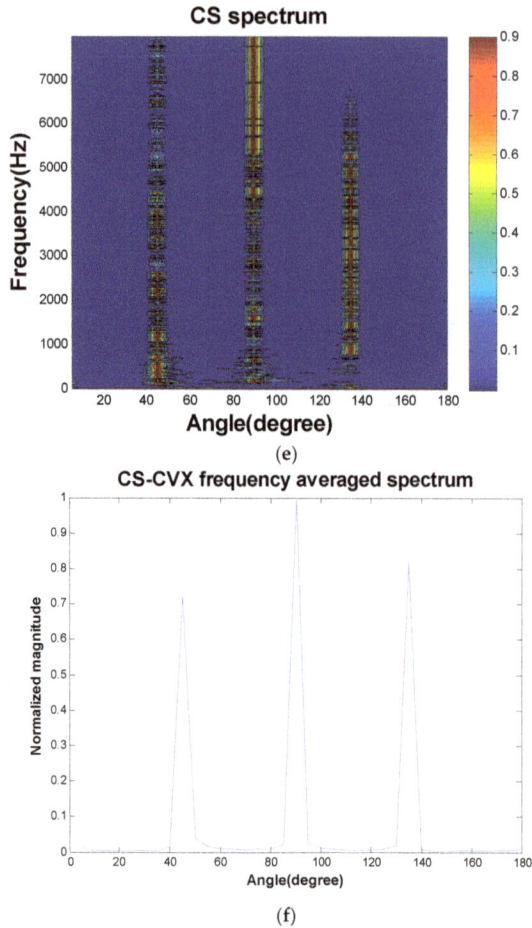

Figure 6. Source localization results obtained using the focal underdetermined system solver (FOCUSS) algorithm and a uniform linear array. (**a**) Tikhonov regularization (TIKR) spectrum (**b**) TIKR averaged spectrum (**c**) FOCUSS spectrum (**d**) FOCUSS frequency-averaged spectrum. (**e**) Compressive sampling-convex optimization (CS-CVX) spectrum (**f**) CS-CVX frequency-averaged spectrum.

The separation results obtained using TIKR, FOCUSS, and CS-CVX are summarized in Table 1. PESQ is an objective test measure for speech quality evaluation. It is a full-reference algorithm and analyzes the speech signal sample-by-sample after a temporal alignment of corresponding excerpts of reference and test signal. The mean opinion score (MOS) is calculated on the basis of PESQ ranging from 1 to 5; MOS signifies the difference in speech quality between the clean and the separated signals, which is affected by separation performance and signal distortion. The segmental SNR (segSNR) is defined as

$$SNR_{seg} = \frac{1}{N} \sum_{k=1}^{N} 10 \log_{10} \left[\frac{\sum\limits_{n \in frame_k} |\mathbf{s}(n)|^2}{\sum\limits_{n \in frame_k} |\hat{\mathbf{s}}(n) - \mathbf{s}(n)|^2} \right] \tag{39}$$

The segSNR correlates with the effect of noise reduction. The FOCUSS-PINV algorithm was observed to achieve the highest score in PESQ and segSNR (Table 1), although it required more computation time than TIKR and FOCUSS-TIKR.

Table 1. Separation performance of the Tikhonov regularization (TIKR) and focal underdetermined system solver (FOCUSS) methods for three sources in the underdetermined system.

Methods		TIKR	FOCUSS-PINV	FOCUSS-TIKR	CS-CVX
	Source 1	2.034	3.966	3.350	3.251
PESQ	Source 2	1.696	3.146	2.879	2.783
	Source 3	1.912	3.818	3.394	3.347
segSNR	Source 1	0.554	11.92	11.27	8.817
CPU time (s)		53	810	487	16302

In our previous simulation, we simulated microphones that had no noise; therefore, our regularization parameter was very close to zero. In the current simulation, our microphones did have white noise with a magnitude equal to the magnitude of the microphone signals divided by 100. Therefore, the potential loss of SNR was 40 dB.

The regularization parameter is chosen by the maximal singular σ_1 value dividing 100 in 100 Hz. In our case, the maximal singular value was equal to 5.32. Next, a coarse search was performed by varying β in orders of 10 and using the GSS algorithm to find the optimal regularization parameter. In this case, the optimal regularization parameter was 0.0174, and the FOCUSS-PINV of the robustness to the noise was very ineffective because PINV artifacts caused discontinuities in regularization. CS-CVX was sensitive to the noise, and the segSNR and PESQ achieved lower scores than FOCUSS-TIKR did, as listed in Tables 2 and 3 (20dB). Therefore, noise was present, and FOCUSS-TIKR was the best choice in this case. It outperformed PESQ and segSNR.

Table 2. Separation performance of the Tikhonov regularization (TIKR), focal underdetermined system solver (FOCUSS), and compressive sampling-convex optimization (CS-CVX) methods for three sources with additive white noises (40dB SNR).

Methods		TIKR	FOCUSS-PINV	FOCUSS-TIKR	CS-CVX
	Source 1	2.135	1.170	2.941	2.412
PESQ	Source 2	1.810	1.121	2.234	1.321
	Source 3	1.723	1.168	2.841	2.241
segSNR	Source 1	0.52	−10.00	1.558	0.946
CPU time (s)		58	817	497	16431

Table 3. Separation performance of the TIKR, FOCUSS and CS-CVX methods for three sources with additive white noises (20dB SNR).

Methods		TIKR	FOCUSS-PINV	FOCUSS-TIKR	CS-cvx
	Source 1	1.512	1.168	2.512	1.436
PESQ	Source 2	1.401	1.118	2.012	1.712
	Source 3	1.489	1.044	2.487	1.324
segSNR	Source 1	−0.12	−13.11	0.12	−1.112
CPU time (s)		57	815	494	16421

3.2. Random Array

A simulation was conducted for localization and separation of two point sources located at $(0, 0, -1 \text{ m})$ and $(0.8, 0.3, -1 \text{ m})$, both of which were emitting clean speech signals, as illustrated in Figure 7. A 30-element random array with aperture dimension 0.48 m × 0.4 m situated at z = 0 was utilized to capture the signals emitted by these two sources. To set up the propagation matrix, 100 (10 × 10) equivalent sources were distributed on the image plane located 1 m away from the array. Figure 8 shows the condition numbers of different frequencies. The problem is ill-conditioned at low frequencies.

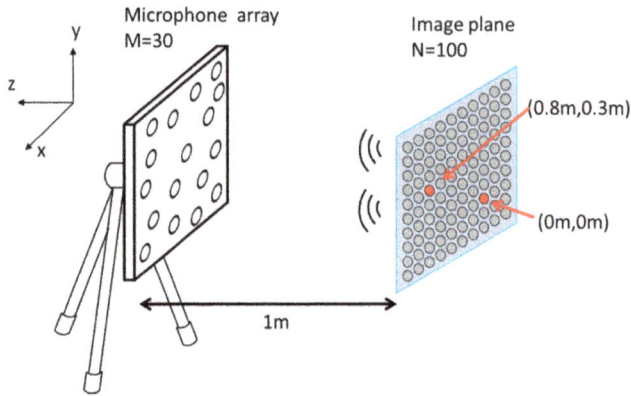

Figure 7. Arrangement for simulation of localization and separation of two sources.

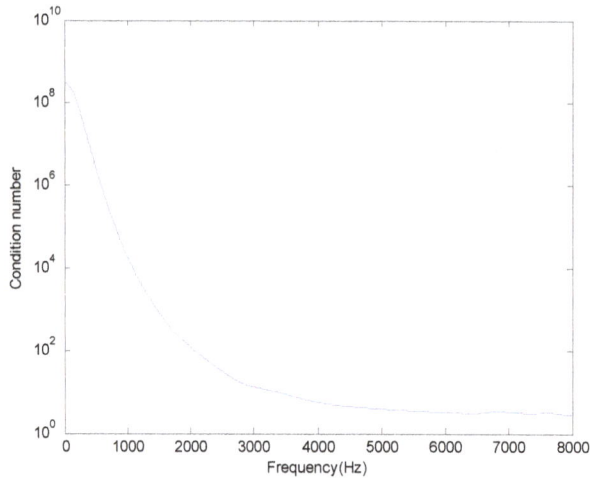

Figure 8. Condition number versus frequency.

Figure 9a–f show the source localization results obtained using six approaches. Two sources were correctly located on the noise map with varying degrees of resolution by all methods. A conventional method delay and sum (DAS) method (a) gave the poorest resolution, whereas the CS-CVX method provided the highest resolution. The SD method (d) and CG method (f) were acceptable but did not perform quite as well as the CS-CVX method. The NT method (e) yielded accurate source locations with a slightly increased sidelobe level, but it was the most computationally efficient (Table 2).

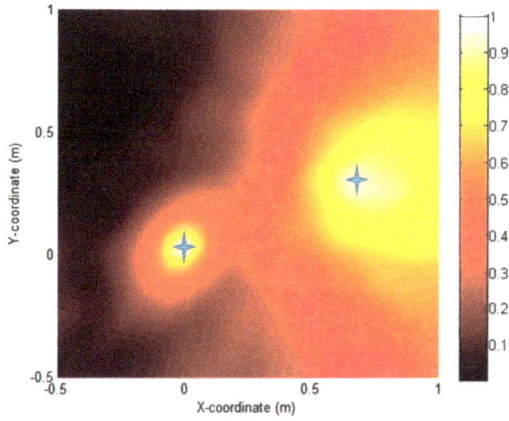

(a)

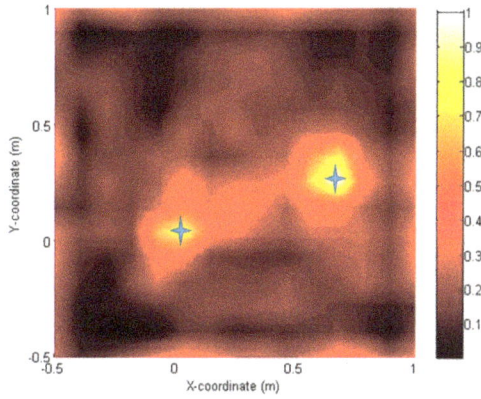

(b)

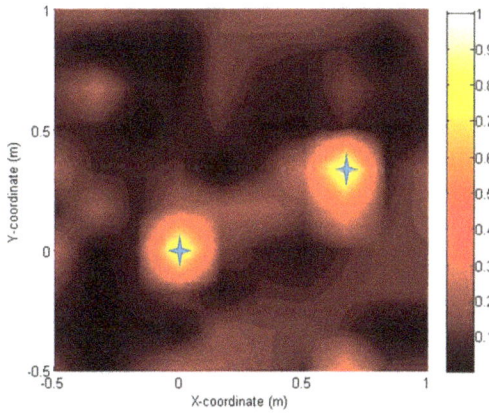

(c)

Figure 9. *Cont.*

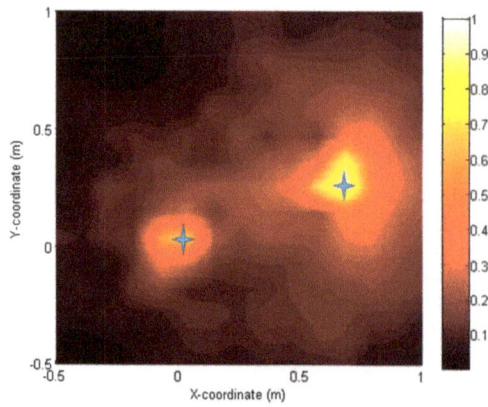

(d)

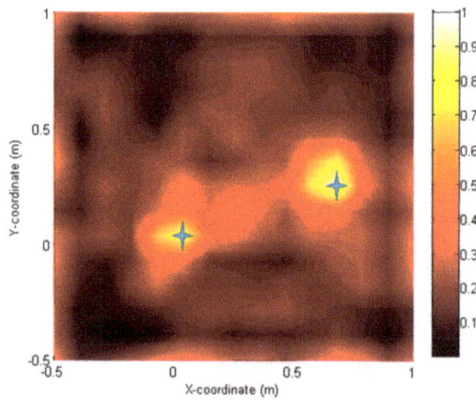

(e)

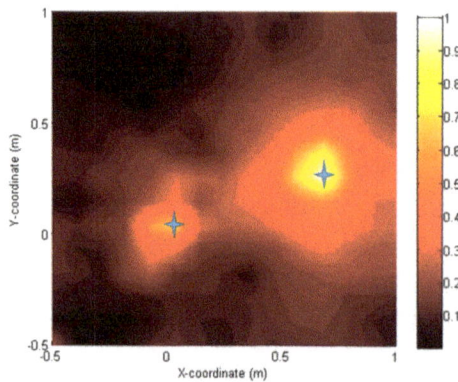

(f)

Figure 9. Localization results of two point sources. (**a**) Delay and sum algorithm, (**b**) Tikhonov regularization algorithm, (**c**) compressive sampling-convex optimization algorithm, (**d**) steepest descent algorithm, (**e**) Newton's algorithm, and (**f**) conjugate gradient algorithm.

Table 4 presents a comparison of the separation results obtained using five methods. CS-CVX displayed the highest scores for PESQ and segSNR, despite being extremely time-consuming. Iterative CS approaches were determined to be far more computationally efficient than the CS-CVX method. The CG method attained the highest PESQ, but the lowest segSNR. This suggests that the favorable separation performance of the CG method comes at the price of signal distortion. The SD and NT methods demonstrated acceptable PESQ and high segSNR. Although signals were not perfectly separated by using these two methods, the incurred distortion was minor. In general, methods present a trade-off between separation performance and signal distortion. Table 5 shows the separation results with additive noise (SNR = 28 dB). All the methods were observed to suffer from the interference of noise; consequently, the values of PESQ and segSNR were notably low. However, all the methods were determined to be robust to noise. The present study also considered mismatches between equivalent sources (dictionaries) and real sources. Table 6 shows the separation results of the NT method with different levels of mismatch. Mismatch means that the real source is not precisely on the deployed source location (dictionary). Extreme mismatch means the real source is exactly on the center of four nearest the deployed source location. More descriptions are added to the revised manuscript. Unless the source was just at the center of the near dictionaries, the separation performance was high and not influenced by noise.

Table 4. Separation performance of five algorithms for two speech sources.

Methods		TIKR	CS-CVX	SD	NT	CG
PESQ	Source 1	1.99	3.12	2.39	2.48	2.76
	Source 2	2.60	3.31	2.77	2.83	3.13
segSNR	Source 1	2.15	7.54	5.08	6.04	0.50
	Source 2	2.73	8.24	7.02	7.26	1.53
CPU time (s)		201	31065	377	296	386

Table 5. Separation performance of five algorithms for two speech sources with additive noise.

Methods		TIKR	CS-CVX	SD	NT	CG
PESQ	Source 1	1.82	2.83	2.22	2.38	2.59
	Source 2	2.12	3.20	2.53	2.53	2.88
segSNR	Source 1	1.40	2.11	2.03	4.13	−0.50
	Source 2	2.30	6.51	2.09	1.72	1.23
CPU time (s)		211	31348	358	290	307

Table 6. Separation performance of Newton's algorithm for two speech sources with different mismatch conditions.

Methods		without Mismatch	with Mismatch	Extreme Mismatch
PESQ	Source 1	2.52	2.48	2.14
	Source 2	2.81	2.83	1.70
segSNR	Source 1	6.90	6.04	−0.40
	Source 2	6.33	7.26	0.18

3.3. SFA and SFS

Depending on the sparsity of the sound sources, the SFA stage can be implemented in several manners [35–38]. For the sparse-source scenario, a two-stage algorithm is utilized; the source bearings are estimated using the minimum power distortionless response (MPDR) [7] and the associated amplitudes of plane waves are estimated using the TIKR algorithm. For the nonsparse-source scenario, a one-stage algorithm based on the CS-CVX algorithm or the FOCUSS algorithm is employed.

The SFS stage is carried out using a loudspeaker array to reconstruct the sound field with the source bearing and amplitude obtained in the SFA stage. Pressure matching was employed for the SFS purpose in this study by sampling a large number of virtual control points in the interior area surrounded by the loudspeakers. The pressure matching procedure can be described as the following optimization problem:

$$\min_{\mathbf{s}_s(\omega)} \| \mathbf{B}(\omega)\mathbf{s}_p(\omega) - \mathbf{H}(\omega)\mathbf{s}_s(\omega) \|, \tag{40}$$

where $\mathbf{s}_p(\omega) = \begin{bmatrix} s_1(\omega) & \cdots & s_P(\omega) \end{bmatrix}^T$ is the amplitude vector of the Pth primary plane-wave component, $\mathbf{s}_s(\omega) = \begin{bmatrix} s_1(\omega) & \cdots & s_L(\omega) \end{bmatrix}^T$ denotes the amplitude vector of the input signals to the L secondary loudspeaker sources, $\mathbf{H}(\omega) \in \mathbb{C}^{K \times L}$ denotes the room response matrix, and $\mathbf{b}_d = \begin{bmatrix} e^{-j\mathbf{k}_d \cdot \mathbf{y}_1} & \cdots & e^{-j\mathbf{k}_d \cdot \mathbf{y}_K} \end{bmatrix}^T$ is the steering vector for the dth primary plane-wave component to the nth control point, $\mathbf{y}_n, n = 1, \ldots, K$. $\mathbf{B}(\omega) = \begin{bmatrix} \mathbf{b}_1 & \cdots & \mathbf{b}_d \end{bmatrix} \in \mathbb{C}^{K \times P}$ is the steering matrix from the plane-wave components obtained in the preceding SFA stage to the control points. Therefore, the optimal solution can be written as

$$\mathbf{s}_s(\omega) = \mathbf{H}^{\#}(\omega)\mathbf{B}(\omega)\mathbf{s}_p(\omega), \tag{41}$$

where "#" symbolizes some type of inverse operation on the matrix $\mathbf{H}(\omega)$. In this study, TIKR was utilized to calculate the input signal amplitudes to the secondary sources. GSS can be used to find the optimal regularization parameter.

Experiments were conducted to validate the proposed audio analysis and synthesis system. In the SFA stage, a 24-element circular microphone array with a radius of 12 cm was utilized to capture and parameterize the sound field in an anechoic chamber (the recording room), as illustrated in Figure 10. In the SFS stage, a rectangular, 32-loudspeaker array was employed to reproduce in a live room (the reproduction room) the sound field previously encoded in the SFA stage. The walls of the room were lined with acoustically reflective boards (Figure 11).

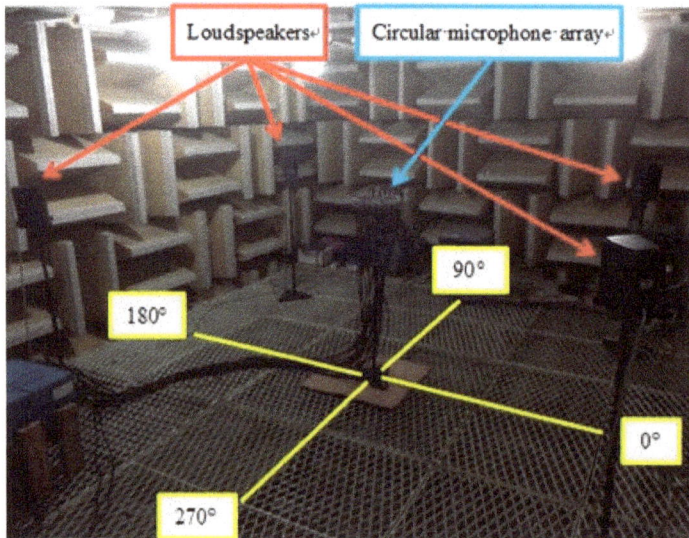

Figure 10. Sound field analysis experimental arrangement in a 5.4 m × 3.5 m × 2 m anechoic room.

Figure 11. Sound field synthesis experimental arrangement in a 3.6 m × 3.6 m × 2 m live room fitted with reflective walls.

To process microphone output and loudspeaker input signals, multichannel analog-to-digital converters (M-32 AD) and digital-to-analog converters (M-32 DA) (RME, Haimhausen, Germany) were used with a sampling frequency of 16 kHz.

An audio codec system involves three inverse problems, namely the SFA stage, room response modeling, and the SFS stage. The condition numbers are plotted against the frequencies of three steering matrices in Figure 12a–c. Figure 12c indicates that the ill-posedness encountered in the room response modeling procedures must be addressed, with the aid of appropriate regularization methods. Large regularization parameters can increase the robustness of the inverse problem. In this study, we set the regularization parameter to 10.

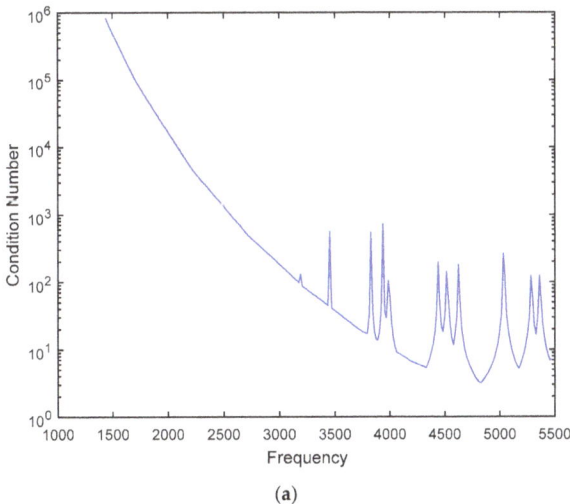

(a)

Figure 12. *Cont.*

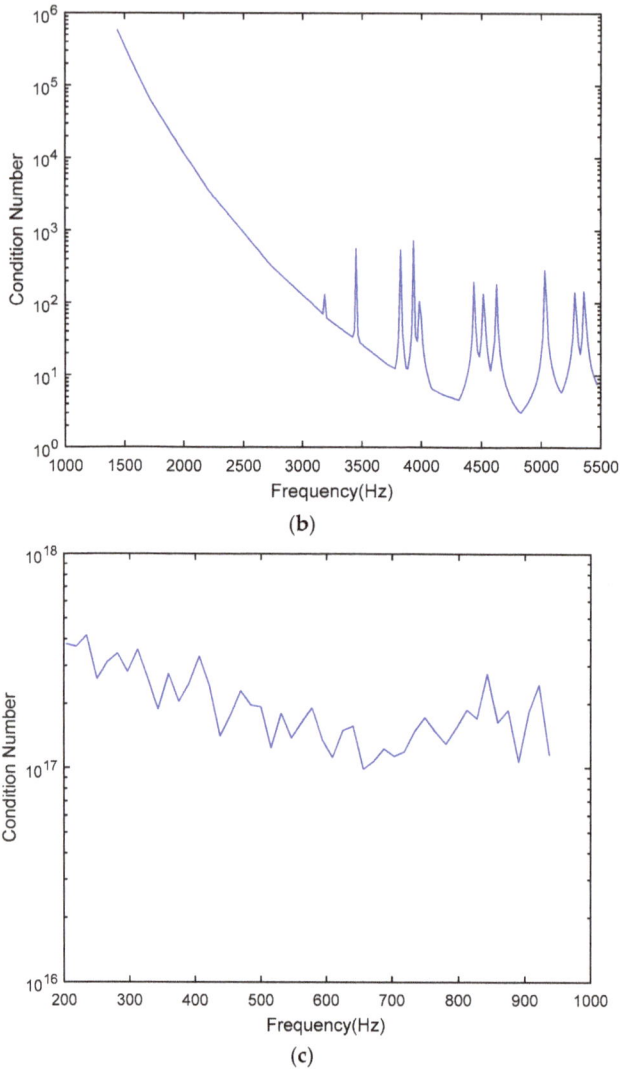

Figure 12. Plots of condition numbers versus frequencies of each steering matrix in three inverse problems: (**a**) sound field analysis stage; (**b**) room response modeling; and (**c**) sound field synthesis stage.

In the SFA experiment, loudspeaker sources positioned at the angles $\theta = 60°, 240°$ played two 10-s speech clips. After recording the source by CMA, we used three algorithms to extract the source signals. First, we applied the two-stage MPDR and TIKR algorithms. The MPDR spectrum is plotted as a function of angle and frequency in Figure 13a. The resulting frequency-averaged and normalized MPDR spectrum is illustrated in Figure 13b, which peaks at the angles $\theta = 60°, 240°$ as desired. The results show that the source was accurately localized using MPDR. Next, the source signals were extracted using the TIKR algorithm. We also applied one-stage CS algorithms and one-stage FOCUSS algorithms, which located sources and separated their amplitudes in a single calculation.

(a)

(b)

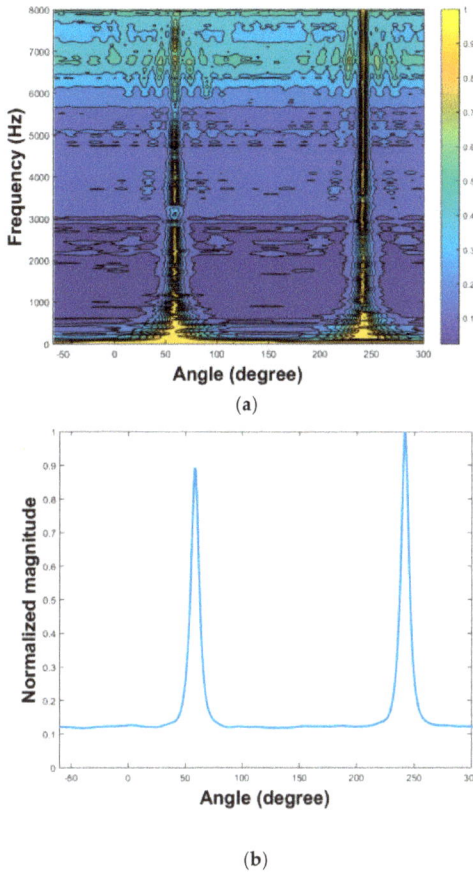

Figure 13. Localization of one music source signal with sampling rate 16 kHz, located at the angles 60°, 240° in an anechoic chamber. By using a uniform circular array with a radius of 12 cm, the source direction can be identified by the peak in the angular spectrum. (a) Minimum power distortionless response (MPDR) spectrum plotted versus angle and frequency. (b) Frequency-averaged and normalized MPDR spectrum.

The signals extracted using different methods were evaluated by using the MOS of the PESQ test. Results confirmed that the TIKR performed well in signal separation with satisfactory audio quality. The results are summarized in Table 7.

Table 7. Mean opinion score of perceptual evaluation of speech quality for source signal separation at the angles 60°, 240° with speech signals using Tikhonov regularization (TIKR), compressive sampling-convex optimization (CS-CVX), and focal underdetermined system solver (FOCUSS).

Methods		Two-Stage TIKR	One-Stagecs-CVX	One-Stagefocuss
PESQ	Source 1	2.84	3.11	1.56
	Source 2	2.79	2.99	1.61
segSNR	Source 1	17.40	20.74	13.54
	Source 2	16.59	17.34	12.68
CPU time (s)		8	27,588	275

One sample coherence function between one loudspeaker and one microphone is shown in Figure 14, indicating the signal quality to be poor below 200 Hz. Therefore, band-limited processing was applied for all frequencies up to 200 Hz in the SFS stage. In this frequency range, pressure matching was used on the basis of the room response model.

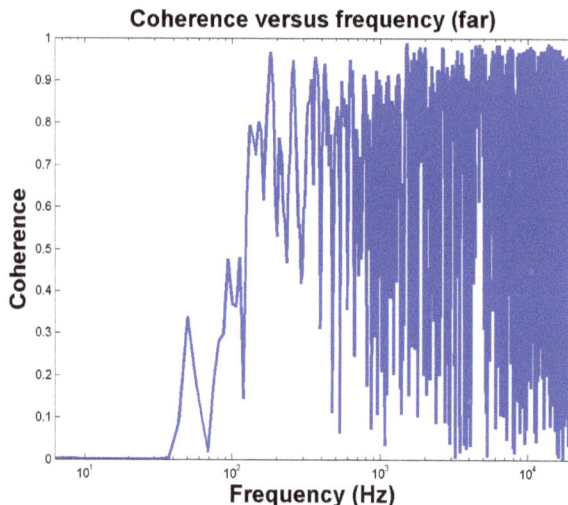

Figure 14. Coherence curve measured from one loudspeaker to one microphone in a live room. A PULSE analysis platform made from Brüel & Kjær was arranged to measure the coherence curve. A white noise signal with sampling rate 16 kHz was used as the driving signal of the loudspeaker.

The SFS stage was conducted for three different methods. The coherence between the loudspeaker and the microphone was poor below 200 Hz; therefore, the signals below 200 Hz were not processed. Method 1, band-limited processing, was applied from 200 Hz to the spatial aliasing frequency, 952 Hz, in the SFS stage. In this frequency range, pressure matching was performed on the basis of the room response model. Below 200 Hz, unprocessed audio signals were fed directly to the loudspeakers. Above 952 Hz, a simple vector panning [39] approach was adopted. The optimal regularization parameter β achieving the highest MOS in room response modeling was calculated using GSS [21] as $\beta = 0.0008634$.

In the second method, instead of a vector panning method, we used DAS to process signals above 952 Hz. In the third method, we used pressure matching to obtain signals above 200 Hz. The use of different regularization parameters in pressure matching results in different levels of localization performance and audio quality.

Figure 15a,c shows the MPDR spectrum and the normalized MPDR spectrum obtained using the third method for $\beta = 0.01$ and $\beta = 10$, respectively. Low values of the regularization parameter β yielded higher localization performance than high values did. These two signals were compared with the clean signal through the PESQ test. The results showed that the high β ensured satisfactory voice quality, whereas the low β impaired voice quality.

The results of three localization methods are presented in Figure 16a–f. The MPDR spectra are plotted as functions of angle and frequency in Figure 16a,c,f. The resulting frequency-averaged and normalized MPDR spectra are shown in Figure 16b,d,f.

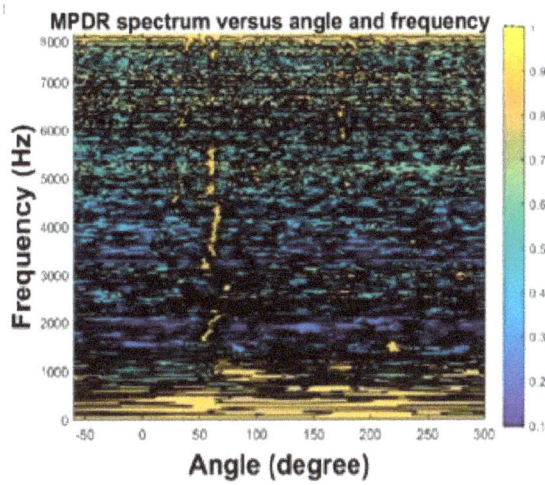

(a)

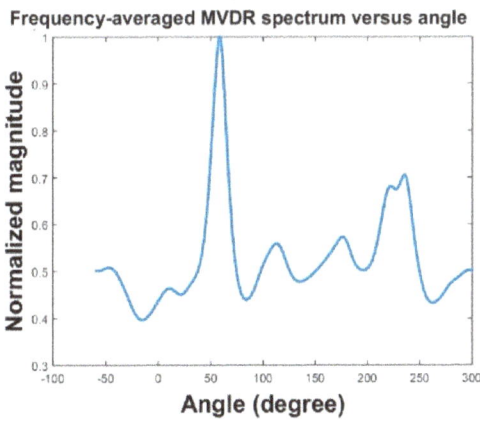

(b)

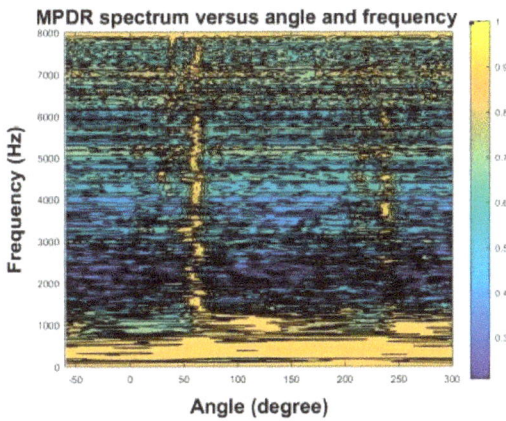

(c)

Figure 15. *Cont.*

(d)

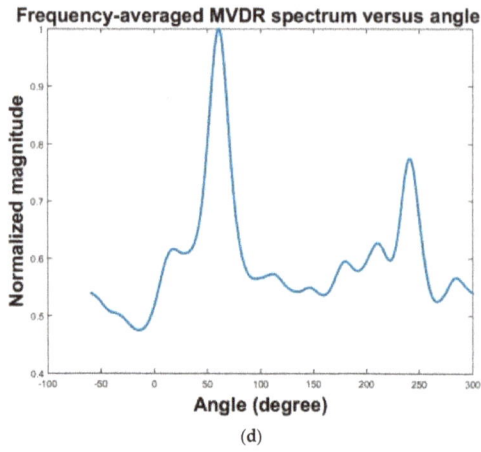

Figure 15. Localization results in the sound field synthesis experiment by the third method with different regularization parameters. $\beta = 10$ (**a,b**) and $\beta = 0.01$ (**c,d**).

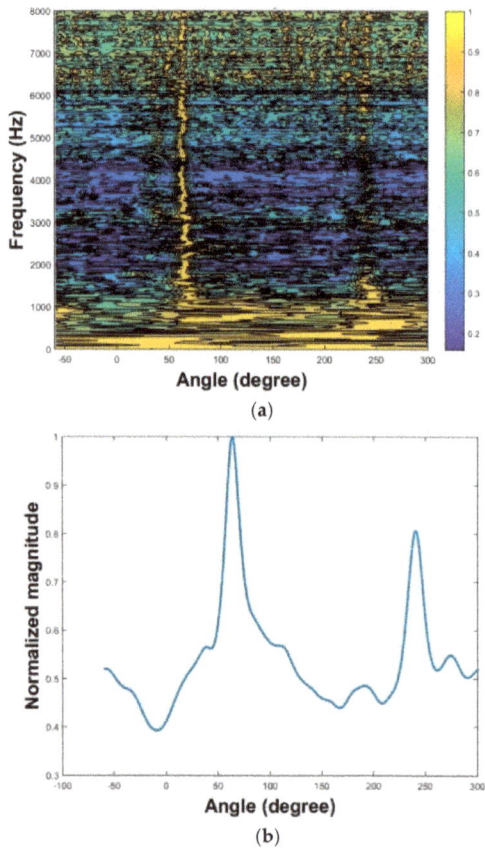

(a)

(b)

Figure 16. *Cont.*

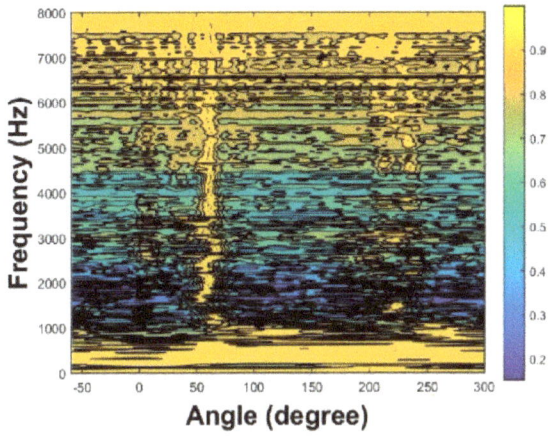

(c)

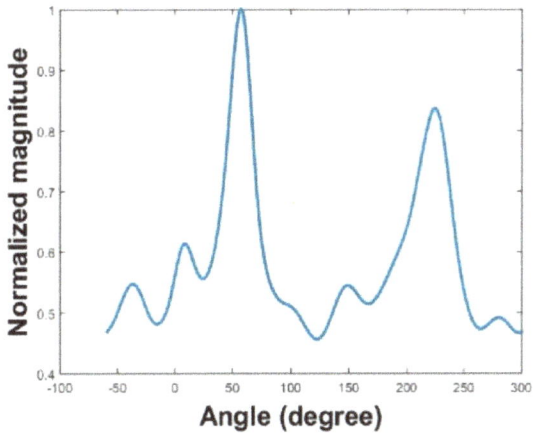

(d)

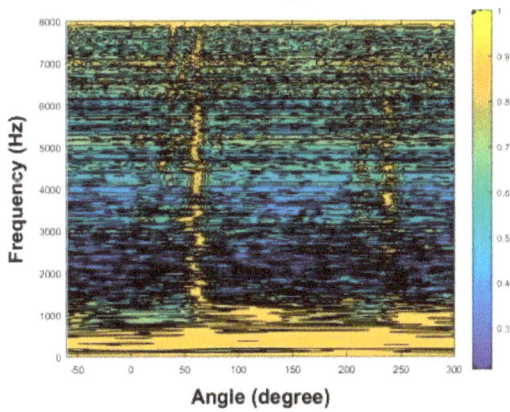

(e)

Figure 16. *Cont.*

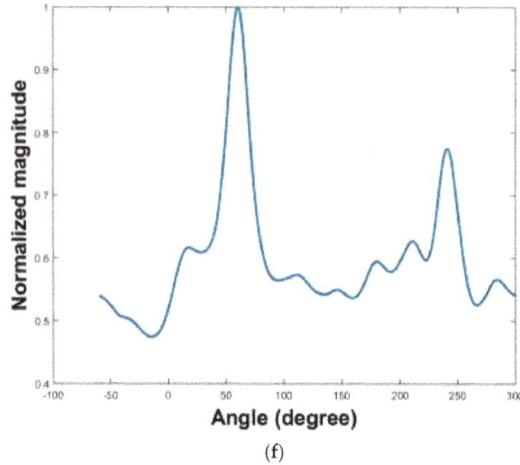

(f)

Figure 16. Localization results in the sound field synthesis experiment with three different approaches. (a,b) first method, (c,d) second method, and (e,f) third method.

4. Conclusions

This study developed algorithms for solving inverse problems generally encountered in spatial audio signal processing. The TIKR algorithm was shown to solve overdetermined problems. However, the regularization parameter in the TIKR method was not effectively chosen. This study thus presents a guideline for choosing the optimal regularization parameter β in the TIKR method. Specifically, choosing the optimal β involves dividing the maximal singular value at low frequency by the threshold and then running the simulation forward and backward for powers of 10 of β. Optimization methods such as GSS can be used by observing the trend of an objective function (such as PESQ). Some trade-offs must be made between localization performance and voice quality. In general, a high β results in a small solution norm with high voice quality, whereas a low β yields a small residual norm with high localization performance.

Inverse problems in noise sound source localization and separation problems can be solved by 1-stage and 2-stage (overdetermined and underdetermined), each of the 1-stage and 2-stage methods has its advantages and disadvantages. In general, the 1-stage methods provide both localization and separation results with good performance. The 2-stage methods give slightly better separation performance than the 1-stage methods. From our experience, PESQ correlates better with separability and segSNR correlates better with distortion.

For 1-stage (underdetermined) problems, iterative CS algorithms have been developed for solving acoustic inverse problems, with applications to localization and separation. The results demonstrate that the CS-CVX method was effective in solving CS problems, despite being computationally expensive. Iterative CS methods achieved comparable performance to the CS-CVX method for CS problems, in far less computation time. The FOCUSS-TIKR and CG methods attained high PESQ, whereas the SD and NT methods attained high segSNR. In general, iterative CS methods were determined to perform better than the TIKR method. For 1-stage methods, FOCUSS-TIKR attains the highest MOS value of PESQ for clean signals, while the Newton method performs the best. Both methods require less CPU time than the CS-CVX.

Inverse solution approaches are also useful in solving SFA and SFS problems. In this study, three inverse problems were solved for implementing an audio codec system. Because of the ill-posed yield at low frequencies, particularly in the room response modeling stage, choosing an appropriate regularization parameter β was crucial. Therefore, this stage required a larger regularization parameter

Appl. Sci. **2017**, *7*, 582

than the SFA and SFS stages required. In the analysis stage, the one-stage CS algorithm was determined to be more computationally expensive than the two-stage TIKR algorithm. In the synthesis stage, the first method performed well in localization, but did not perform well in reproduced voice quality. As compared with the third method, the second method reproduced signals with boosted high-frequency content above 952 Hz with poor localization. The third method had the highest performance in terms of voice quality and localization performance.

Acknowledgments: The work was supported by the Ministry of Science and Technology (MOST) of Taiwan, Republic of China, under project number 102-2221-E-007-029-MY3.

Author Contributions: M.R. Bai and C.-M. Yang conceived and designed the experiments; Y.-H. Chiang and P.-C. Wu performed the experiments; P.-C. Wu and C. Chung analyzed the data; C. Chung and Y.-H. Chinag wrote the paper.

Conflicts of Interest: The authors declare no conflict of interest.

References

1. Kim, Y.; Nelson, P.A. Spatial resolution limits for the reconstruction of acoustic source strength by inverse methods. *J. Sound Vib.* **2003**, *265*, 583–608. [CrossRef]
2. Nelson, P.A.; Yoon, S.H. Estimation of acoustic source strength by inverse methods: Part I, conditioning of the inverse problem. *J. Sound Vib.* **2000**, *233*, 639–664. [CrossRef]
3. Kim, Y.; Nelson, P.A. Optimal regularisation for acoustic source reconstruction by inverse methods. *J. Sound Vib.* **2004**, *275*, 463–487. [CrossRef]
4. Maynard, J.D.; Williams, E.G.; Lee, Y. Nearfield acoustic holography: I. Theory of generalized holography and the development of NAH. *J. Acoust. Soc. Am.* **1985**, *78*, 1395–1413. [CrossRef]
5. Hald, J. Fast wideband acoustical holography. *J. Acoust. Soc. Am.* **2016**, *139*, 1508–1517. [CrossRef] [PubMed]
6. Brooks, T.F.; Humphreys, W.M. A deconvolution approach for the mapping of acoustic sources (DAMAS) determined from phased microphone arrays. *J. Sound Vib.* **2006**, *294*, 856–879. [CrossRef]
7. Bai, M.R.; Kuo, C.H. Deconvolution-based acoustic source localization and separation algorithms. *J. Acoust. Soc. Am.* **2014**, *135*, 2358. [CrossRef]
8. Bai, M.R.; Hua, Y.H.; Kuo, C.H.; Hsieh, Y.H. An integrated analysis-synthesis array system for spatial sound fields. *J. Acoust. Soc. Am.* **2015**, *137*, 1366–1376. [CrossRef] [PubMed]
9. Bai, M.R.; Hsu, H.S.; Wen, J.C. Spatial sound field synthesis and upmixing based on the equivalent source method. *J. Acoust. Soc. Am.* **2014**, *135*, 269–282. [CrossRef] [PubMed]
10. Elliott, S.J.; Cheer, J.; Murfet, H.; Holland, K.R. Minimally radiating sources for personal audio. *J. Acoust. Soc. Am.* **2010**, *128*, 1721–1728. [CrossRef] [PubMed]
11. Bai, M.R.; Hsieh, Y.H. Point focusing using loudspeaker arrays from the perspective of optimal beamforming. *J. Acoust. Soc. Am.* **2015**, *137*, 3393–3410. [CrossRef] [PubMed]
12. Bai, M.R.; Tung, C.W.; Lee, C.C. Optimal design of loudspeaker arrays for robust cross-talk cancellation using the Taguchi method and the genetic algorithm. *J. Acoust. Soc. Am.* **2005**, *117*, 2802–2813. [CrossRef] [PubMed]
13. Loizou, P.C. *Speech Enhancement: Theory and Practice*; Taylor & Francis: Park Drive, UK; Abingdon, UK, 2007.
14. Shabtai, N.R. Optimization of the directivity in binaural sound reproduction beamforming. *J. Acoust. Soc. Am.* **2015**, *138*, 3118–3128. [CrossRef] [PubMed]
15. Miyoshi, M.; Kaneda, Y. Inverse filtering of room acoustic. *IEEE Transac. Acoust. Speech Signal Process.* **1998**, *36*, 145–152. [CrossRef]
16. Groetsch, C.W. *The Theory of Tikhonov Regularization for Fredholm Equations of the First Kind*; Pitman Advanced Pub. Program: Boston, MA, USA, 1984.
17. Hansen, P.C. *Rank-Deficient and Discrete Ill-Posed Problems*; Society for Industrial and Applied Mathematics: Philadelphia, PA, USA, 1998.
18. Bertero, M.; Poggio, T.; Torre, V. Ill-Posed Problems in Early Vision. *Proc. IEEE* **1988**, *76*, 869–889. [CrossRef]
19. Hansen, P.C. *Analysis of Discrete Ill-Posed Problems by Means of the L-Curve*; Society for Industrial and Applied Mathematics: Philadelphia, PA, USA, 1992.

20. Hansen, P.C.; O'leary, D.P. *The Use of the L-Curve in the Regularization of Discrete Ill-Posed Problems*; Society for Industrial and Applied Mathematics: Philadelphia, PA, USA, 1993.

21. Brent, R.P. *Algorithms for Minimization without Derivatives*; Prentice-Hall, Inc.: Englewood Cliffs, HJ, USA, 1973; pp. 48–75.

22. Candes, J.; Wakin, M.B. An introduction to compressive sampling. *IEEE Signal Process. Mag.* **2008**, *25*, 21–30. [CrossRef]

23. Edelmann, G.F.; Gaumond, C.F. Beamforming using compressive sensing. *J. Acoust. Soc. Am.* **2011**, *130*, 232–237. [CrossRef] [PubMed]

24. Boyd, S.; Vandenberghe, L. *Convex Optimization*; Cambridge University Press: New York, NY, USA, 2004; Chapters 1–7.

25. Bai, M.R.; Chen, C.C. Application of Convex Optimization to Acoustical Array Signal Processing. *J. Sound Vib.* **2013**, *332*, 6596–6616. [CrossRef]

26. Grant, M.; Boyd, S. CVX, Version 1.21 MATLAB Software for Disciplined Convex Programming. Available online: http://cvxr.com/cvx (accessed on 14 June 2013).

27. Gorodnitsky, I.F.; Rao, B.D. Sparse Signal Reconstruction from Limited Data Using FOCUSS: A Re-weighted Minimum Norm Algorithm. *IEEE Trans. Signal Process.* **1997**, *45*, 600–616. [CrossRef]

28. ITU-T Recommendation P.862. *Perceptual Evaluation of Speech Quality (Pesq): An Objective Method for End-to-End Speech Quality Assessment of Narrow-Band Telephone Networks and Speech Codecs*; International Telecommunication Union: Geneva, Switzerland, 2001; p. 21.

29. Bai, M.R.; Ih, J.G.; Benesty, J. *Acoustic Array Systems: Theory, Implementation, and Application*, 1st ed.; Wiley-IEEE Press: Singapore, 2013; Chapters 3–4.

30. Golub, G.H.; van Loan, C.F. *Matrix Computations*, 3rd ed.; Johns Hopkins University Press: Baltimore, MD, USA, 1989; Chapter 12.

31. Noble, B.; Daniel, J.W. *Applied Linear Algebra*; Prentice Hall: Englewood, NJ, USA, 1977.

32. Hestenes, M.R.; Stiefel, E. Methods of conjugate gradients for solving linear systems. *J. Res. Natl. Bur. Stand.* **1952**, *49*, 409–436. [CrossRef]

33. Fraysse, V.; Giraud, L. *A Set of Conjugate Gradient Routines for Real and Complex Arithmetics*; CERFACS Technical Report TR/PA/00/47; Cedex: Toulouse, France, 2000.

34. Ginn, K.B.; Haddad, K. Noise source identification techniques: Simple to advanced applications. *Proc. Acoust.* **2012**, *2012*, 1781–1786.

35. Capon, J. High-Resolution Frequency-Wavenumber Spectrum Analysis. *Proc. IEEE* **1969**, *57*, 1408–1418. [CrossRef]

36. Gomes, J.; Hald, J.; Juhl, P.; Jacobsen, F. On the Applicability of the Spherical Wave Expansion with a Single Origin for Near-Field Acoustical Holography. *J. Acoust. Soc. Am.* **2009**, *125*, 1529–1537. [CrossRef] [PubMed]

37. Candes, J.; Romberg, J.; Tao, T. Stable Signal Recovery from Incomplete and Inaccurate Measurements. *Commun. Pure Appl. Math.* **2006**, *59*, 1207–1223. [CrossRef]

38. Candes, J.; Romberg, J.; Tao, T. Robust Uncertainty Principles: Exact Signal Reconstruction Form Highly Incomplete Frequency Information. *IEEE Trans. Inf. Theory* **2006**, *52*, 489–509. [CrossRef]

39. Kim, Y.H.; Choi, J.W. *Sound Visualization and Manipulation*; Wiley: Singapore, 2013.

applied sciences

MDPI

Article

Low Frequency Interactive Auralization Based on a Plane Wave Expansion

Diego Mauricio Murillo Gómez *,†, **Jeremy Astley and Filippo Maria Fazi**

Institute of Sound and Vibration Research, University of Southampton, SO17 1BJ, UK; rja@isvr.soton.ac.uk (J.A.); filippo.fazi@soton.ac.uk (F.M.F.)
* Correspondence: diego.murillo@usbmed.edu.co; Tel.: +57-312-505-554
† Current address: Faculty of Engineering, Universidad de San Buenaventura Medellín, Cra 56C No 51-110, 050010 Medellín, Colombia.

Academic Editors: Woon-Seng Gan and Jung-Woo Choi
Received: 2 March 2017; Accepted: 23 May 2017; Published: 27 May 2017

Abstract: This paper addresses the problem of interactive auralization of enclosures based on a finite superposition of plane waves. For this, room acoustic simulations are performed using the Finite Element (FE) method. From the FE solution, a virtual microphone array is created and an inverse method is implemented to estimate the complex amplitudes of the plane waves. The effects of Tikhonov regularization are also considered in the formulation of the inverse problem, which leads to a more efficient solution in terms of the energy used to reconstruct the acoustic field. Based on this sound field representation, translation and rotation operators are derived enabling the listener to move within the enclosure and listen to the changes in the acoustic field. An implementation of an auralization system based on the proposed methodology is presented. The results suggest that the plane wave expansion is a suitable approach to synthesize sound fields. Its advantage lies in the possibility that it offers to implement several sound reproduction techniques for auralization applications. Furthermore, features such as translation and rotation of the acoustic field make it convenient for interactive acoustic renderings.

Keywords: interactive auralization; plane wave expansion; inverse method; finite element method

1. Introduction

Auralization is a subject of great interest in different areas because it enables the generation of an audible perception of the acoustic properties of a specific environment [1]. It is a powerful technique because it allows the sound field to be rendered according to the characteristics of the medium, which has applications in the evaluation and understanding of the physical phenomenon under consideration. For room acoustics, auralization provides a convenient tool for experimental tests, subjective evaluations, virtual reality and architectural design.

A significant feature that enhances the auralization technique is the generation of interactive environments in which the listener can move within the enclosure. This is achieved by synthesizing the acoustic field in real time according to the properties of the room and the source-receiver paths. Several approaches have been proposed in the scientific literature to generate interactive auralizations based on Geometrical Acoustics (GA) [2–6]. Nevertheless, at low frequencies, the assumptions required for GA are not generally satisfied, which requires the use of different techniques, such as the numerical solution of the wave equation. The Finite Element Method (FEM) [7], the Boundary Element Method (BEM) [8] and the Finite Difference Time Domain (FDTD) method [9] are some of the techniques commonly used to estimate room impulse responses. However, the computational cost required by these approaches to predict the solution constrains their use for real-time applications.

Despite the significant computational cost of the above methods, some alternatives have been formulated to generate interactive environments based on wave propagation. Mehra et al. proposed the use of the equivalent source method for the rendering of acoustic fields of large and open scenes in real time [10]. Although the approach allows for the reconstruction of the sound field in real time, limitations related to static sound sources or the inclusion of the Doppler effect have to be overcome. The inclusion of the directivity of sound sources and listeners using a spherical harmonic representation has also been proposed by the same author to extend the versatility of the methodology [11]. Another approach proposed by Raghuvanshi [12] simulates the synthesis and propagation of sound. The solution method, denoted as Adaptive Rectangular Decomposition (ARD), permits real-time computation providing a platform for interactive auralizations. The main advantages of the method are the use of dynamic listener/sources and its ability to simulate large complex 3D scenes. Alternatively, Savioja presented a different strategy to predict acoustic fields in real time [13]. The numerical solver corresponds to an FDTD model running over GPUs using parallel computing techniques. The results indicate that this methodology allows the simulation up to 1.5 kHz in moderate size domains. Dynamic listener and multiple sources are also possible based on this processing scheme.

A different approach to interactive auralizations based on the numerical solution of the wave equation is to encode spatial information from the predicted acoustic pressure data. Translation and rotation of the acoustic field can be then achieved by the application of mathematical operators. Southern et al. [14] proposed a method to obtain a spherical harmonic representation of FDTD simulations. The approach uses the the Blumlein difference technique to create a higher order directivity pattern from two adjacent lower orders. This is achieved by approximating the gradient of the pressure as the difference between two neighbouring pressure points on the grid where the solution was computed. The rotation of the acoustic field can be easily computed by a rotation matrix, whereas the translation can be recreated by an interpolation process between spatial impulse responses [15]. Sheaffer et al. [16] suggested the formulation of an inverse problem to generate a spherical harmonic representation of acoustic fields predicted using FDTD. A binaural rendering is achieved based on the relation between spherical harmonics, plane waves and HRTFs.

An implementation of a Plane Wave Expansion (PWE) is carried out in the current study as an alternative methodology to generate an interactive auralization from predicted acoustic numerical data. The methodology has been evaluated by using FE results, but can readily be implemented using other numerical methods. The approach is based on the concept that the acoustic pressure at each node of the mesh can be understood as the output of a virtual omnidirectional microphone. By using the data from the mesh, it is possible to create a virtual microphone array, which, with the implementation of an inverse method, allows for the estimation of complex amplitudes of a set of plane waves that synthesize the desired acoustic field. The use of an inverse method to generate a plane wave expansion from predicted pressure data was previously studied by Støfringsdal and Svensson for 2D cases [17]. An extension of their work is presented in this paper to the 3D case.

Based on a plane wave representation, mathematical operators can be implemented to enable interactive features for auralization applications. The translational movement of the listener can be generated by the translation of the plane wave expansion [18,19]. In terms of the listener's rotation, a spherical harmonic transformation can be used to rotate the acoustic field [20]. A rotation in the plane wave domain is achieved by the implementation of a VBAP algorithm. Nevertheless, experiments conducted by Murillo [21] suggest that the spherical harmonic transformation is more accurate for this specific application.

A complete framework for an interactive auralization based on a plane wave representation is presented in this article. The processing chain involves the use of an inverse method to extract spatial information from FE simulations, the use of translation and rotation operators, the combination of real-time audio processing with a visual interface and binaural rendering to reproduce the acoustic field. The proposed approach is evaluated by testing it within a real-time auralization system as a reference case. This auralization system allows us to emulate interactively wave phenomena, such as the modal behaviour of the enclosure or acoustic diffraction. The remaining parts of the paper are organized as follows: The mathematical foundations of the plane wave expansion and the derivation of the translation and rotation operators are reviewed in Section 2. The implementation of an inverse method to generate a plane wave representation of the predicted acoustic fields is presented in Section 3. In Section 4, a real-time auralization system is developed, which allows for the evaluation of the proposed approach. The discussion and evaluation of the results are considered in Section 5. Finally, conclusions of the current work are presented in Section 6.

2. Mathematical Foundations

A particular solution of the homogeneous Helmholtz equation for a singular radian frequency ω is given by a complex pressure field $q(\omega)e^{jk\mathbf{x}\cdot\hat{\mathbf{y}}}$, which corresponds to a plane wave arriving in the direction of a unit vector $\hat{\mathbf{y}}$ with an arbitrary complex amplitude q and wavenumber k. The vector $\mathbf{x}$ identifies the position where the acoustic pressure is evaluated, and the symbol "·" represents the scalar product operation. The plane wave approximation is appropriate at a large distance from the acoustic source, where the curvature of the wave can be ignored. In a similar way, an acoustic field that satisfies the homogeneous Helmholtz equation can be represented by means of a Plane Wave Expansion (PWE) as:

$$p(\mathbf{x},\omega) = \int_{\hat{\mathbf{y}}\in\Omega} e^{jk\mathbf{x}\cdot\hat{\mathbf{y}}} q(\hat{\mathbf{y}},\omega)d\Omega(\hat{\mathbf{y}}), \tag{1}$$

in which $\mathbf{x}$ is the evaluation point, $\hat{\mathbf{y}}$ indicates the different incoming directions of the plane waves, $q(\hat{\mathbf{y}},\omega)$ is the amplitude density function and Ω is the unitary sphere [19]. The synthesis of acoustic fields based on a plane wave representation is a common approach [16,17] being adaptable to several audio reproduction techniques. Binaural reproduction is performed by the convolution of the plane waves with the HRTFs according to the direction of arrival [22]. The relation and transformation between the PWE, Ambisonics and wave field synthesis are presented in [23]. Furthermore, mathematical operators to translate [18,19] and rotate [20] the acoustic field are available, which makes this sound field representation convenient for interactive applications. The disadvantage of this method is that the assumption of plane waves makes the approach suitable for sound fields generated by sources located at a large distance from the listener. In addition, the implementation of an infinite number of plane waves is not feasible, and the use of discrete wave directions generates artefacts in the sound field reconstruction.

2.1. Translation of the Acoustic Field

Figure 1 shows the vector $\mathbf{x}'$, which identifies the origin of a relative coordinate system corresponding to the centre of the listener's head. The vector $\mathbf{x}_{rel}$ defines the same point in space in the relative coordinate system as is identified by the vector $\mathbf{x}$ in absolute coordinates.

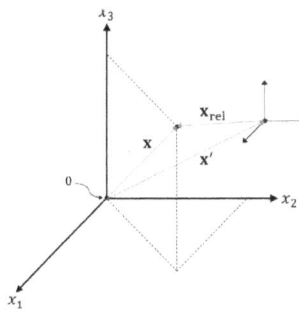

Figure 1. Vector $\mathbf{x}$ is represented as $\mathbf{x}_{rel}$ in the relative coordinate system $\mathbf{x}'$.

The sound field translation operator is derived by considering two plane wave expansions of the same sound field, but centred at different points in space, specifically at the origin of the absolute and relative coordinate systems. In this case, the difference between the two plane wave expansions is given by the plane wave amplitude densities $q(\hat{y}, \omega)$ and $q_{rel}(\hat{y}, \omega)$, respectively. Therefore, the objective is to express one density in terms of the other. This is achieved by expanding x_{rel} as $x - x'$, leading to:

$$q_{rel}(\hat{y}, \omega) = q(\hat{y}, \omega)e^{jkx' \cdot \hat{y}}. \tag{2}$$

Equation (2) indicates that $e^{jkx' \cdot \hat{y}}$ is the translation operator for the plane wave expansion from the origin to x'. Its equivalence in the time domain can be easily found by using the shifting property of the Fourier transform:

$$Q_{rel}(\hat{y}, t) = Q\left(\hat{y}, t - \frac{x' \cdot \hat{y}}{c}\right). \tag{3}$$

2.2. Rotation of the Acoustic Field

A spherical harmonic transformation of the plane wave expansion can be performed based on the Jacobi–Anger relation [24]. From the spherical harmonic representation, the rotation can be generated by the implementation of a rotation matrix. The derivation of a sound field rotation operator in the spherical harmonic domain proceeds as follows. A rotation in the azimuthal plane by ϕ_0 can be expressed as:

$$p(r, \theta, \phi - \phi_0, \omega) = \sum_{n=0}^{\infty} \sum_{m=-n}^{n} A_{nm}(\omega) j_n(kr) Y_n^m(\theta, \phi - \phi_0). \tag{4}$$

Expanding the right-hand side of Equation (4) gives:

$$p(r, \theta, \phi - \phi_0, \omega) = \sum_{n=0}^{\infty} \sum_{m=-n}^{n} A_{nm}(\omega) j_n(kr) \sqrt{\frac{(2n+1)}{4\pi} \frac{(n-m)!}{(n+m)!}} P_n^m(\cos\theta) e^{jm\phi} e^{-jm\phi_0}, \tag{5}$$

which yields:

$$p(r, \theta, \phi - \phi_0, \omega) = \sum_{n=0}^{\infty} \sum_{m=-n}^{n} j_n(kr) Y_n^m(\theta, \phi) A_{\phi_0 nm}(\omega), \tag{6}$$

in which:

$$A_{\phi_0 nm}(\omega) = A_{nm}(\omega) e^{-jm\phi_0}. \tag{7}$$

Equation (7) indicates that the azimuthal rotation of the sound field can be performed by taking the product of the complex spherical harmonic coefficients and a complex exponential whose argument depends on the angle of rotation. An inverse spherical harmonic transformation can be implemented to return to the plane wave domain after the rotation has been carried out.

3. Plane Wave Expansion from Finite Element Data

Although the plane wave expansion is an integral representation, for the implementation of the method, Equation (1) is discretized into a finite number of L plane waves, namely:

$$p(x, \omega) = \sum_{l=1}^{L} e^{jkx \cdot \hat{y}_l} q_l(\omega) \Delta\Omega_l, \tag{8}$$

where $\Delta\Omega_l$ corresponds to the portion of the unit sphere that is associated with the plane wave l. The discretization of Equation (1) is performed by using a predefined uniform distribution of L plane waves over a unit sphere [25]. Based on a finite plane wave expansion, an inverse method can be implemented to estimate a discrete set of plane waves whose complex amplitudes synthesize the target acoustic field. To that end, the acoustic pressure calculated with the FEM at a specific location of the domain can be understood as corresponding to the output of an omnidirectional microphone. The combination of acoustic pressures at discrete points generates a virtual microphone array that is used to extract spatial information of the sound field. Based on that information, the amplitude q_l of each plane wave is determined by the inversion of the transfer function matrix between the microphones and the plane waves [26]. This principle is explained as follows: the complex acoustic pressure predicted with the FE model at M virtual microphone positions is denoted in vector notation as:

$$\mathbf{p}(\omega) = [p_1(\omega), p_m(\omega), \dots, p_M(\omega)]^T, \tag{9}$$

where p_m is the acoustic pressure at the m-th virtual microphone. Likewise, the complex amplitudes of L plane waves used to reconstruct the sound field are represented by the vector:

$$\mathbf{q}(\omega) = [q_1(\omega), q_l(\omega), \dots, q_L(\omega)]^T. \tag{10}$$

Finally, the transfer function that describes the sound propagation from each plane wave to each virtual microphone is arranged in matrix notation as:

$$\mathbf{H}(\omega) = \begin{vmatrix} h_{11}(\omega) & \cdots & h_{1L}(\omega) \\ \vdots & h_{ml}(\omega) & \vdots \\ h_{M1}(\omega) & \cdots & h_{ML}(\omega) \end{vmatrix}$$

in which $h_{ml} = e^{jk\mathbf{x}_m \cdot \hat{\mathbf{y}}_l}$. Consequently, the relationship between the plane wave amplitudes and the virtual microphone signals is:

$$\mathbf{p}(\omega) = \mathbf{H}(\omega)\mathbf{q}(\omega). \tag{11}$$

The amplitude of the plane waves is calculated by solving Equation (11) for $\mathbf{q}(\omega)$. This is carried out in terms of a least squares solution, which minimizes the sum of the squared errors between the reconstructed and the target sound field [26]. In the case of an overdetermined problem (more virtual microphones than plane waves), the error vector can be expressed as:

$$\mathbf{e}(\omega) = \tilde{\mathbf{p}}(\omega) - \mathbf{p}(\omega), \tag{12}$$

where $\tilde{\mathbf{p}}(\omega)$ is the pressure reconstructed by the plane wave expansion and $\mathbf{p}(\omega)$ is the target pressure from the FE model. The least squares solution is achieved by the minimization of a cost function $J(\omega) = \mathbf{e}^H(\omega)\mathbf{e}(\omega)$ in which $(\cdot)^H$ indicates the Hermitian transpose. The minimization of the cost function $J(\omega)$ is given by [26]:

$$\mathbf{q}(\omega) = \mathbf{H}^\dagger(\omega)\mathbf{p}(\omega), \tag{13}$$

in which $\mathbf{H}^\dagger(\omega)$ is the Moore–Penrose pseudo-inverse of the propagation matrix $\mathbf{H}(\omega)$ [26].

The use of a finite number of plane waves leads to artefacts in the sound field reconstruction. Ward and Abhayapala [27] proposed the following relation between the area of accurate reconstruction, the number of plane waves and the frequency of the field:

$$L = \left(\left\lceil 2\pi \frac{R}{\lambda} \right\rceil + 1 \right)^2, \tag{14}$$

in which L is the number of plane waves, $\lceil \cdot \rceil$ indicates the round operator, λ is the wavelength and R is the radius of a sphere within which the reconstruction is accurate. Numerical simulations have been performed to evaluate the effects of discretizing the plane wave expansion. Figure 2 shows the real part of the target and the reconstructed acoustic pressure (Pa) in a cross-section of the domain using 36, 64 and 144 plane waves in the expansion. The frequency of the field corresponds to 250 Hz.

The black circle displayed in the figure represents the area of accurate reconstruction predicted by solving Equation (14) for R.

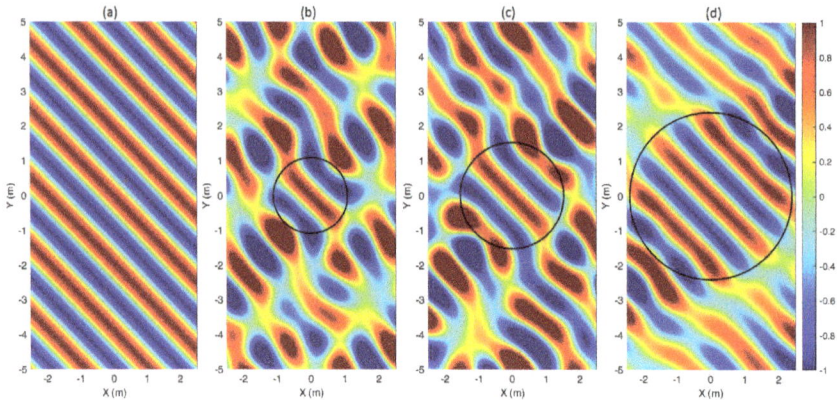

Figure 2. Reconstructed acoustic field using different numbers of plane waves in the expansion. (a) target field; (b) reconstructed field $L = 36$; (c) reconstructed field $L = 64$; (d) reconstructed field $L = 144$.

Figure 2 indicates that the region in which the reconstruction is accurate increases when a higher number of uniformly-distributed plane waves is considered in the expansion. Good agreement was found between the radius predicted by Equation (14) and the area where the reconstruction is correct. A preliminary analysis was performed to establish the size of the microphone array and the number of plane waves required to generate an inverse matrix whose condition number is smaller than 10^6. The condition number is defined as the ratio between the largest and the smallest singular value of the propagation matrix $\mathbf{H}(\omega)$. It has been shown in [26] that the stability of the solution provided by the inverse method is determined by the condition number; therefore, high values of this parameter indicate that errors in the model, such as noise or non-linearity of the system, will affect the result for $\mathbf{q}(\omega)$ significantly. The criterion of a condition number smaller than 10^6 was motivated due to the fact that the data come from numerical simulations, which are free from measurement noise. Although the model is still affected by numerical inaccuracies, the level of this type of noise is expected to be much lower than in the case of measured noise.

Firstly, a simple incoming plane wave of 63 Hz ($\theta = 90°$, $\phi = 45°$) in free field was selected as a target. The sound field was analytically calculated in a rectangular domain of dimensions (5 m, 10 m, 3 m) and captured by four different virtual cube arrays with linear dimensions of 1.2 m, 1.6 m, 2 m and 2.4 m, respectively. The spatial resolution between microphones corresponded to 0.2 m. A frequency of 63 Hz was selected as a reference since the condition number decreases with frequency, and 63 Hz therefore provides a reasonable lower threshold. Table 1 shows the condition number of matrix $\mathbf{H}(\omega)$ for different sizes of the array and numbers of planes waves uniformly distributed over a unit sphere [25].

Table 1. Condition number of the matrix $\mathbf{H}(\omega)$ as a function of the size of the microphone array and the number of plane waves.

Length of the Array	$L = 64$	$L = 144$	$L = 324$
1.2 m (343 mics)	2.39×10^7	2.86×10^{13}	2.13×10^{18}
1.6 m (729 mics)	3.09×10^6	8.88×10^{11}	1.32×10^{17}
2 m (1331 mics)	6.11×10^5	6.13×10^{10}	7.65×10^{16}
2.4 m (2197 mics)	1.56×10^5	6.89×10^9	3.01×10^{16}

As expected, the condition number decreases as the size of the array increases. The reason for that may be attributed to the wavelength of the plane wave, which is approximately 5.4 m at this frequency, and a larger array captures more information of the sound field. Regarding the number of plane waves, the results suggest that increasing its number does not improve the situation, instead it increases the condition number significantly. This can be explained because at 63 Hz, there is not enough information in the sound field that is captured by the virtual microphone array; therefore, additional plane waves only make the inversion of the propagation matrix $\mathbf{H}(\omega)$ more difficult. In addition, the use of a higher number of plane waves makes the method computational expensive due to the number of convolutions that must be performed in real time. An optimal relation of 64 plane waves, a microphone array of 1.6 m in length with virtual microphone spacing of 0.2 m (729 microphone positions) was found at 63 Hz. This spatial resolution (0.2 m) yields an aliasing frequency of ≈850 Hz, which is sufficient for the range of the FE simulations. This high frequency limit is calculated based on the Nyquist theorem for sampling signals [28].

3.1. Reference Case

A typical meeting room (No. 4079 in Building 13 at the Highfield Campus of the University of Southampton) was selected as a reference case. It is an L-shaped room with a volume of approximately 88 m³. FE simulations were conducted using the commercial package COMSOL v5.1 (COMSOL Inc., Stockholm, Sweden). The reader is referred to [21] for a detailed description of the simulation procedure, i.e., the characterization of the acoustic source, the geometric model of the enclosure, the boundary conditions and the measurements carried out to validate the predictions. Figure 3 shows a model of the enclosure identifying the location of the virtual microphone array.

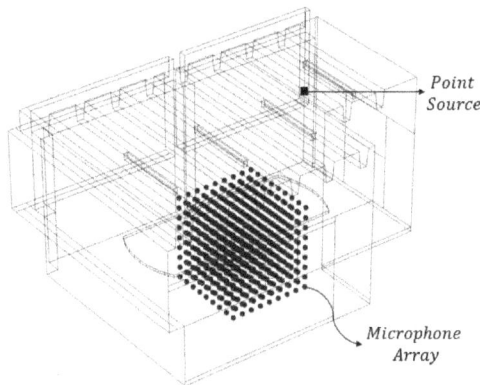

Figure 3. Model of the reference room model.

Two types of figures are presented to assess the performance of the inverse method. The first shows a comparison between the real parts of the target (numerical) and reconstructed acoustic pressures (Pa) over a cross-section of the domain (1.6 m). The second type of figure plots the absolute and phase components of the error between the target and reconstructed pressure fields, defined as:
amplitude error:

$$E_{pa}(\mathbf{x}) = 20\log_{10}\left(\frac{|\tilde{p}(\mathbf{x})|}{|p(\mathbf{x})|}\right). \tag{15}$$

phase error:

$$E_{pp}(\mathbf{x}) = \angle\left(p(\mathbf{x})\tilde{p}(\mathbf{x})^{*}\right), \tag{16}$$

where $\tilde{p}(\mathbf{x})$ is the reconstructed pressure, $p(\mathbf{x})$ is the target pressure, $(\cdot)^{*}$ indicates the complex conjugate operator and $\angle$ represents the phase of a complex number. The amplitude error gives

189

insight about whether the reconstructed acoustic field is louder or quieter compared to the target one. The phase error indicates the phase differences between the reconstructed and target acoustic fields. Figures 4 and 5 show the synthesized acoustic pressure at two frequencies (63 Hz and 250 Hz). The dotted black square represents the position of the microphone array.

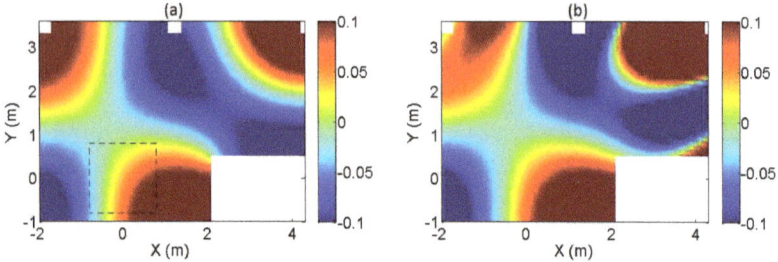

Figure 4. Target (**a**) and reconstructed (**b**) field of the reference room, 63 Hz.

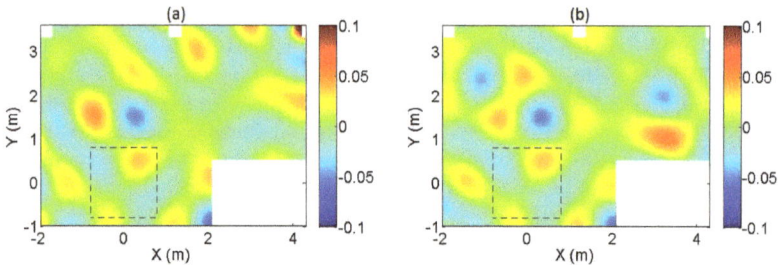

Figure 5. Target (**a**) and reconstructed (**b**) field of the reference room, 250 Hz.

Figures 4 and 5 indicate that the inverse method is able to predict quite well a plane wave expansion whose complex amplitudes synthesize the computed sound field even in small rooms and at low frequencies where the plane wave propagation assumption is not completely satisfied. As expected, the reconstruction is accurate around the virtual microphone array. The corresponding acoustic errors are presented in Figures 6 and 7, respectively.

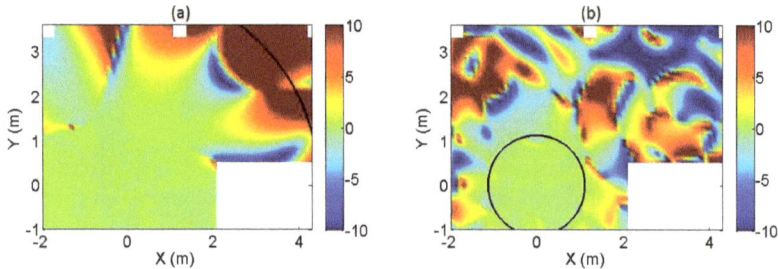

Figure 6. Amplitude error of the reference room at 63 Hz (**a**) and 250 Hz (**b**).

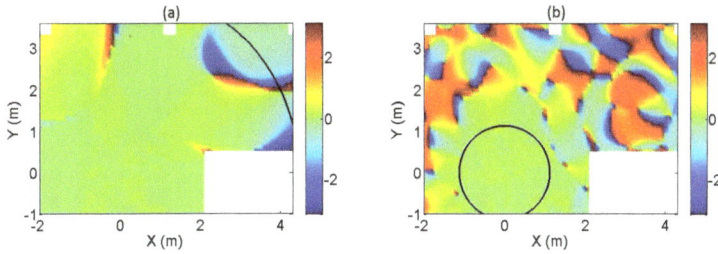

Figure 7. Phase error of the reference room at 63 Hz (**a**) and 250 Hz (**b**).

These indicate that the area of accurate reconstruction depends on the frequency of the acoustic field, being more extensive at low frequencies. In terms of defining a radius within which the reconstruction is accurate, the acoustic errors show good agreement with Equation (14), represented by the circular arc in Figures 6 and 7, only for the higher frequency of 250 Hz. At the lower frequency, 63 Hz, the area is overestimated. This relates to the presence of the walls. At the lower frequency, the homogeneous Helmholtz equation is not satisfied within the radius predicted by Equation (14) since the walls of the room intrude into this region acting as reflective boundaries and playing a role similar to acoustic sources.

3.1.1. Regularization in the Formulation of the Inverse Problem

A well-established technique to improve the stability of the solutions of inverse methods is the use of regularization in the inversion of the propagation matrix $\mathbf{H}(\omega)$ [29,30]. Tikhonov regularization is used for this purpose. It is based on the concept of changing the cost function $J(\omega)$ by the inclusion of an additional term [26], that is:

$$J(\omega) = \mathbf{e}^H(\omega)\mathbf{e}(\omega) + \beta(\omega)\mathbf{q}^H(\omega)\mathbf{q}(\omega), \tag{17}$$

where $\beta(\omega)$ is the regularization parameter. The minimization of the cost function $J(\omega)$ of Equation (17) is given by [26]:

$$\mathbf{q}(\omega) = \left[\mathbf{H}^H(\omega)\mathbf{H}(\omega) + \beta(\omega)\mathbf{I}\right]^{-1}\mathbf{H}^H(\omega)\mathbf{p}(\omega). \tag{18}$$

Equation (17) indicates that the minimization of the cost function takes into account the sum of the squared errors between the reconstructed and target acoustic pressure and, in addition, the sum of the squared norm of plane wave amplitude vector. Figures 8–11 show the results obtained for the reference problem when Tikhonov regularization is applied. The value of β, which is given in each case, was calculated as:

$$\beta = \|\mathbf{H}(\omega)\|^2 \Gamma, \tag{19}$$

in which $\|\cdot\|$ is the spectral norm (the largest singular value) of the propagation matrix $\mathbf{H}(\omega)$ and Γ is an arbitrary constant whose value is selected between 1×10^{-3} and 1×10^{-6}.

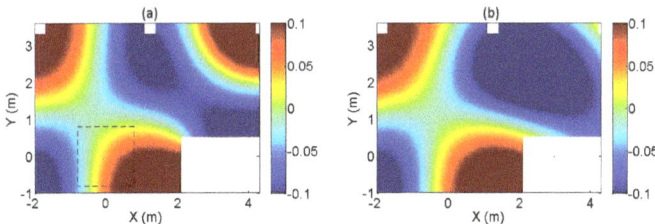

Figure 8. Target (**a**) and reconstructed (**b**) field of the reference room, regularized $\beta = 0.33$.

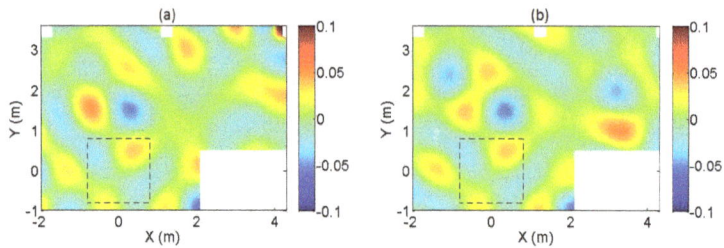

Figure 9. Target (**a**) and reconstructed (**b**) field of the reference room, regularized $\beta = 0.33$.

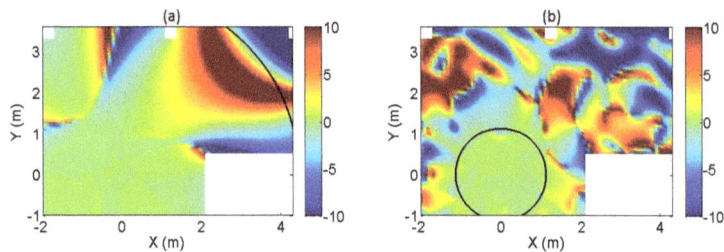

Figure 10. Amplitude error of the reference room at 63 Hz (**a**) and 250 Hz (**b**), regularized $\beta = 0.33$.

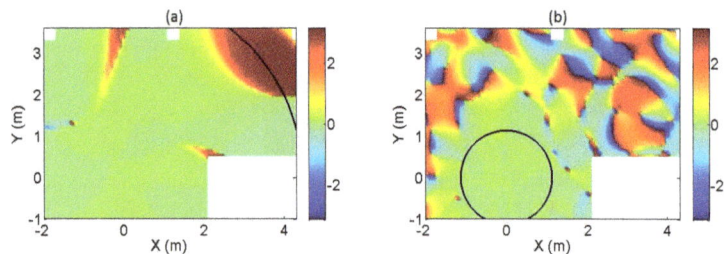

Figure 11. Phase error of the reference room at 63 Hz (**a**) and 250 Hz (**b**), regularized $\beta = 0.33$.

Figures 8–11 indicate that the implementation of regularization reduces the area of accurate reconstruction at 63 Hz compared to the non-regularized case. In contrast, regularization does not have a significant effect in the sound field reconstruction at 250 Hz. This can be explained because the condition number at this frequency is low (45.9), and regularization therefore has little effect on the inversion of the matrix $\mathbf{H}(\omega)$. Nevertheless, according to Figures 6 and 10, the amplitude of the acoustic field tends to be quieter for the regularized case at 63 Hz. This particular result is convenient for an interactive auralization system because the translation operator can lead to a zone with high acoustic pressure if no regularization is applied, which affects the stability and robustness of the implementation.

An additional analysis of the energy distribution of the plane wave density q_l was conducted to evaluate the effects of regularization in the formulation of the inverse problem. Figures 12 and 13 show an interpolated distribution of the complex amplitude of the plane waves. This is plotted in two dimensions by unwrapping the unit sphere onto a 2D plane whose axes represent the elevation and azimuth angle in degrees. The total energy of the plane wave expansion is calculated from the expression:

$$q_{total}(\omega) = \sum_{l=1}^{L} |q_l(\omega)|^2 \tag{20}$$

in which q_l is the complex amplitude of the l-th plane wave. This value is noted in the figure captions.

These figures indicate that regularization has an important effect on the spatial distribution of the energy of the plane wave density at 63 Hz. For this frequency, a more concentrated directional representation was found when regularization is implemented. Figure 14 shows the energy of each plane wave component for the regularized and non-regularized solution. A significant reduction in the energy of the plane waves is evident in the regularized case (up to four orders of magnitude). This result suggest that the use of regularization leads to a more efficient solution in terms of the energy used to reconstruct the acoustic field.

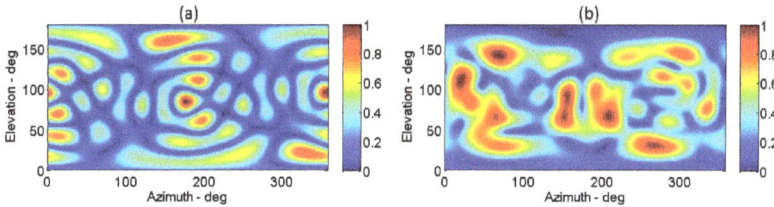

Figure 12. Normalized amplitude PWE at 63 Hz (**a**) and 250 Hz (**b**), non-regularized, $q_{total}(63) = 26.11$, $q_{total}(250) = 7 \times 10^{-4}$.

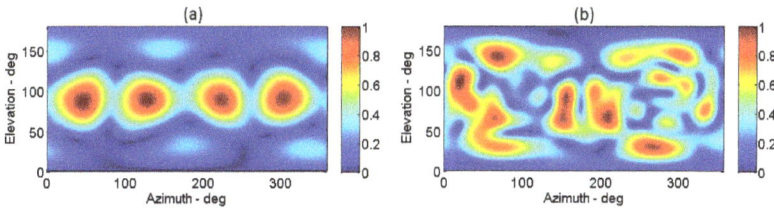

Figure 13. Normalized amplitude PWE at 63 Hz (**a**) and 250 Hz (**b**), regularized $\beta = 0.33$, $q_{total}(63) = 2.6 \times 10^{-3}$, $q_{total}(250) = 7 \times 10^{-4}$.

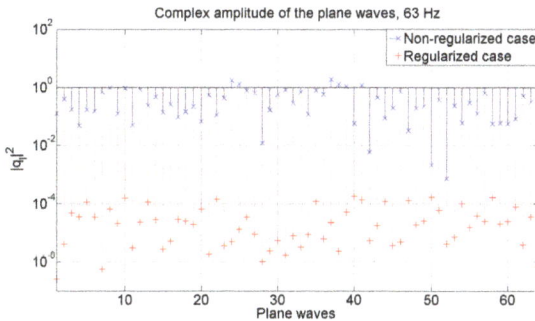

Figure 14. Comparison of the complex amplitude of the plane wave expansion.

4. Real-Time Implementation of an Auralization System

An interactive auralization system based on the plane wave expansion was developed. The system allows for real-time acoustic rendering of enclosures by using synthesized directional impulse responses calculated in advance. Due to the pre-computation of the PWE, the proposed auralization system enables interactive features such as translation and rotation of the listener within the enclosure. However, changes in the boundary conditions or modifications of the acoustic source in terms of its directivity and spatial location are not included at this stage without recalculating the PWE.

The maximum frequency simulated was 447 Hz, which is sufficient to auralize the modal behaviour of the enclosure. The time required to recalculate the PWE depends on the volume of the enclosure and the maximum frequency computed in the FE solution. For this reference case, the computation time is about one day using six nodes per wavelength on a standard desktop computer with 32 GB of RAM and Intel i7 processor. A high-pass filter (cut-off at 20 Hz) and a low-pass filter (cut-off 355 Hz) were applied to reduce the ripples produced by the truncation of the data. The proposed auralization system combines a real-time acoustic rendering with a graphical interface based on a video game environment.

Figure 15 shows the general architecture of the signal processing chain. It is composed of four main blocks. The first block is the convolution of anechoic audio material with a plane wave expansion. The second module refers to the implementation of the translation operator in the plane wave domain. The third block corresponds to the application of a rotation operator in the spherical harmonic domain, and finally, the last stage is the sound reproduction using a headphone-based binaural system with non-individual equalization. The implementation has been made using the commercial packages Max v.7.2 and Unity v.5.0. Max is a visual programming language oriented toward audio and video processing. In contrast, Unity is a programming language dedicated to video game development. It was used in the current research to create a graphical interface for the interactive auralization.

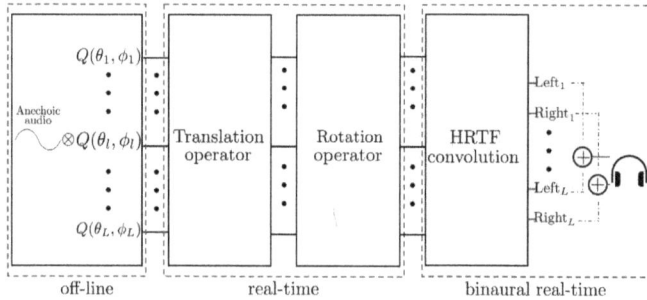

Figure 15. General architecture for a real-time auralization based on the plane wave expansion.

4.1. Translation of the Acoustic Field

It is important to point out that the reconstructed field will be exact for the target field if an integral plane wave expansion (Equation (1)) is used for the synthesis. This outcome means that the translation operator will lead to the correct field regardless of the location where the translation is intended. However, if the plane wave expansion is approximated by a finite sum, as in Equation (8), the reconstructed field will contain errors, and the translation operator will lead to the correct field only in the area where the discretized plane wave expansion matches the target field. An indication of the amount of translation that can be applied before noticeable sound artefacts occur can be estimated from Equation (14) as:

$$r_t = \left(\frac{(\sqrt{L} - 1)\lambda}{2\pi} \right) - r_l.$$ (21)

where r_t is the translation distance, L is the number of plane waves used for the synthesis, λ is the wavelength and r_l is the radius of the listener's head (e.g., 0.1 m) where it is desirable that the sound field is reproduced accurately. Indeed, r_l must be taken into consideration in order to preserve the binaural cues. An example of the translation radius is given in Figure 16, in which $R(\omega) = \frac{(\sqrt{L}-1)\lambda}{2\pi}$.

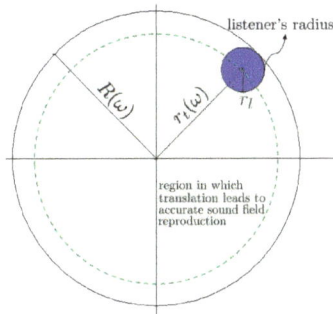

Figure 16. Region of accurate translation given by the PWE.

4.2. Rotation of the Acoustic Field

Although sound field rotation can be generated by interpolating HRTFs, this methodology has the limitation that it is restricted to binaural reproduction only. The use of a spherical harmonic representation to rotate the acoustic field is suitable for several audio reproduction techniques, which increases the flexibility of the proposed approach. In general, the implementation of the rotation in the auralization system using a spherical harmonic transformation can be divided into three main steps: the encoding, rotation and decoding stages, respectively. The general concept can be defined as the encoding of each of the directional impulse responses into a finite number of spherical harmonic coefficients, the application of the rotation and, finally, the return to the plane wave domain again through a decoding process. Nevertheless, to reduce the computational cost required by the generation of the rotation in the acoustic field, the encoding stage is computed by using as an input the difference between the angles of the directional impulse responses $\hat{y}_l(\theta_l, \phi_l)$ and the rotation angle, rather than by multiplying the spherical harmonic coefficients by a rotation operator. Figure 17 shows the signal processing chain for the rotation operator.

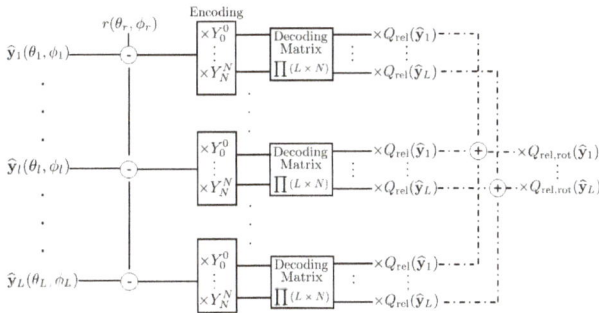

Figure 17. Rotation scheme.

One relevant consideration on the use of this approach is the number of audio files to be processed in real time $(L \times N)$. Due to the very large amount of operations, the number of spherical harmonic coefficients was limited to the fifth order (36 coefficients) for the encoding and decoding stages. The encoding is performed using real-valued spherical harmonics [31], which are calculated from their complex pairs. As a consequence of the reduced order used for rotation, the area of accurate reconstruction is reduced following the relation $N = kr$. In order to preserve the translation area, the translation operator is applied before the rotation. In this case, the area reduction given by the lower spherical harmonic order does not reduce the region where the translation operator accurately reconstructs the target field. This means that the outcome of the rotation operation has only to be accurate in the radius corresponding to the listener's head (r_l). The truncation of the spherical harmonic series up to fifth order is sufficient to cover the listener's radius for the frequency range considered.

4.3. Graphical Interfaces

A virtual environment was created in Unity to generate a platform where the listener can move using a first-person avatar and hear the changes in the acoustic field based on its relative position with respect to the enclosure. This is achieved by sending from Unity to Max the location and orientation of the avatar. The interaction between these two software packages was achieved using the Max-Unity Interoperability Toolkit [32]. Figure 18 illustrates the model made in Unity to generate the interactive auralizations.

Figure 18. Room model created in Unity. Exterior (**a**) and interior (**b**) view of enclosure.

5. Evaluation of the Auralization System

A series of experiments to assess the accuracy of the sound field reconstruction given by the proposed method is presented. For this, the real-time implementation in Max was used to record the synthesized acoustic field at different positions of the enclosure. Two types of analysis were performed: monaural (based on omnidirectional signals) and a spatial evaluation based on first order B-format signals. The procedure consisted of recording the output signals from the real-time auralization system implemented in Max and comparing them to numerical references from the FE model.

5.1. Monaural Analysis

A comparison of the predicted omnidirectional frequency responses at different receiver positions was carried out. This was performed by rendering the sound field in real time using the auralization system developed in Max. The omnidirectional frequency responses from the auralization system were obtained by adding all of the directional impulse responses corresponding to the different L plane waves used to represent the field and recording the total output after the rotation stage. This information is compared to omnidirectional frequency responses that were synthesized individually at the receiver locations. These omnidirectional references do not use the directional information of the plane wave expansion. They correspond to the frequency response of omnidirectional receivers obtained directly from the FE solution. The use of the numerical information as a reference is due to the lack of measurements and spatial information across the enclosure to evaluate different positions.

Five receivers' positions were selected as shown in Figure 19. The central point of the expansion corresponds to Location 01. The predicted frequency responses in narrow band and in 1/3 octave band resolution for each receiver are illustrated in Figures 20–24. The vertical cyan line indicates the cut-off frequency (355 Hz) of the low-pass filter, and the vertical black line indicates the maximum frequency below which the translation is expected to provide accurate results. The maximum frequency was estimated by solving Equation (14) for R. This was done taking into account the distance between the central point of the expansion and each receiver's position. The mean error displayed in the figures was selected as a metric, and it is defined as:

$$\text{ME(dB)} = \frac{1}{n} \sum_{i=1}^{n} \left| 10 \log_{10}(|\tilde{p}_i|^2) - 10 \log_{10}(|p_i|^2) \right|, \tag{22}$$

in which n is the number of 1/3 octave frequency bands and $|\tilde{p}_i|^2$ and $|p_i|^2$ are the predicted and reference energy of the acoustic pressure in the 1/3 octave band i, respectively. This error is based on an equal contribution from all of the 1/3 octave bands, being analogous to a model in which pink noise is used as the input signal. It was created to provide insight into how dissimilar on average the reconstructed field is from the reference one. A summary of the mean errors according to the receiver location is presented in Table 2.

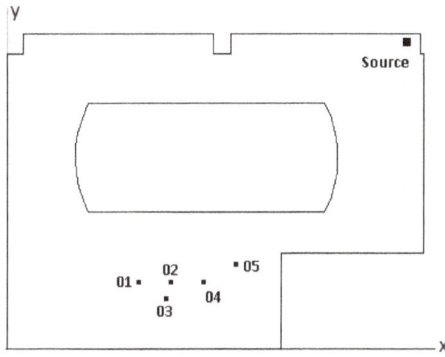

Figure 19. Listener positions selected to evaluate the auralization system.

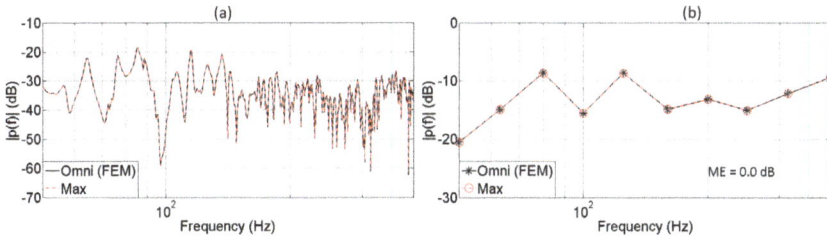

Figure 20. Comparison of full (**a**) and 1/3 octave band (**b**) frequency responses. PWE and omnidirectional room impulse response (FEM) at reference Position 1.

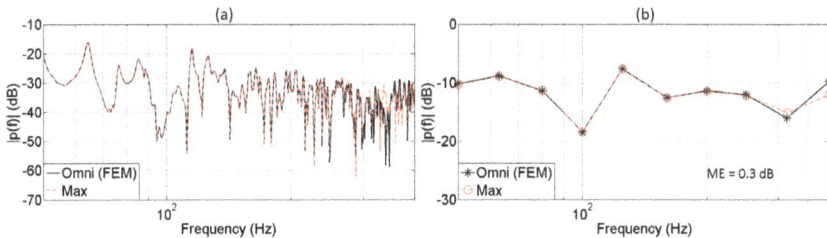

Figure 21. Comparison of full (**a**) and 1/3 octave band (**b**) frequency responses. Translated PWE and omnidirectional room impulse response (FEM) at translated Position 2.

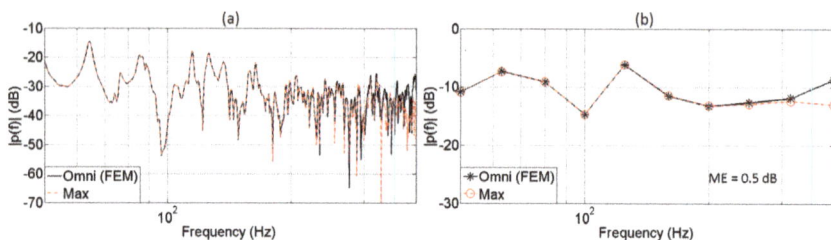

Figure 22. Comparison of full (**a**) and 1/3 octave band (**b**) frequency responses. Translated PWE and omnidirectional room impulse response (FEM) at translated Position 3.

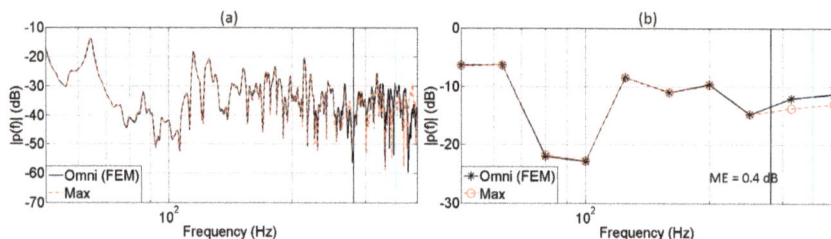

Figure 23. Comparison of full (**a**) and 1/3 octave band (**b**) frequency responses. Translated PWE and omnidirectional room impulse response (FEM) at translated Position 4.

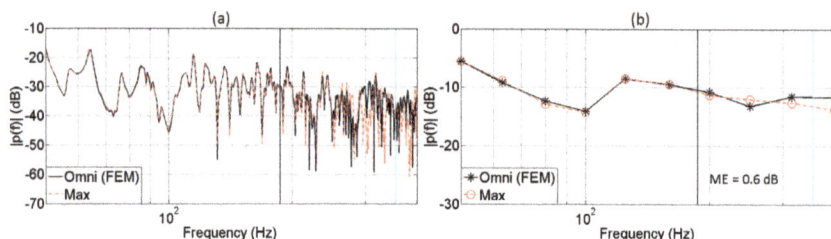

Figure 24. Comparison of full (**a**) and 1/3 octave band (**b**) frequency responses. Translated PWE and omnidirectional room impulse response (FEM) at translated Position 5.

Table 2. Mean errors at different receiver locations. The distance to the central point of the expansion and the maximum frequency at which achieving an accurate reconstruction is expected are reported.

Receiver	Distance (m)	Frequency (Hz)	ME (dB)
2	0.5	$\approx$562	1.5
3	0.5	$\approx$562	1.7
4	1	$\approx$281	1.9
5	1.5	$\approx$187	2.0

The results indicate that the changes in the modal response of the enclosure are correctly predicted by the translation operator. Furthermore, the figures show good accuracy in the sound field reconstruction up to the frequencies predicted by Equation (14) as long as these frequencies are below the cut-off frequency. The differences found at this frequency range may arise from three causes: the implementation of integer delays in Max, the numerical accuracy used by Max to perform mathematical operations (summing the directional impulse responses) and the application of regularization in the inverse problem, which decreases the matching between the radius of validity and the effective area where the reconstruction is accurate.

5.2. Spatial Analysis

In the previous section, the auralization system was evaluated in terms of its accuracy in reconstructing the acoustic pressure at different reference locations. Although this provides useful insight into the performance of the method, it does not give any information about the spatial characteristics of the synthesized sound field. This aspect was investigated by assessing the ability of the system to accurately reconstruct the zero and first order terms in the spherical harmonic expansion of the sound field, as described by Equation (4), often referred to as Ambisonics B-format signals. An accurate reconstruction of the zero and first order sound fields at the listening position implies an accurate reproduction of binaural localization cues at low frequencies [33]. Nevertheless, this approach is a preliminary analysis, and further investigation is required.

The B-format reference signals were estimated by the implementation of an inverse method according the formulation proposed by the authors [34]. For this, virtual microphone arrays were used to sample the FE data at the positions where the B-format signals were intended to be synthesized. The B-format signals from the auralization system were obtained by recording the output of the rotation module (see Figure 15), which is based on a spherical harmonic transformation. The zero and first order components were recorded and compared to the numerical references.

The B-format consists of four signals, which correspond to an omnidirectional (zero order) and three orthogonal dipoles (first order). They are refereed to in the Ambisonics literature as W, X, Y and Z, respectively. The analysis of the B-format signals was carried out at Receivers 2 and 5. These were chosen because they corresponded to the locations where the best and worst agreement was found in terms of the mean errors for the monaural evaluation. The frequency responses in the narrow band and in the 1/3 octave band resolution of the reference and synthesized B-format signals are illustrated in Figures 25–32. A summary of the mean errors according to the B-format component is presented in Table 3.

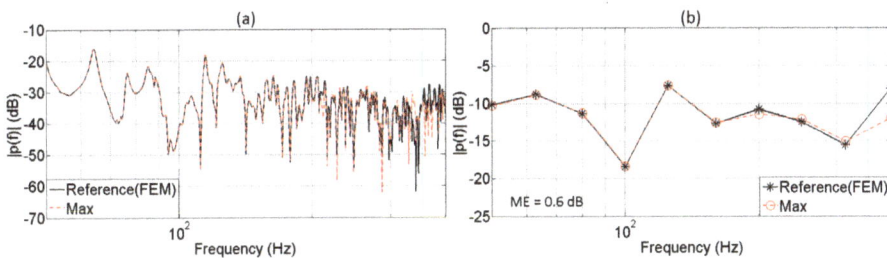

Figure 25. Comparison of full (**a**) and 1/3 octave band (**b**) frequency responses. Translated PWE (W) and reference (W) at translated Position 2.

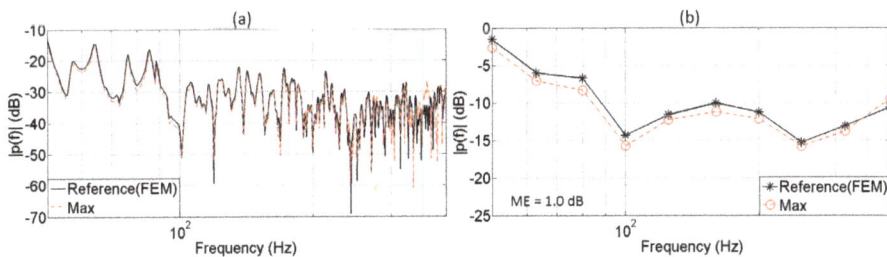

Figure 26. Comparison of full (**a**) and 1/3 octave band (**b**) frequency responses. Translated PWE (X) and reference (X) at translated Position 2.

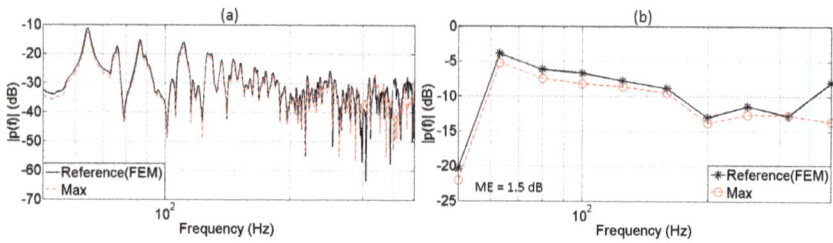

Figure 27. Comparison of full (**a**) and 1/3 octave band (**b**) frequency responses. Translated PWE (Y) and reference (Y) at translated Position 2.

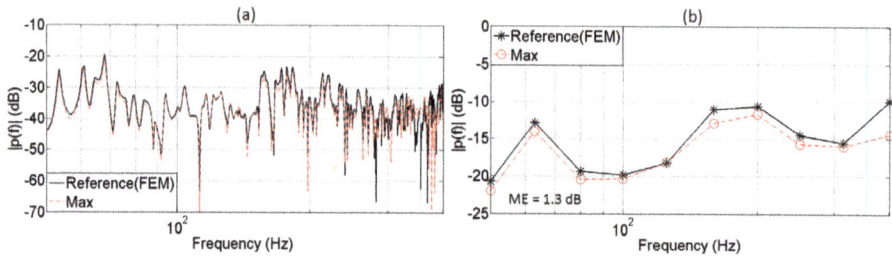

Figure 28. Comparison of full (**a**) and 1/3 octave band (**b**) frequency responses. Translated PWE (Z) and reference (Z) at translated Position 2.

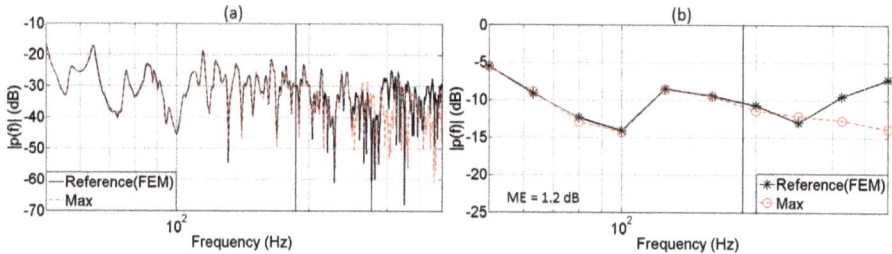

Figure 29. Comparison of full (**a**) and 1/3 octave band (**b**) frequency responses. Translated PWE (W) and reference (W) at translated Position 5.

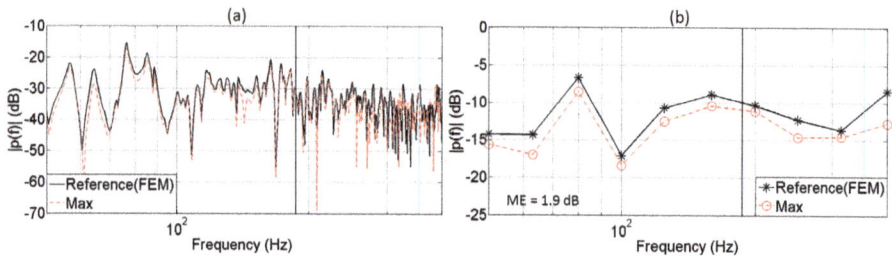

Figure 30. Comparison of full (**a**) and 1/3 octave band (**b**) frequency responses. Translated PWE (X) and reference (X) at translated Position 5.

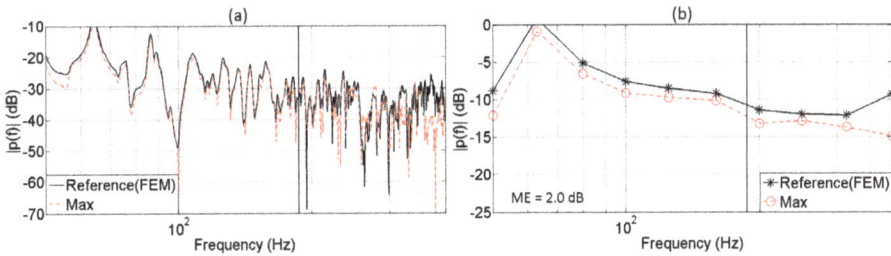

Figure 31. Comparison of full (**a**) and 1/3 octave band (**b**) frequency responses. Translated PWE (Y) and reference (Y) at translated Position 5.

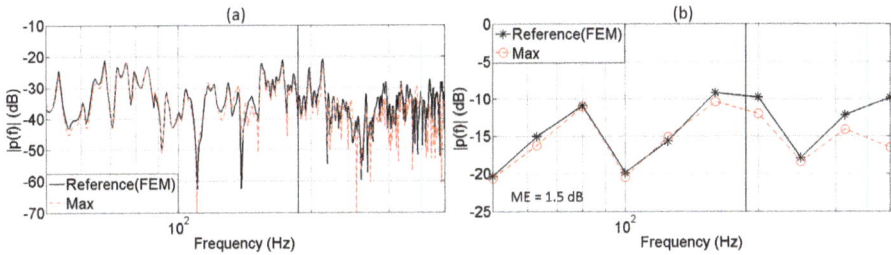

Figure 32. Comparison of full (**a**) and 1/3 octave band (**b**) frequency responses. Translated PWE (Z) and reference (Z) at translated Position 5.

Table 3. Mean errors in the B-format signals for the interactive auralization at different receiver locations.

Receiver	W (dB)	X (dB)	Y (dB)	Z (dB)	Average (dB)
2 (0.5 m)	0.6	1.0	1.5	1.3	1.1
5 (1.5 m)	1.2	1.9	2.0	1.5	1.7

A comparison of the mean errors indicates that the field at Receiver 2 has smaller errors values than at Receiver 5. This is expected as the distance to the central point of the expansion of Receiver 2 is smaller. Regarding the B-format signals, the outcomes show that the reconstruction is more accurately performed for the W signals. In this case, a very good agreement between the frequency response synthesized by the auralization system and the reference signal was found up to the frequency established by Equation (14). For the remaining coefficients, the match is not as good as the zero order, but with good agreement in terms of the envelope of the frequency response.

6. Conclusions

A framework for the generation of an interactive auralization of an enclosure based on a plane wave expansion has been presented. This acoustic representation not only allows for interactive features, such as the translation and rotation of sound fields, but is also compatible with several sound reproduction techniques, such as binaural rendering, Ambisonics, WFS and VBAP. The directional impulse responses corresponding to this plane wave representation were predicted by means of the finite element method.

An analysis of the reconstruction of the sound field in terms of monaural and B-format signals indicates that the interactive auralization system based on a plane wave representation is able to synthesize the acoustic field at low frequencies correctly, making it suitable for the auralization of enclosures, whose sound field is characterized by a modal behaviour.

The suitability of inverse methods to estimate the amplitude of a set of plane waves that reconstruct a target field has been proven. This technique is useful to extract directional information from data obtained from FE simulations. However, the discretization of the integral representation into a finite number of plane waves limits the spatial accuracy of the sound field reconstruction. The extent of the region in which the synthesis is accurate depends on the number of plane waves and the frequency of the field.

The use of Tikhonov regularization has three main effects on the sound field representation. The first is that the energy of the plane wave density used for the synthesis of the sound fields is considerably lower than for the non-regularized solution. The second consequence is that the energy distribution of the plane wave density is much more directionally concentrated. The last effect is a reduction of the area where the sound field reconstruction is accurate compared to the non-regularized solution. Nevertheless, the implementation of regularization is convenient for an interactive auralization system because the translation operator can generate zones with high acoustic pressure if no regularization is applied.

Future work will include the combination of finite element and geometrical acoustic results to extend the method outlined here to middle and high frequencies.

The code developed for the interactive system can be download from "https://drive.google.com/file/d/0BwuuNpQpY5UKZ2htSW9JbXN5OVU/view?usp=sharing".

Author Contributions: The current paper is the result of the research conducted by Diego Mauricio Murillo Gómez for the degree of doctor of philosophy at the Institute of Sound and Vibration Research, University of Southampton. Filippo Maria Fazi and Jeremy Astley were the supervisors; their contributions correspond to the advice, support and review at all stages of the research.

Conflicts of Interest: The authors declare no conflict of interest.

Abbreviations

The following abbreviations are used in this manuscript:

GA	Geometrical Acoustics
FEM	Finite Element Method
BEM	Boundary Element Method
FDTD	Finite Difference Time Domain
GPU	Graphics Processor Unit
PWE	Plane Wave Expansion
WFS	Wave Field Synthesis
VBAP	Vector-Based Amplitude Panning

References

1. Vorländer, M. *Auralization*, 1st ed.; Springer: Berlin, Germany, 2010.
2. Savioja, L.; Huopaniemi, T.; Lokki, T.; Vaananen, R. Creating Interactive Virtual Acoustic Environments. *J. Audio Eng. Soc.* **1999**, *47*, 675–705.
3. Funkhouser, T.; Tsingos, N.; Carlbom, I.; Elko, G.; Sondhi, M.; West, J.; Pingali, G.; Min, P.; Ngan, A. A beam tracing method for interactive architectural acoustics. *J. Acoust. Soc. Am.* **2004**, *115*, 739–756.
4. Noisternig, M.; Katz, B.; Siltanen, S.; Savioja, L. Framework for Real-Time Auralization in Architectural Acoustics. *Acta Acust. United Acust.* **2008**, *94*, 1000–1015.
5. Chandak, A.; Lauterbach, C.; Taylor, M.; Ren, Z.; Manocha, D. AD-Frustum: Adaptive Frustum Tracing for Interactive Sound Propagation. *IEEE Trans. Vis. Comput. Graph.* **2008**, *14*, 1707–1714.
6. Taylor, M. RESound: Interactive Sound Rendering for Dynamic. In Proceedings of the 17th International ACM Conference on Multimedia 2009, Beijing, China, 19–24 October 2009; pp. 271–280.
7. Astley, J. Numerical Acoustical Modeling (Finite Element Modeling). In *Handbook of Noise and Vibration Control*, 1st ed.; Crocker, M., Ed.; John Wiley & Sons: Hoboken, NJ, USA, 2007; Chapter 7, pp. 101–115.

8. Herrin, D.; Wu, T.; Seybert, A. Boundary Element Method. In *Handbook of Noise and Vibration Control*, 1st ed.; Crocker, M., Ed.; John Wiley & Sons: Hoboken, NJ, USA, 2007; Chapter 8, pp. 116–127.

9. Botteldooren, D. Finite-Difference Time-Domain Simulation of Low-Frequency Room Acoustic Problems. *J. Acoust. Soc. Am.* **1995**, *98*, 3302–3308.

10. Mehra, R.; Raghuvanshi, N.; Antani, L.; Chandak, A.; Curtis, S.; Manocha, D. Wave-Based Sound Propagation in Large Open Scenes using an Equivalent Source Formulation. *ACM Trans. Graph.* **2013**, *32*, 19.

11. Mehra, R.; Antani, L.; Kim, S.; Manocha, D. Source and Listener Directivity for Interactive Wave-Based Sound Propagation. *IEEE Trans. Vis. Comput. Graph.* **2014**, *20*, 495–503.

12. Raghuvanshi, N. Interactive Physically-Based Sound Simulation. Ph.D. Thesis, University of North Carolina, Chapel Hill, NC, USA, 2010.

13. Savioja, L. Real-Time 3D Finite-Difference Time-Domain Simulation of Low and Mid-Frequency Room Acoustics. In Proceedings of the 13th Conference on Digital Audio Effects, Graz, Austria, 6–10 September 2010.

14. Southern, A.; Murphy, D.; Savioja, L. Spatial Encoding of Finite Difference Time Domain Acoustic Models for Auralization. *IEEE Trans. Audio Speech Lang. Process.* **2012**, *20*, 2420–2432.

15. Southern, A.; Wells, J.; Murphy, D. Rendering walk-through auralisations using wave-based acoustical models. In Proceedings of the 17th European Signal Processing Conference, Glasgow, UK, 24–28 August 2009; pp. 715–716.

16. Sheaffer, J.; Maarten, W.; Rafaely, B. Binaural Reproduction of Finite Difference Simulation Using Spherical Array Processing. *IEEE Trans. Audio Speech Lang. Process.* **2015**, *23*, 2125–2135.

17. Støfringsdal, B.; Svensson, P. Conversion of Discretely Sampled Sound Field Data to Auralization Formats. *J. Audio Eng. Soc.* **2006**, *54*, 380–400.

18. Menzies, D.; Al-Akaidi, M. Nearfiled binaural synthesis and ambisonics. *J. Acoust. Soc. Am.* **2006**, *121*, 1559–1563.

19. Winter, F.; Schultz, F.; Spors, S. Localization Properties of Data-based Binaural Synthesis including Translatory Head-Movements. In Proceedings of the Forum Acusticum, Krakow, Poland, 7–12 September 2014.

20. Zotter, F. Analysis and Synthesis of Sound-Radiation with Spherical Arrays. Ph.D. Thesis, University of Music and Performing Arts, Graz, Austria, 2009.

21. Murillo, D. Interactive Auralization Based on Hybrid Simulation Methods and Plane Wave Expansion. Ph.D. Thesis, Southampton University, Southampton, UK, 2016.

22. Duraiswami, R.; Zotkin, D.; Li, Z.; Grassi, E.; Gumerov, N.; Davis, L. High Order Spatial Audio Capture and Its Binaural Head-Tracked Playback Over Headphones with HRTF Cues. In Proceedings of the 119th Convention of the Audio Engineering Society, New York, NY, USA, 7–10 October 2005.

23. Fazi, F.; Noisternig, M.; Warusfel, O. Representation of Sound Fields for Audio Recording and Reproduction. In Proceedings of the Acoustics 2012, Nantes, France, 23–27 April 2012; pp. 1–6.

24. Williams, E. *Fourier Acoustics*, 1st ed.; Academic Press: London, UK, 1999.

25. Fliege, J. *Sampled Sphere*; Technical Report; University of Dortmund: Dortmund, Germany, 1999.

26. Nelson, P.; Yoon, S. Estimation of Acoustic Source Strength By Inverse Methods: Part I, Conditioning of the Inverse Problem. *J. Sound Vib.* **2000**, *233*, 639–664.

27. Ward, D.; Abhayapala, T. Reproduction of a Plane-Wave Sound Field Using an Array of Loudspeakers. *IEEE Trans. Audio Speech Lang. Process.* **2001**, *9*, 697–707.

28. Herlufsen, H.; Gade, S.; Zaveri, H. Analyzers and Signal Generators. In *Handbook of Noise and Vibration Control*, 1st ed.; Crocker, M., Ed.; John Wiley & Sons: Hoboken, NJ, USA, 2007; Chapter 40, pp. 101–115.

29. Kim, Y.; Nelson, P. Optimal Regularisation for Acoustic Source Reconstruction by Inverse Methods. *J. Sound Vib.* **2004**, *275*, 463–487.

30. Yoon, S.; Nelson, P. Estimation of Acoustic Source Strength By Inverse Methods: Part II, Experimental Investigation of Methods for Choosing Regularization Parameters. *J. Sound Vib.* **2000**, *233*, 665–701.

31. Poletti, M. Unified description of Ambisonics using real and complex spherical harmonics. In Proceedings of the Ambisonics Symposium 2009, Graz, Austria, 25–27 June 2009; pp. 1–10.

32. Department of Music. Virginia Tech-School of Performing Arts. 2016. Available online: http://disis.music.vt.edu/main/index.php (accessed on 25 February of 2017).

33. Menzies, D.; Fazi, F. A Theoretical Analysis of Sound Localisation, with Application to Amplitude Panning. In Proceedings of the 138th Convention of Audio Engineering Society, Warsaw, Poland, 7–10 May 2015; pp. 1–5.
34. Murillo, D.; Fazi, F.; Astley, J. Spherical Harmonic Representation of the Sound Field in a Room Based on Finite Element Simulations. In Proceedings of the 46th Iberoamerican Congress of Acoustics 2015, Valencia, Spain, 21–23 September 2015; pp. 1007–1018.

MDPI AG

St. Alban-Anlage 66

4052 Basel, Switzerland

Tel. +41 61 683 77 34

Fax +41 61 302 89 18

http://www.mdpi.com

Applied Sciences Editorial Office

E-mail: applsci@mdpi.com

http://www.mdpi.com/journal/applsci